David E. Rowe

# Bernhard Riemann: His Life and Wondrous Mathematical Legacy

David E. Rowe
Institut für Mathematik
Johannes Gutenberg University Mainz
Mainz, Germany

ISBN 978-3-032-25456-6 ISBN 978-3-032-25457-3 (eBook)
https://doi.org/10.1007/978-3-032-25457-3

Mathematics Subject Classification: 01-02



This Springer imprint is published by the registered company Springer Nature Switzerland AG
The registered company address is: Gewerbestrasse 11, 6330 Cham, Switzerland

## Preface: Riemann's Life and Legacy

When considering famous names in the history of mathematics, Bernhard Riemann's inevitably appears high on any list. At the time of his death in 1866, he had already attained a modicum of fame, but nothing like what would come afterward. Some might argue that Riemann would have achieved lasting fame during his lifetime had he not already died shortly before reaching his fortieth birthday. Against that argument, it should be noted that some of Riemann's most famous and influential papers were first published immediately after his death. Had he lived, there is good reason to believe those works might never have surfaced at all. Just imagine how diminished Riemann's fame would have been if mathematicians today had never heard of the Riemann integral or Riemannian manifolds! This book attempts to lift the veil on the life of this brilliant but mysterious figure. A great deal of the mystery will nevertheless remain, however, as Riemann was an elusive personality even for those who knew him best.

The gist of Riemann's life story was first told in 1876, ten years after his death, by his friend Richard Dedekind. Since then, little has been added to Dedekind's rather dry account, as others have mainly repeated information he imparted, much of which he obtained from Riemann's widow, Elise Riemann née Koch.[1] English readers from the last century were treated to a more colorful telling of the tale in Eric Temple Bell's *Men of Mathematics*, but even he could not really enliven it all that much. Readers were thus left to contemplate the stark contrast between Riemann's rich inner life – the boldness of his remarkable mathematical world – and his shy, awkward manner as an unremarkable personality. Most would say that his life was short and uneventful. The present account should easily dispel that rather facile picture, an impression in some part due to a dearth of knowledge about the circumstances that shaped Riemann's unusual life and career.

Many commentators have written about Riemann's ideas without paying much attention to when and how these first surfaced. My approach, by contrast, aims to anchor his intellectual life in the events of his time, while never losing sight of the chronology. This hardly need be emphasized for those who study history, but here the larger methodological implications concern the nature of mathematical knowledge as a product of human intellectual and social activity. In the present case, I maintain that Riemann's novel ideas were not uniquely his own, as if discovered in a vacuum. On the other hand, I doubt that anyone could give a complete and convincing explanation for what, when, and why Riemann produced his wondrous achievements, especially given the dearth of source material that sheds light on his motivations. One can always argue

[1] Dedekind was, of course, in no position to write about Riemann's early life, nor his later married years in Italy, where he died in July 1866. For the former, Elise Riemann could consult with Ida Riemann, the mathematician's older sister, with whom she lived after her husband's death. One cannot claim that Dedekind's essay was ghostwritten by these two women, but one can clearly gauge the importance of their input from the lengthy letter and biographical sketch Elise sent him on 1 May 1875 [Scheel 2015, 309–317].

that Riemann was a mathematical genius, so trying to account for his sources and inspiration can never really explain what he accomplished. In only partial deference to that viewpoint, I have tried to resist the temptation to explain Riemann's mathematics, other than to make a few hopefully suggestive remarks. Instead, I want to let Riemann speak for himself, especially when his words clearly bring out the larger import of his ideas.

The structure I have adopted for this book attempts to sort out important themes associated with Riemann's life. Part I unfolds in four phases, leaving more detailed discussions of his major works for Part II. As background for this biographical section, the Introduction provides a thumbnail sketch of mathematics during the first half of the nineteenth century, but with special emphasis on Riemann's *alma mater*, Göttingen University. This ends with a timeline chronicling some of the main events in his life. In several respects, Riemann's career followed in the footsteps of three legendary Göttingen figures: Carl Friedrich Gauss, Gustav Lejeune Dirichlet, and Wilhelm Weber, though his relations with Gauss were more distant than with the other two. All three play key roles in the biographical chapters, but Gauss's larger intellectual influence on Riemann receives special attention in Chapter 10, as a way of summing up Part II. Part I already underscores how Riemann's theory of algebraic functions and their integrals arose from ideas in potential theory and partial differential equations. Riemann learned these subjects very early on: first, by reading Gauss's works on geomagnetism and conformal mappings and, secondly, by following Dirichlet's lecture courses in Berlin.

In Part II, the focus shifts to Riemann's better known works, though with due attention to their publication history and subsequent reception. Chapter 5 discusses his new geometric approach to the theory of functions of a single complex variable, as presented in his doctoral dissertation of 1851. Most mathematicians – Gauss being an exception – considered this work nearly unreadable, and yet it planted the seeds for almost everything Riemann thereafter contributed to this field. Around two years later, he wrote the two papers discussed in Chapter 6. Both were written to meet the faculty's requirements for Habilitation, i.e., admission as an unsalaried postdoc. This formal procedure began in 1853, when Riemann submitted his postdoctoral thesis, which dealt with representations of real functions by trigonometric series, a topic studied earlier by Dirichlet. Along the way to generalizing Dirichlet's findings, Riemann reformed Cauchy's concept of integration by introducing the nowadays ubiquitous Riemann integral. His second paper, written in preparation for his final qualifying lecture, dealt with a famous philosophical problem of considerable interest to Gauss, namely, to account for the principles underlying geometrical knowledge. During Riemann's lifetime, however, these now classic works never appeared in print and were thus altogether forgotten. Luckily, Richard Dedekind resurrected both texts soon after his friend's death, part of the complex story surrounding Riemann's legacy.

In Chapter 7, I draw on Riemann's lecture courses on elliptic and Abelian functions to show how Riemann went on to explore the terrain he had sketched in his dissertation. This led to a famous sequel, his "Theory of Abelian Functions," published in 1857 in "Crelle," Germany's most prestigious mathematical periodical. He wrote this at the request of Carl Wilhelm Borchardt, its editor. Borchardt was well aware that his Berlin colleague, Karl Weierstrass, was hard at work trying to solve the Jacobi inversion problem for general Abelian integrals. In fact, Weierstrass's latest paper was already in press when Riemann's study came out. In it, Riemann presented a new method for solving this famous problem once and for all. After studying it, Weierstrass decided to withdraw his own paper from publication, creating a minor sensation. Although Riemann's paper was the most influential work to appear during his lifetime, it was, in many respects, even harder to read than his dissertation.

Two other shorter papers are presented in Chapter 8: Riemann's note on the zeta function and his posthumously published Paris prize paper. Both attained considerable posthumous fame but for very different reasons. Dedekind and others saw the importance of the prize paper in connection with Riemann's unpublished lecture on the foundations of geometry. Later, with the advent of Einstein's general theory of relativity, Riemannian curvature came to occupy a central place in modern differential geometry. Riemann's approach to the zeta function had a similarly profound impact on analytic number theory, though today nearly all mathematicians identify this with one profoundly difficult and still unsolved problem: the Riemann hypothesis.

For an expert, like S.J. Patterson, Riemann's paper may seem readable enough, but for those of us who are less fortunate, we can be grateful for the lucid explanations given by H.M. Edwards in his by now classic study *Riemann's Zeta Function*. Edwards would have rejected that appellation, I'm sure, having been one of the foremost advocates for the view that mathematicians should read the classical *primary* literature, including Riemann's paper.[2] He ended his Preface by writing:

> Read the classics, not just Riemann, but all the major contributions to analytic number theory that I discuss in this book. The purpose of a secondary source is to make the primary sources available to you. If you can read and understand the primary sources without reading this book, more power to you. If you read this book without reading the primary sources you are like a man who carries a sack lunch to a banquet. [Edwards 1974, xi]

Without wishing to raise the obvious question – what is a mathematical classic anyway? – the present study hopefully makes the case that Riemann's genius was nurtured from an early age by reading the works of other authors. His taste for classic works was surely impeccable. How many were able to read *him* with comparable understanding? Who can pretend to know?

Riemann also wrote a number of papers on special topics in physics, though these left no major imprint on subsequent research. Nevertheless, he attached great importance to several of these works, which are essential for understanding the central role of physical ideas in his research program. In Chapter 9, a selection of Riemann's papers on physical topics is taken up from that vantage

[2] For a taste of his passionate opposition to what he saw as mainstream views among historians of modern mathematics, see [Edwards 2020].

point. Part II then concludes with the aforementioned account of Gaussian influences in Chapter 10. Although Gauss never served as a mentor to the young Riemann, his works in both mathematics and physics strongly influenced Riemann's distinctive approach to complex analysis.

Like Part II, which gives a selective survey of Riemann's principal works, Part III offers an even more selective overview of his legacy. Chapter 11 discusses seven mathematicians who made noteworthy contributions to the evolution of Riemannian ideas. Three of them – Friedrich Prym, Gustav Roch, and Karl Hattendorff – had the opportunity to attend lecture courses taught by Riemann. In Italy, Riemann's work in complex analysis was strongly received by Enrico Betti and Felice Casorati, whereas in Germany it met with a mixed reception. Alfred Clebsch was the first to exploit the fertility of Riemannian methods in algebraic geometry, even if his work was viewed skeptically by Roch, who was a true disciple of Riemann. Nor should one overlook Clebsch's influence on the young Felix Klein, whose road to Riemann passed through his association with Clebsch and his school. Klein's approach to Riemann surfaces met with considerable resistance during the 1880s. Indeed, by that time Riemann's intellectual legacy had become a battleground for various groups of (mainly German) mathematicians. Several were prominent analysts who tried to salvage Riemann's main results by developing new methods to replace Riemann's arguments based on what he called Dirichlet's principle.

A full account of the reception history stemming from Riemann's work would go far beyond the confines of this book, which merely aims to describe some of its more important features.[3] My version of this complex story ends in Chapter 12 with an account of the fruitful collaboration between Richard Dedekind and Heinrich Weber (no relation to Wilhelm Weber), who together produced the first edition of *Riemanns Werke* [Riemann 1876/1892/1902]. This chapter takes up only a few of the mathematical topics that came to light thanks to the efforts of Dedekind and Weber, whose lifelong friendship began with their work on this project. Much turns on the circumstances that led to this unusual collaboration, which took place outside of Göttingen, where work on the Gauss edition had just begun. Their physical separation forced Dedekind and Weber to communicate via letters, some of extraordinary length, a treasure trove for documenting their mutual contributions to the edition.

A key figure, contributing to their work from behind the scenes, was Elise Riemann. As the mathematician's widow, she clearly had a singular and intensely personal interest in his legacy. Our story thus picks up with events from the end of Chapter 4, recounting how, after returning from Italy in 1866, Elise Riemann persuaded Dedekind to go through her husband's scientific estate to determine what might still be publishable. Dedekind was immediately fascinated by the materials she gave him, despite their chaotic, often barely legible state. His enthusiasm, coupled with Wilhelm Weber's desire to promote Riemann's legacy alongside that of Gauss, were key factors that led to the idea of

[3] Walter Purkert listed over 550 papers influenced by Riemann, which appeared in the 25 years after his death [Riemann 1990, 870 ff.].

an expanded edition that would incorporate material from Riemann's estate. Alfred Clebsch threw himself headlong into this project, but he died before it could be brought to fruition. With Dedekind's help, however, Heinrich Weber took up and completed this task, which led to publication of the first edition of *Bernhard Riemanns Gesammelte Mathematische Werke und wissenschaftlicher Nachlass*. Heinrich Weber also suggested the idea of adding a short biography of Riemann's life, written by someone who knew him personally. That suggestion, too, came to fruition, despite Dedekind's reluctance to write about Riemann's personal life. Chapter 12 describes how this came about, while underscoring how Elise Riemann made her influence felt on the content as well as the spirit of Dedekind's biographical sketch.

## Acknowledgments

It is my pleasure to thank some of the many people who have contributed in various ways to the making of this book, beginning with the friendly staff at the Niedersächsische Staats- und Universitätsbibliothek Göttingen (hereinafter SUB), where Riemann's scientific estate is held. As usual, Bärbel Mund answered all my queries very promptly and was otherwise very helpful. Several people sent comments on earlier versions of this book, including two anonymous referees who contributed useful feedback. Two other readers, Donal O'Shea and Gert Schubring, imparted extremely helpful advice for improving the first part of the text; I'm very grateful for their invaluable input. A special thanks goes to Erhard Scholz, whose expertise on Riemann shows up in many places in this book. Erhard noted a number of errors and infelicitous phrases, which I never would have caught otherwise. This book benefited immensely from scholarly work on Riemann's personal and academic life undertaken over the last fifty years, in particular the meritorious efforts of Erwin Neuenschwander, who turned up much new source material. Others whom I'd like to thank for their support include Tom Archibald, Leo Corry, Olivier Darrigol, Moritz Epple, Jeremy Gray, Lizhen Ji, John McCleary, Stefan Müller-Stach, Volker Remmert, Wayne Rowe, Norbert Schappacher, Reinhard Siegmund-Schultze, Renate Tobies, Klaus Volkert, Scott A. Walter, Michael Wiescher, and Anita Zieher. I'm also very pleased to acknowledge that this book profited from the efforts of Remi Lodh, senior editor at Springer Heidelberg. He and I planned that it might come out in 2026 – the "Riemann Year" – and thanks to his help, we succeeded. I'm also happy to announce a forthcoming event I initiated while writing this book: Sandra Schüddekopf will direct a new play dealing with the life of Riemann and which will be performed at several institutions throughout Germany. Thanks to the enthusiastic support of Walter Purkert and others, "Wer war Bernhard Riemann?" will open in November 2026 at Bonn University. Last of all, following the usual custom, my deep thanks go to the special person in my life, my dear wife Hilde. This time she even volunteered to share in the proofreading, which definitely helped to reduce the number of typos.

# Contents

## Part III Riemann's Legacy

# Introduction: Riemann's Times

## The Mystery of Riemann's Fame

Richard Dedekind was only one among many prominent mathematicians who knew Riemann personally. A small number had studied under him and many more had read his published work. Yet to understand how Riemann's ideas came into the German, Italian, and French mathematical worlds, one must first realize that only a relatively small part of his work was actually known during his lifetime. Indeed, some of his most influential ideas only surfaced after his death, sometimes even many years later. When he died in Italy in 1866, few if any still remembered the topic of his now famous lecture on the hypotheses underlying geometry. This was one of several manuscripts his widow, Elise Riemann, gave to Dedekind, who then prepared it for publication in [Riemann 1868/1876]. Dedekind also decided to publish Riemann's still unknown postdoctoral dissertation [Riemann 1868/1876] as well as a short paper on electrodynamics [Riemann 1867a/1876]. Riemann's lecture attracted almost immediate attention, whereas his thesis – the second paper, in which he introduced the Riemann integral – eventually became a classic in the history of series representations for real-valued functions. All three of these papers became far more easily accessible in 1876, when they were reprinted in the first edition of *Bernhard Riemanns Gesammelte Mathematische Werke und wissenschaftlicher Nachlass* together with Dedekind's biography [Dedekind 1876a].

Two things should be emphasized in this connection. First, Dedekind played an important, but secondary role in producing this and subsequent editions of *Riemanns Werke*. It was his friend Heinrich Weber who took responsibility for the lion's share of this work. Second, a quick glance at the table of contents reveals that Riemann published only nine papers during his lifetime. Moreover, just five of those publications were contributions to pure mathematics; the other four dealt with special problems in physics. The key paper that at once established his fame and reputation was [Riemann 1857b], a very difficult two-part study devoted to the theory of Abelian functions. It met with a mixed reception, in large part because of technical difficulties connected with



D. E. Rowe, *Bernhard Riemann: His Life and Wondrous Mathematical Legacy*,
https://doi.org/10.1007/978-3-032-25457-3_1

Riemann's methodology, in particular his use of Riemann surfaces and what he called Dirichlet's principle. Riemann had long before outlined this new approach to complex analysis, having laid the groundwork for it in his doctoral dissertation [Riemann 1851/1876]. However, few outside Göttingen ever saw this earlier publication, so one can well imagine the shock Riemann's contemporaries experienced when they tried to read [Riemann 1857b] cold, as it were. Several years later, he elaborated on part of this theory in his paper on the zeros of $\theta$-functions [Riemann 1866/1876], the last of the nine publications he saw into print.

Riemann's life did not unfold on a grand stage, nor did he call much attention to himself when interacting with others. As a Gymnasium pupil, his teachers described him as well-behaved and diligent (*brav und fleißig*), traits that remained pronounced as he grew older. Less obvious, but no less important, was his intellectual ambition and the wide range of interests he pursued. Some of his philosophical views and physical investigations have been discussed elsewhere, but usually with little heed to the circumstances in his life during which that work was undertaken. It has also been remarked that Riemann was a voracious reader, even though it has been hard to pin down what and when he became acquainted with various works. Nothing is known, for example, about the books in his private library, though we can in some instances make intelligent guesses about works he probably owned. Thanks to the recently published list of books and journals he borrowed from Göttingen's well-endowed University Library, it becomes possible to form a quite detailed picture of Riemann's varied intellectual interests during the years he lived there [Neuenschwander 2022]. Even a cursory glance through this list immediately reveals that Riemann was deeply immersed in ongoing research in physics, beginning no later than 1849, when he began studying under Wilhelm Weber. Naturally, he also paid close attention to publications in the organs of the Paris and Berlin academies, but mathematics was only one of his many interests.

Riemann's intellectual interests were very broad and overlapped with those of Carl Friedrich Gauss, who was first and foremost a professional astronomer and geodesist. Both subjects dominated Gauss's teaching activity and a great deal of his research as well. Those who chose to do doctoral work under him received very little direct guidance, and Riemann's case was no different in this respect. This was not so much a personality quirk as a reflection of the fact that Gauss came from an era that preceded the emergence of research schools in mathematics at the German universities. In this sense, Riemann was a transitional figure. During his very brief career, he attracted a handful of disciples who promoted his approach to function theory, though with limited success. It was only during the modern period, dominated by competing research schools at the Prussian universities, that Felix Klein and David Hilbert succeeded in creating a strong Riemannian tradition in complex analysis. By doing so, they added much luster to Göttingen mathematics.

For Riemann, as for other leading German mathematicians from the mid-nineteenth century onward, teaching and research were inseparable activities. Thus, it should come as no surprise that Riemann's lecture courses – in particular, the many different extrapolations of them that appeared over several decades after his death – provide important insights into his key motivating ideas. Trying to understand his mathematics as presented in his (few) published works – loaded with terse explanations and abstract formulations – can seem like a hopeless undertaking. Those who had the opportunity to attend his lectures often found them difficult to follow as well, but he made every effort to make himself understood and succeeded in improving his presentations over time. Only in rare instances did Riemann teach a course that was part of the canonical Göttingen curriculum. Moreover, aside from such occasional offerings, his entire teaching career centered on two quite closely related areas: 1) complex analysis and 2) mathematical physics. The former was largely his own unique invention, though he drew heavily on the works of Augustin-Louis Cauchy as well as on Gaussian ideas. The latter area was represented by two courses, one on partial differential equations with applications to physics, the other on potential theory in connection with gravity, electricity, and magnetism; both were closely akin to courses also taught by Gustav Dirichlet.

Thanks to the diligent efforts of several historians of mathematics, I have been able to draw on earlier studies, while taking advantage of little known source material. Many of the papers in Riemann's scientific estate have also been digitized in recent years, making for easy access to sources supplementing those found in *Riemann's Werke*. From his letters to family members and other third party correspondence, we can easily see that Riemann, despite some lonely phases, was anything but a loner type. He had many acquaintances and a few key supporters, but there were also others whom he viewed as potential rivals. Was he a typical hypochondriac, as many thought, or just prone to sickness like many in his family? Richard Dedekind had to face that question when he wrote his oft-cited biographical essay about the events in Riemann's life. For Riemann, the deaths of his father and younger brother in the mid-1850s were heavy blows that led him to withdraw socially and become increasingly morbid. Outward recognition could never overcome the suffering he felt after losing his parents and eventually all but one of his five siblings. One can truly say that Riemann was acquainted with grief.

Money was always a problem until he finally gained a full professorship in Göttingen in 1859. Honors and recognition swiftly followed, as leading mathematicians in Berlin and Paris feted him. In 1862, he married his older sister Ida's good friend, Elise Koch, who comforted him throughout the years when his health problems dominated all else. Riemann had been very close to his father, a Lutheran pastor, and his wife's background was identical to his own. Their shared religious faith thus served to strengthen their marriage. As we shall see, without Heinrich Weber's prodding, Dedekind would surely never have written his short biography of Riemann, which relied heavily on information he received from Ida and Elise Riemann.

## Göttingen in the Era of Gauss

Riemann's life and legacy have long been connected with Göttingen University, a special place in the annals of modern mathematics. In popular lore, this Göttingen mathematical tradition began with Gauss, passed through Dirichlet and Riemann, then came into full bloom in the era of Felix Klein and David Hilbert [Rowe 2004], and ended during the Weimar period, when Richard Courant reigned over a vibrant international community up until the Nazis destroyed it in 1933 [Rowe 2018, Part 5]. Courant, like Klein before him, devoted a great deal of effort to forging links between modern Göttingen mathematics and the works of Bernhard Riemann, whose ideas inspired a great number of distinguished mathematicians. Yet, however strong those intellectual links may have been, the gulf separating Riemann's world from what came afterward can hardly be exaggerated.

In order to grasp his life and times, we should recall some of the key institutions and cultural norms as well as leading personalities in Göttingen during the first half of the nineteenth century.[4] Göttingen's Georgia Augusta was not yet a modern research university in that era, which is not to say that scholarly research was held in low esteem. On the contrary, from the very beginning some of its faculty members were among the most productive scholars in the German states. They formed a small elite, however, recognized by their membership in the local Royal Society (*Königliche Gesellschaft der Wissenschaften*), which normally convened on Saturdays at 3 o'clock. These meetings were open only to its members, who were required to present papers in Latin – often short announcements of works they intended to publish later – in at least six sessions each year [Oesterley 1838, 94–95]. Public lectures only took place on rare occasions, in particular to honor a recently deceased former member, as was the case after Riemann's death in 1866 [Schering 1909/1990].

Göttingen's Royal Society had three sections: two larger ones representing, on the one hand, medicine and the natural sciences, on the other, the historical-philological disciplines, alongside a smaller mathematical section, which included astronomy and physics [Gierl 2004, 155–160]. Members were elected for life and they conducted the Royal Society's business on the basis of seniority; they also were expected to publish in the organs of the society, the *Göttingische Gelehrte Anzeigen* and the volumes of the *Commentationes societatis regiae scientiarum Gottingensis* (later superceded by the *Abhandlungen* series), which carried longer studies. Aside from the regular full members, younger scholars could be appointed as *Assessoren*, who had publication rights, but no others.

Riemann became an *Assessor* in 1856 and a full member in 1860. In that year, he joined Wilhelm Weber and Justus Ulrich in the mathematics section. At that time, there were eight members in the historical-philological section and ten representing the natural sciences. Like other similar institutions, the Göttingen Society also had foreign and corresponding members as well as sixteen honorary memberships (Goethe once held this title). After the mid-century, the

[4] For an overview of mathematics in Germany, see [Scharlau 1990] and [Schubring 1990].

importance of such societies and academies would decline rapidly as new forms of publication for specialized disciplines grew. Yet in Riemann's day, membership in such an institution was a sure sign one had arrived at the pinnacle of academic success.

Visitors who came to Göttingen had the opportunity to visit some of the university facilities: the astronomical observatory, botanical garden, museum, portrait gallery, cabinet of machines and models, collection of physical instruments, and the chemical laboratory. Among the university's varied attractions, its library stood at the top of the list; in Riemann's time it housed well over 300,000 volumes. Not that it was easy gaining access to them: official opening hours were from 1 to 2 during the week and from 2 to 4 on Saturdays and Sundays. Some books had to be studied on the premises, whereas all others could only be borrowed by submitting a permission form signed by a professor. According to the records of books Riemann borrowed from the university library, he was especially interested in works by Gauss – no surprise there – but also publications by French mathematicians. These included classics such as Cauchy's *Cours d'analyse* and Legendre's *Traité des fonctions elliptiques* [Neuenschwander 1981a, 86].

Göttingen University, which opened in 1737, has traditionally been regarded as one of the earliest modern institutions of higher education. It thus behooves us to consider its overall structure more than one hundred years later, if only to recognize its deep roots in the past. As elsewhere in the German states, the university had four faculties, a structure dating back to the Middle Ages. These were the traditional higher faculties – theology, jurisprudence, and medicine – plus the philosophical faculty, which in Göttingen had been founded with a status comparable to the other three. This innovation came about through the ambitions of the Hanoverian Monarch Georg August (George II of Britain), who wanted to establish a prestigious university which could attract talented sons of the country's nobility. Previously, many went to Halle University in Prussia to take their degrees, before returning to Hanover to pursue civil service careers. Members of Göttingen's four faculties were also civil servants, which meant their academic freedoms were circumscribed by the fact that they served at the pleasure of the King.

Only those who were full professors (*Ordinarien*) in the four faculties were entitled to manage their internal affairs. In many cases, these were older men who held the honorific title of *Hofrat*. Among those who taught mathematics courses, Gauss and Bernhard Friedrich Thibaut were *Ordinarien*, whereas Justus Ulrich taught as an associate professor up until his promotion to *Ordinarius* in 1831, the year before Thibaut's death. These lesser-known names were important, since they carried the brunt of the teaching load for decades. Gauss's courses were confined to standard topics in astronomy and geodesy. As director of the astronomical observatory, he acted as its unofficial theoretician; whereas the observational astronomer, Ludwig Harding, carried out the practical work. This division of labor was formalized much later, but during Riemann's lifetime the ambiguous status of Gauss's professorship as mathematician/astronomer led to conflicts and confusion. Following the death of Gauss in 1855, the physicist

Wilhelm Weber tried to coordinate scientific activities at the observatory, but with limited success. Matters finally came to rest in 1869 when Ernst Schering was promoted to full professorship and made director of the theoretical division of the observatory. In attaining that status, he effectively assumed Gauss's former position.

At the time Riemann entered Göttingen University in 1846, one could easily gain an overview of those in charge: its theology faculty had five *Ordinarien*, the law faculty six, the medical faculty nine, and the philosophical faculty, which was by far the largest, had seventeen members. Their names, given by order of seniority, were listed at the back end of the course catalogs, along with the names of other instructors with lesser titles. These catalogs, which contain much important information about the course offerings at the Georgia Augusta, were issued each semester from the mid-eighteenth century onward. If we turn to the philosophical faculty, the courses fell into the following fields of study (omitting the ubiquitous term *Wissenschaft*, roughly research field): philosophy, politics and economics, mathematics, natural sciences, history, literary history, fine arts, ancient cultures, oriental and ancient languages, and modern languages and literature. Naturally, some of these fields were more strongly represented than others, but one might well be astounded by the heavy emphasis on humanistic studies at the expense of the hard sciences. This apparent distortion, however, will come as no surprise to those familiar with the curriculum taught at the German Gymnasien during the nineteenth century and long afterward.

Carl Friedrich Gauss (Fig. 0.1) was nearly seventy by the time Riemann met him. Although still mentally sharp, his days as a creative mathematician were now over. Gauss first came to Göttingen as a student during the mid-1790s. Before then, he lived in his native Braunschweig (Brunswick) under the protection of its ruler, Duke Karl Wilhelm Ferdinand. He might have chosen to attend Braunschweig's local university in Helmstedt, but did not, and for a simple reason: books [Küssner 1979, 48]. Already as a pupil at the Collegium Carolinum in Brunswick, Gauss was reading Newton's *Principia*, a book he bought in 1794. Thanks to a stipend from the Duke, he gained the opportunity to spend three years pursuing his intellectual interests, which were by no means confined to mathematics [Goldstein et al. 2007]. Still, he knew that to become an educated and versatile mathematician he would need ready access to the current scientific literature. Helmstedt, though a much older university than Göttingen's Georgia Augusta, simply lacked the resources of the newer institution.

So Gauss had good reason to believe he would be pleased with the scholarly apparatus housed in Göttingen's university library. Traveling the 90 km by foot, he arrived in October 1795 for the winter semester. A short time later, he wrote to his former teacher, Eberhard A. W. Zimmermann, that he was impressed with the library holdings. In the meantime, he had already begun studying several volumes from the Proceedings of the Petersburg Academy, which contained numerous articles written by the ultra-prolific Leonhard Euler. Gauss thereafter made regular use of these and other works by Euler [Küssner 1979].

Gauss returned to Brunswick in 1798 and one year later took his doctorate in Helmstedt. His dissertation presented the first of his four proofs of the funda-

Fig. 0.1: Carl Friedrich Gauss (1777–1855) painted in 1840 by Christian Albrecht Jensen.

mental theorem of algebra [Gauss 1799]. He followed this with his magnum opus in number theory, *Disquisitiones Arithmeticae* [Gauss 1801/1966],[5] though his best-known accomplishments were in astronomy. Gauss broke into that field at the same time he invented the statistical method of least squares ([Stigler 1981], [Stigler 1986]); he then relocated the asteroid Ceres on the basis of just three observations [Gauss 1809/1857]. In Brunswick, Gauss continued to enjoy the protection of its ruler, though not for long.

As a field marshal in the Prussian army, Duke Karl Wilhelm Ferdinand commanded the troops at the decisive Battle of Jena-Auerstedt in 1806. There his forces were trounced and he himself mortally wounded by a musket ball. Napoleon thereafter dictated politics throughout this region in Germany. Having lost royal support, Gauss had to look for a regular academic position during the brief era when large parts of the German states were under French rule. He returned to Göttingen in 1807, where he was appointed full professor and director of the astronomical observatory, the position he held up until his death in February 1855. The city was by this time part of the newly established Kingdom of Westphalia, whose government demanded that Gauss pay two thousand francs as a war contribution; eventually the sum was paid by an anonymous donor.

Gauss's career spanned a period during which scientific academies were gradually losing their dominant position in the academic world. Their slow decline was accompanied by the rising importance of the grand écoles in France and the research-oriented German universities. The curriculum at the École Polytechnique under Gaspard Monge was dominated by higher mathematics courses, and several of its graduates became prominent figures in French mathematics. Gauss belonged to the mathematical sciences division of the Göttingen Royal Society of Sciences, joined in 1831 by the physicist Wilhelm Weber. During Gauss's early career, he published in Latin, often in the *Commentationes* series, which ended in 1841. Although Latin retained considerable importance during Riemann's time, the growing acceptance of German as a scholarly language led the Royal Society to replace this traditional series with a new one, its *Abhandlungen*. Gauss published in nearly every issue, which lent this publication great prestige. The *Abhandlungen* only appeared irregularly, usually in intervals of two to three years. Although largely a compendium of "local knowledge," it remained highly prestigious even in Riemann's days.

During his first ten years in Göttingen, Gauss worked in the old observatory, a converted fortification tower on the wall surrounding the city. This had been the working quarters for Göttingen's first astronomer, Tobias Mayer the elder, who was especially known for his detailed lunar maps and tables. As Stigler has shown, the latter were compiled using a statistical method similar to linear regression analysis [Stigler 1986]. Mayer's work addressed the problem of lunar libration, but it was also motivated by the more pressing technical problem of determining longitude at sea. For this, one needed either a reliable clock on

[5] For a long-range view of the importance of *Disquisitiones Arithmeticae* in the emergence of modern number theory, see [Goldstein et al. 2007].

Fig. 0.2: The New Astronomical Observatory in 1817, watercolor by Christian Andreas Besemann.

board or precise astronomical data, such as the position of the moon. Mayer's lunar tables, first published in 1752, were sufficiently accurate for determining longitude to within half a degree, as confirmed by the British astronomer Royal James Bradley. Great Britain, as the leading sea-power nation, had a huge stake in solving this technical problem, which naturally preoccupied Mayer as an employee of the House of Hanover.

Gauss was none too pleased with the state of the astronomical instruments he inherited from his Göttingen predecessors. After a long delay, though, he took up quarters in 1816 in the new classically designed observatory, where Riemann would later live (see Fig. 0.2). Throughout his career, Gauss was mainly known for his accomplishments as an applied mathematician. As a professional astronomer, he corresponded with other leading practitioners, including Wilhelm Olbers (Bremen), Friedrich Wilhelm Bessel (Königsberg), and Christian Schumacher (Altona). Bessel was on close terms with the physicist Franz Neumann, who together with Carl Gustav Jacob Jacobi founded the Königsberg mathematical-physical seminar. One of the hallmarks of Neumann's seminar was based on Bessel's methods of measurement and data reduction [Olesko 1991].

Like many other astronomers from this time, Gauss also worked on geodetic surveys. In this connection, he gained fame for his invention of a new instrument, the heliotrope. As he later described in a letter to Wilhelm Olbers, the inspiration for building it came to him one day in 1818 when he was in Lüneburg.

Looking northward, he saw a reflected beam of sunlight which formed a bright spot in a window at the top of the Michaelis Church tower in Hamburg more than 40 kilometers away. An unrelated, but nevertheless ironically parallel incident occurred more than twenty years later in May 1842, shortly after Riemann first began attending school in Lüneburg. As Dedekind reported, the young lad was deeply moved to see the rising clouds of smoke to the north during the great fire that destroyed much of the old city of Hamburg [Dedekind 1876a, 510].

As Erhard Scholz has shown, Gauss's surveying work paved the way for higher geodesy and his closely related, but purely mathematical research on the differential geometry of surfaces [Scholz 2004]. Using the heliotrope, Gauss could chart out large geodetic triangles, including the one with vertices at Hoher Hagen – Brocken – Inselberg. In speaking of precise measurement – a theme of great importance during the era of Gauss[6] – one should not neglect to mention the important role played by the makers of scientific instruments. The heliotrope and other geodetic devices were mainly built for Gauss by the firm of J. Philipp Rumpf, whose workshop was located near the city's northern gate, the Weender Tor. Rumpf did contract work and held the title of University Mechanic up until his death in 1833. His firm was then purchased by Moritz Meyerstein, who as a young man had done his apprenticeship under Rumpf [Hentschel 2005]. For the next thirty years, Meyerstein would deliver a steady stream of sophisticated scientific instruments, fulfilling orders presented to him by Gauss and Wilhelm Weber. These included an instrument first designed by Gauss, who played a major role in Alexander von Humboldt's project to study global fluctuations in the earth's magnetic field (about which, see below). Riemann would come to know the theory and praxis of this instrument – the bifilar magnetometer – very well (see Section 9.2).

At the time of Gauss's appointment, Göttingen belonged to the Kingdom of Westphalia, ruled by Jérôme Bonaparte, the youngest brother of Napoleon. This was the town and university Riemann's father came to know as a theology student; no doubt he was the first to tell his son about the legendary figure of Carl Friedrich Gauss. This period ended abruptly, however, with the War of Liberation of 1813 [Bühler 1987, 54–55]. Napoleon's defeat ushered in an era of conservative reaction, as the Congress of Vienna largely restored the earlier monarchies. The Confederation of German states that emerged in 1815 was, however, in significant ways very different from the former Holy Roman Empire that Napoleon had dissolved. All the tiny Catholic principalities in the Rhineland were gone and the largely Lutheran state of Prussia now acquired control of these substantial territories in the West. Hanover was restored and became a Kingdom, but its geographic situation had become precarious, wedged between the two parts of Prussia. In Riemann's time, its position weakened further, as after 1837 Hanover was no longer linked in a personal union with Great Britain.

In the meantime, pressure grew for expanded political rights in the form of constitutional monarchies, a movement Austria's Prince Metternich hoped

[6] See the essays in [Wise 1997], in particular [Olesko 1997].

to suppress. Since a good deal of this liberal agitation flourished at German universities, Metternich and other conservatives countered these developments in 1819 by enacting the Carlsbad Decrees. Despite this repression, however, calls for democratization and liberal constitutions could be heard in some of the German states after 1830, when the Bourbon King Charles X was toppled from power in France. The July Revolution put Louis Philippe, the "Citizen King," on the throne for the next eighteen years. His supporters, the Orléanists, continued to collide with Legitimists, who backed the House of Bourbon, as well as with Bonapartists. In short, cracks in the Metternichian had become ever more visible.

On the other side of the Rhine, in the Kingdom of Hanover things remained quiet up until January 1831, when violent protests broke out in the town of Göttingen. One of the ringleaders was Dr. Johann Hermann von Rauschenplat, a *Privatdozent* in the law faculty who initially protested against the attempted censure of a dissertation. This case and other grievances led to an armed takeover of the local government and the formation of a new municipal council run by students and local citizens. Von Rauschenplat was made head of its "National Guard," consisting of around 500 armed students. This short-lived affair ended when Hanoverian troops arrived to restore order, but the events of that time left a lasting memory. As a staunch conservative, Gauss was one of many in Göttingen who were appalled by these developments.

## Berlin in the Era of Humboldt and Dirichlet

Even if few in Germany looked upon France as a harbinger for Europe's political future, no one could overlook the fact that its scientific culture was the envy of the world. Alexander von Humboldt spent some twenty years in Paris as a member of the Institut de France, during which time he worked on assembling data he had gathered during his travels throughout the Western Hemisphere. This material would later serve as the basis for the lectures he delivered at the University of Berlin, beginning in 1827. Afterward, he elaborated on this material and produced his masterpiece, *Kosmos*, published in five volumes between 1845 and 1862. Riemann familiarized himself with at least parts of it.

Berlin's Friedrich Wilhelm University had opened in 1810 during the reform era that began after Prussia's loss of territories to Napoleon. One of its founding figures was Wilhelm von Humboldt, Alexander's older brother, who helped inaugurate its neo-humanist tradition. This went hand-in-hand with a broader reform of the elite secondary schools in Prussia, the Gymnasien, which imparted large dosages of Latin and Greek. Humboldt's neo-humanistic model sought to instill character by adopting educational ideals rooted in classical antiquity. Throughout much of the nineteenth century, Prussia's vaunted educational system remained somewhat singular, as no single model prevailed among the various states of the German Confederation.

Even among the Protestant northern states – our sole focus here – one can detect striking differences between their respective educational systems (on mathematics education, see [Schubring 2012]). The Prussian model, spearheaded by its new university in Berlin, became particularly famous for promoting a fusion of teaching with scholarly research. This dynamic, which R. Steven Turner dubbed the Prussian "research imperative" [Turner 1980], spread throughout the leading German universities after the founding of the Second Empire in 1871.[7] One of the key factors that led to the rise of research-oriented universities throughout the German states was the excellence of the elite secondary schools when it came to preparing young men for further study.

Riemann benefited greatly from the two years he spent studying in Berlin, which contrasted sharply from the time he spent as a student in Göttingen. Some of these differences reflected the fact that Hanover had not implemented the same type of state examination for teachers as the one introduced in Prussia in 1810. Those who, having passed this exam, became mathematics teachers at the Prussian Gymnasien obtained a status comparable with their colleagues, who taught ancient Greek and Latin. In Hanover, which introduced a state examination for teachers in 1831, even those who were qualified to teach the highest level courses in mathematics and natural sciences at the Gymnasium still had a lesser status, as those subjects were not viewed as part of the core curriculum [Schubring 2015, 7–8]. Riemann apparently never took this exam, as he had set his mind on a university career, a risky decision as shall be seen.

In 1834, the Vienna Ministerial Conference proposed new articles that became federal law in all member states. This required those states that had not yet instituted a university entrance qualification (*Abitur*) to do so immediately. The *Abitur* had, in fact, already been introduced in Hanover in 1829, but henceforth the certificate of academic preparation required for enrollment was to include an assessment of the student's moral conduct. This background suggests that state authorities adopted these measures not only to raise educational standards but also to prevent politically undesirable elements from gaining admission [Wolter 1987]. These circumstances serve as a reminder that the *Studiosus* Bernhard Riemann lived during an era when authorities kept a watchful eye on political activities at some German universities.

Unlike the natural sciences, mathematics instruction fared relatively well within the reformed Gymnasium curriculum. The subject was taught, however, in a manner sharply opposed to the prevailing trend in France. Rather than cultivating utilitarian methods, the German schools emphasized formal procedures and logic, both of which figured prominently in philosophical studies. Even those Germans whose background and interests favored applied mathematics, like the civil engineer August Leopold Crelle, found themselves unable to resist the trend toward purism. Thus, Crelle's *Journal für die reine und angewandte Mathematik*, which he founded in 1826, was dominated from the outset by pure mathematics. Its first volumes carried works by Abel, Jacobi, and Steiner that

[7] The impact of the German universities on higher mathematics in the United States was documented in [Parshall/Rowe 1994].

set the tone for what followed [Rowe 1998]. This eventually led some to joke that Crelle should have called his publication *Journal für die reine unangewandte Mathematik* (Journal for pure *unapplied* Mathematics) [Klein 1926, 95].

A noteworthy feature of "Crelle's Journal" during the years 1826–1855, when it appeared under his name, was the division of its contents into the two main categories of pure and applied works. Initially, Crelle adopted the curious convention of subdividing the far larger first category into three fields: 1. analysis, 2. geometry, and 3. mechanics. Since relatively few submissions fell into the third area, however, Crelle soon dropped it; afterward papers in mechanics appeared within the much smaller second category of applied mathematics. This more conventional division of pure mathematics into just two fields was rooted in a tradition that drew a sharp epistemic line between analysis and geometry, the former being rooted in the concept of number, the latter deriving from the study of figures. One finds this same classification in the thirteen books of Euclid's *Elements*. Still, by the time Carl Wilhelm Borchardt took over as editor, Crelle's rubric looked increasingly antiquated. Indeed, by the time Riemann published his famous four-part article in Volume 54 of the journal this classification had disappeared altogether; Riemann's study appeared simply as numbers 11 to 14 on a running list with all the other contributions.

Despite the long-term trend toward purism, during the 1820s German mathematicians could ill afford to ignore the vital new impulses that issued from Paris, without which the sudden ascendance of higher mathematics east of the Rhine would have been unthinkable. Alexander von Humboldt spent nearly two decades living in Paris before he returned to Berlin in 1826. As one of the most socially active scientists of his time, von Humboldt also cultivated ties with some of the era's leading mathematicians. Among his most important Parisian discoveries was a young German, Gustav Lejeune Dirichlet, who spent the years from 1822 to 1826 studying there. Over the course of that time, Dirichlet enjoyed the support of Joseph Fourier and Siméon-Denis Poisson at the Collège de France and the Faculté des Sciences of the Sorbonne. By 1825, Humboldt hoped to use his influence with ministerial officials to create a position for Dirichlet at a Prussian university. Despite his many years abroad, Humboldt still enjoyed much prestige at the Prussian court of King Friedrich Wilhelm III, as this case soon revealed. Given Dirichlet's importance for Riemann's development, we have good reason to follow his teacher's early career a bit further.[8]

Johann Peter Gustav Lejeune Dirichlet (Fig. 0.3) was born in 1805 in Düren, a town in the Rhineland near Aachen. Although he spent his formative years in Paris, nothing excited Dirichlet's interest so strongly as Gauss's great work on number theory, *Disquisitiones Arithmeticae*. Dirichlet's later successor in Berlin, Ernst Eduard Kummer, aptly described that attraction:

> This [work] exercised a much more significant influence on his whole mathematical education and development than his other Paris studies. Rather than having merely read through it once or even several times, his whole life long, over and again, he never stopped repeatedly studying the wealth of deep mathematical thoughts it contained.

[8] On archival sources relating to Dirichlet's career, see [Schubring 1986].

> That is why it was never placed on his bookshelf, but rather always lay on the table at which he was working. One can well imagine the exertion it must have cost him in analyzing this extraordinary work, considering that more than twenty years after it had appeared there was still no one alive at that time who had studied all of it and understood it completely. ... Dirichlet was the first not only to understand this work completely, but also to have made it accessible to others, in that he made its rigid methods, behind which the deepest thoughts lay hidden, fluid and transparent, while replacing many of the main points by simpler more genetic ones, without compromising the complete rigor of the proofs in the slightest. He was also the first to go beyond it and reveal the rich treasures and still deeper secrets of number theory. [Kummer 1860, 315–316]

Humboldt, who was about to return to Prussia after his prolonged stay in Paris, had learned that leading French mathematicians held this young Rhinelander in high esteem. Still, nothing would carry as much weight in Berlin as a letter of recommendation from Gauss, who was well aware of Dirichlet's work. On May 21, 1826, Humboldt wrote him as follows:

> As you know, I cannot pretend to have a serious opinion when it comes to the higher regions of mathematics, but I do know through the great mathematicians that Paris possesses, and especially through my oldest friends Fourier and Poisson, that Herr Dirichlet has by nature the most brilliant talent, that he is progressing along the best Eulerian paths, and that one day Prussia will have in him (he is barely 21 years old!) an outstanding professor and academician. Grant my young friend, whose fortune interests me dearly, the protection of your great name. [Biermann 1959, 13]

After receiving Humboldt's entreaty, Gauss wrote to his former student, the Berlin astronomer Franz Encke. The latter then urged the Prussian Ministry to create a position for Dirichlet before the French won him for themselves. Speaking of Gauss's high regard for Dirichlet's work, Encke added:

> In my eyes what gives this opinion of Gauss such high value is that this unique man has always sharply distinguished between works that are worthy through their diligence and those of true genius, and that so long as I have had the fortune to know him, he has never spoken of anyone with such warmth, however respectfully he may have spoken of the accomplishments of others. [Biermann 1959, 14]

Soon afterward, in September 1826, Dirichlet received a very friendly letter from Gauss. Reading Dirichlet's work had clearly given him pleasure, as he recalled his own special love for number theory:

> It is all the more pleasing to me that you have a great attachment to that part of mathematics that has always been my favorite field of study, however seldom I have pursued it. I dearly wish you a situation in which you will have as much control over your time and the choice of your work as possible. Immediately after the appearance of my *Disquisitiones*, I myself was very much hindered by other business, and later by my external circumstances, from following my inclinations to the degree that I would have wished. [Dirichlet 1889/1897, 2: 375]

Gauss went on to say that he had originally planned to write a sequel to the *Disquisitiones*, but had now given up that idea. In the future he would only publish occasional memoirs on number theory, including a three-part study dealing with biquadratic residues, the first of which would appear shortly. "The main materials for the rest," he added, "as well as for the related theory of cubic residues, are essentially finished, although little of it has yet been written up adequately" (*ibid.*). In fact, Gauss had already announced the first of these works, his *Theoria Residuorum Biquadraticorum, Commentatio prima* [Gauss 1828a], in the April 11, 1825 issue of the *Göttingische Gelehrte Anzeigen* [Gauss 1825b]. Dirichlet eventually found and read that notice, which spurred him to undertake his own investigation of the main problem Gauss had described therein.[9] The full article would only appear in 1828, so Dirichlet could only work from Gauss's brief remarks in [Gauss 1825b]. In [Gauss 1831], he finally announced his sequel study, though this time with unusually forthcoming remarks that would later attract Riemann's attention (see Section 10.2).

Although Dirichlet would later become famous for his contributions to many fields, his greatest and most influential work was in number theory. In 1837 he published his first paper on analytic number theory, followed by an important sequel two years later [Merzbach 2018, 105–108]. The latter presented his proof that every arithmetic progression $an + b$ (with $a$ and $b$ relatively prime) contains infinitely many prime numbers (see Section 8.2). The following year he published the first installment of a three-part article entitled "Recherches sur diverses applications de l'analyse infinitésimale à la théorie des nombres," a major contribution to the field of analytic number theory [Merzbach 2018, 145–148].

One might be inclined to assume that, with the backing of von Humboldt and Gauss, Dirichlet encountered smooth sailing during his early career; but such was not the case.[10] For one thing, he had attended the Jesuit College in Cologne only up to age 16, which meant that his knowledge of Latin was somewhat deficient. He spoke excellent French, of course, but to teach at a German university required Habilitation. Although Dirichlet could easily submit a thesis, to do so he needed to hold a doctorate. Minister Altenstein was prepared to appoint Dirichlet in Breslau with a modest salary of 400 Thaler. Altenstein apparently assumed that the young mathematician lacked a doctoral degree, as he advised him to request this title from the faculty at the newly founded University of Bonn. Dirichlet's application met with resistance, however, since he was not prepared to hold forth in a Latin disputation. Such traditions died hard, as Riemann would later learn when he applied for dispensation from this requirement: to his dismay the faculty rejected his petition (see Section 2.5).

To circumvent this requirement in Dirichlet's case, the Bonn faculty instead awarded him the title *doctor honoris causa*, thereby opening the way for his Habilitation in Breslau. Once there, however, Dirichlet's problems only grew more complicated, as the faculty demanded that he not only submit a Latin thesis but also acquit himself in a public Latin disputation. When doubts arose that Dirichlet would meet these requirements in a timely fashion, the Dean decided

[9] For details on what happened afterward, see my article "Gauss, Dirichlet, and the Law of Biquadratic Reciprocity, *The Mathematical Intelligencer* 10(2)(1988): 13–25; reprinted in [Rowe 2018, 29–40].

[10] The events recounted here were first described in [Schubring 1984]; see also [Merzbach 2018, 28–36].

Fig. 0.3: Gustav Lejeune Dirichlet (1805–1859)

to formally admonish him for shirking his duty. This action provoked a response from the natural philosopher Heinrich Steffens, a native Norwegian who wrote in Dirchlet's defense: "No teacher at a university, no philologist will awaken so strange an image before students as when one may be allowed to evaluate a mathematician according to his competence in speaking Latin" [Merzbach 2018, 36]. Dirichlet's ordeal in Breslau reflects the kinds of obstacles that could stand in the way of even the most talented young candidates. His benefactor, Alexander von Humboldt, had considerable influence, but neither he nor the Prussian Ministry of Culture had any say over a faculty's handling of a Habilitation procedure. Luckily, Dirichlet's stint in Breslau was brief, as he took a leave of absence in 1829 to begin teaching in Berlin, first as a private lecturer, then advancing to associate professor in 1831, and finally to Ordinarius (full professor) in 1839.

Similarly arcane regulations were still in place during Riemann's day, so a few general remarks about the granting of academic degrees will not be out of place. If we step back in time, it becomes easier to appreciate how some of these academic traditions evolved. By the beginning of the nineteenth century, the traditional masters degree (M. Art.) had largely fallen to the wayside, replaced by the Dr. Phil. (forerunner of the modern Ph.D.). The doctoral degree had long been conferred by the higher faculties – theology, law, and medicine – whereas the propaedeutic arts and philosophy faculty had earlier only issued the masters title.[11] To receive the title of Dr. Phil., students normally submitted a dissertation, usually around 50 pages, containing original research. If this work was deemed acceptable by the faculty, then the candidate was admitted to the next stage of the procedure.

During the early decades of the nineteenth century, the number of doctoral degrees grew, though some were conferred without a dissertation. When this occurred, the candidate was promoted on the basis of his performance in a public dispute, in which he defended certain theses, submitted in advance, against the counterarguments set forth by (usually) two opponents. The disputation had long played a central role during the Middle Ages and beyond, and even during the nineteenth century candidates were often required to debate in Latin. Those who submitted a dissertation were also normally required to hold forth in a disputation, as did Riemann (see Section 2.5).

Humboldt's involvement in Dirichlet's early career coincided with another initiative that would have profound importance for the mathematical sciences in Göttingen. As a leading natural scientist, he had long been interested in the phenomenon of terrestrial magnetism, and for many years he took measurements of magnetic intensity during his far-flung travels in the Western Hemisphere as well as in Europe [Chapman 1962]. These findings led him to correctly conclude that magnetic intensity increased at higher latitudes, a finding Gauss was later able to account for in [Gauss 1839]. Humboldt also studied regular and irregular magnetic fluctuations, including large scale magnetic storms. When he returned to Berlin in 1827, he began making plans to expand these studies as part of an international scientific project.

His timing could not have been better, as he had been charged with organizing the next meeting of the Society of German Scientists and Physicians to be held in Berlin in 1828. Since its founding in 1822 in Leipzig, this society met each year in September, but in a different city each time. Humboldt thus set about making plans for the Berlin meeting to be a true spectacle that would long be remembered by the 400 scientists who took part in it. To ensure that key participants would get to know one another personally, Humboldt invited Gauss and several others to arrive a week before the official meeting took place.[12] It was in connection with this scientific conference that Gauss and Dirichlet both got to know Wilhelm Weber, who afterward spent several weeks in Berlin attending

[11] In 1810, the higher faculties in Leipzig petitioned the Saxon Ministry in an effort to block the issuance of doctoral degrees by the philosophical faculty, and they won! [Clark 2006, 195].

[12] Daniel Kehlmann re-imagined Gauss's journey from Göttingen to Berlin in the opening chapter of his novel *Measuring the World*. Ute Merzbach gave a detailed account of this meeting in her biography of Dirichlet [Merzbach 2018, 49–53].

Dirichlet's lectures. In effect, the die was hereby cast for what would soon unfold in Göttingen, including developments that would profoundly shape Riemann's early career.

In the meantime, Humboldt had persuaded his friend, the banker Abraham Mendelssohn Bartholdy, to allow him to build a magnetic observatory on the premises of the latter's mansion, located at Leipziger Strasse 3 in Berlin. This building was constructed completely in copper by the famous Berlin architect Karl Friedrich Schinkel. Since observations needed to be made nearly round the clock, Humboldt already faced a major challenge, namely to organize a scientific team of volunteers willing to sacrifice their time for this pursuit. One of its members was Dirichlet, who at first assisted Humboldt in monitoring the magnetic readings; this went on from February 1829 to November 1830.

Dirichlet was throughout this time still on official leave from Breslau. He supplemented his income by teaching mathematics to future officers at the Military Academy, where his services were appreciated. Over the course of this time, he had the good luck to receive frequent invitations from the Mendelssohn Bartholdy family, one of the most socially active in the city. Their two oldest children, Fanny and Felix, were now young adults on the verge of fame in the world of music.[13] The younger daughter, Rebecka, had her share of admirers, too, including the brilliant philosopher Eduard Gans, a celebrated professor of law in Berlin. In 1827, Gans had helped found the *Berliner Jahrbücher für wissenschaftliche Kritik*, official organ of the like-named Hegelian Society, which Dirichlet joined in that year while still teaching in Breslau. Whether or not he ever met the great Georg Wilhelm Friedrich Hegel himself, who died in 1831, is unknown. He most definitely knew Gans, however, and eventually they both vied to win over Rebecka, who surprised many by choosing the modest and soft-spoken Dirichlet over the more charismatic Gans. Their happy marriage began when they wed on 22 May, 1832 [Merzbach 2018, 61–64].

Dirichlet's arrival in the Prussian capital marked the onset of a new era in which Berlin University began to play a conspicuous part in the flowering of a distinctly German mathematical culture. Dirichlet was joined in 1834 by the Swiss geometer Jacob Steiner and then, ten years later, by the dynamic Carl Gustav Jacob Jacobi, who had previously founded a productive school in Königsberg. In Berlin, he became King Friedrich Wilhelm IV's unofficial "court mathematicus" through his appointment to the Prussian Academy. Together this trio of brilliant individualists quickly brought Berlin to the attention of the

[13] Gert Schubring informed me of a family letter from February 1833, written to Julius Schubring (1806–1889), who was a close friend of Felix Mendelssohn Bartholdy; Schubring wrote the libretto for his Oratorio Paulus. Schubring and his bride-to-be were then living in Berlin, and while he was away she attended a masquerade ball thrown by the Mendelssohn family. In this letter, she described the costumes of various prominent participants: Felix dressed as a harlequin and Rebecka as a sack. She was also introduced to a mathematician who was dressed like a telegraph pole, being tall and thin. Since she had no idea how to spell his name, she wrote it phonetically as "Diryklee." This little anecdote confirms that Dirichlet's name was pronounced in Berlin with 'ik' rather than an 'ich' sound. Julius Schubring's son would later become a noteworthy expert on antique sites in Italy, where he met the Riemanns in 1862/63; see Section 4.5.

larger mathematical world. Young Bernhard Riemann got wind of this, too, and thus left Göttingen after only one year to study in the Prussian capital (see Section 1.4 and Chapter 2).

## Collaboration of Gauss and Weber

Fig. 0.4: Wilhelm Weber (1804–1891).

Gauss shared Humboldt's interest in geomagnetism, but not until 1831 did he take up this work in earnest. It was in that year that the much younger Wilhelm Weber (Fig. 0.4) joined him as professor of physics in Göttingen. Soon after Weber's arrival – a journey he undertook by foot from Halle – their shared interests in terrestrial magnetism began to bear fruit. Gauss had previously used spherical harmonic analysis in celestial mechanics, but he now found a way to adapt these same techniques to geomagnetism. On 15 December 1832, he delivered a lecture to members of the Royal Society explaining how one could measure the intensity of a magnetic field in such a way that this did not depend on the properties of the measuring instrument [Gauss 1832b]. Although this marked an important breakthrough, it came at the beginning of a period during which Gauss and Weber faced serious difficulties trying to coordinate their respective measurements.

Identifying changes in magnetic declination required nearly simultaneous observations and involved numerous comparisons of the results. To communicate these, they often sent messengers who had to scamper between their two institutes. It took them roughly fifteen minutes to cover the distance from Gauss's observatory to Weber's physical cabinet, which was then located on the grounds of what is today the old university library. This situation greatly hindered their work up until the spring of 1833. In the meantime, Weber overcame a number of practical problems before he succeeded in constructing a telegraph line that used galvanic currents to transmit messages. Others, including Gustav Fechner, had earlier suggested this possibility, but Weber was the first not only to build such an electric telegraph but to demonstrate how it could be used to send and receive signals over a long distance. Gauss described this invention in one of his early reports on terrestrial magnetism:

> We must not leave unmentioned ... a fantastic new device that we owe to Professor Weber. Already last year he connected the physical cabinet with the observatory by two wires spanned over the roofs of the town, a connection now extended to the magnetic observatory. If one includes the galvanometers at both ends, the galvanic current runs a distance of almost nine thousand feet.[14] [Gauss 1863–1933, 5: 524]

Installing this telegraphic line turned out to be a frustrating task due to constant breakages. It extended from the physics institute to a tower of the Johannes Church and then to the roof of a hospital before reaching the observatory (Fig. 0.5). Gauss and Weber then found a way to synchronize measurements of the magnetic field taken in their respective workplaces by utilizing a binary code for this telegraphic device [H. Weber 1893].

Gauss enjoyed a spectacularly productive collaboration with Wilhelm Weber. In 1834, they founded the Magnetic Union to coordinate measurements of the Earth's magnetic field at different localities, some of which employed devices modeled on those introduced in Göttingen. Beginning in that year they began collecting data on the earth's magnetic field based on observations made at several European stations. Alexander von Humboldt later helped make this into a truly international undertaking by linking the Magnetic Union with similar efforts in Britain and Russia. Before the British took the lead in this work, Göttingen served as the hub for this ambitious activity. Starting in 1836, Gauss and Weber documented these coordinated measurements of terrestrial magnetism in the *Resultate* series [Gauss/Weber 1837–43]. Riemann often studied these reports during the early 1850s when he was a postdoc in Göttingen. By that time, the British scientific community had set up stations in Greenwich, Dublin, Toronto, St. Helena, Cape of Good Hope, and Tasmania, with the British East

[14] One foot corresponded to 1/4 meter at this time.

Fig. 0.5: From a postcard commemorating the Gauss-Weber telegraph.

India Company adding four more in India and Singapore. Humboldt also persuaded the Russian Czar to build observatories across his vast territory, making it possible to draw up world-wide magnetic charts.

Although Göttingen quickly lost its former importance as a center for geomagnetic research, this resulted partly from political as well as scientific developments. The events of 1830–31 temporarily emboldened the liberal movement in Hanover, whose followers rejoiced when King Wilhelm IV decided in 1833 to introduce a new constitution that effectively ended censorship, while safeguarding academic freedoms. This proved to be a short-lived victory, however, as Wilhelm IV died in 1837 without leaving a male heir, making him the last Hanoverian to sit on the British throne. The new King was Ernst August, Wilhelm's brother and a notorious autocrat.

In November 1837, King Ernst August annulled the new constitution, a major setback for the liberal movement. As civil servants, the members of Göttingen's faculty had sworn their allegiance to it, and several felt bound by their oaths. Seven of the more prominent among them openly protested the King's decision, including Gauss's collaborator and his son-in-law Heinrich Ewald (Fig. 0.6). The King responded by dismissing all seven from their academic positions and expelling the three he regarded as the ringleaders of the "Göttingen Seven"

Fig. 0.6: The Göttingen Seven: above, the Brothers Grimm, below left to right, Wilhelm Weber and Heinrich Ewald.

from his land.[15] Sympathy for the victims of this purge spread throughout the German states and eventually all seven gained appointments elsewhere. The damage to the university's reputation would take much longer to repair, however.

Many wondered why Gauss remained silent, as insiders knew he was deeply disappointed over the dismissals of Weber and Ewald. Behind the scenes, he tried to negotiate a compromise, but to no avail [H. Weber 1893]. In the end, he refused to bite the hand that fed him, knowing that even he was a mere *Hofrat* – a court adviser, whose advice was rarely ever sought and certainly not during a contentious political confrontation. In the wake of this incident, enrollments fell dramatically. Throughout the 1820s, the student body had typically averaged around 1400 per year. By the early 1830s, however, enrollments were declining dramatically, in part due to the opening of Berlin's new university in 1810. By 1837 enrollments had fallen below 1,000, but in the wake of the "Göttingen Seven" incident, they fell even more dramatically.

Wilhelm Weber remained in Göttingen as a private citizen until 1843, when he gained a professorship in physics in Leipzig. There he continued to pursue research on terrestrial magnetism, whereas Gauss became increasingly despondent. In Weber's absence, research on geomagnetism was mainly pursued by Gauss's former assistant, Benjamin Goldschmidt. On May 21 1843, Gauss wrote this to his sorely missed friend:

> In the last two months I was quite busy with my own mathematical speculations. This took a lot of time even though I have not yet reached my first goal. I always got distracted by one direction or another, sometimes even by a will-of-the-wisp, as is often the case with mathematical speculations. I am very sorry that this activity and the need to write up a treatise for the Royal Society prevent me from contributing to another volume of the *Resultate*, in case you actually plan to continue with it.[16] Generally after your separation and the loss of your help, I have been losing interest in magnetism. [H. Weber 1893, 77-78]

These were the prevailing scientific circumstances in Göttingen when, on 25 April 1846, Riemann enrolled as a *Studiosus* in philology and theology. By this time, the number of students was only around 560, and Riemann would not remain long either. His family was well aware that his real ambition was to study mathematics, and his father thankfully gave him permission to continue pursuing that subject at the University of Berlin in 1847. Following the tumultuous political events of 1848, Weber was reappointed to his physics chair in Göttingen. His return practically coincided with Riemann's own after spending two years in Berlin, where he absorbed a great deal of higher mathematics from both Dirichlet and Jacobi. Soon thereafter, Riemann attached himself to Wilhelm Weber and his circle, while continuing his already strong interest in research on terrestrial magnetism.

---

[15] One of the three was Jakob Grimm, who settled in Kassel and was later joined there by his younger brother Wilhelm. The Brothers Grimm were the most famous members of the "Göttingen Seven."

[16] Weber would, in fact, cease publishing this series [Gauss/Weber 1837–43], which Riemann later diligently studied as a student.

For ease of reference, the timeline below summarizes the main events in Riemann's life.

## Timeline of Riemann's Life

1737 – Opening of Göttingen University under auspices of Hanoverian Prince-Elector Georg August, otherwise known as George II, King of Great Britain

1807 – Carl Friedrich Gauss appointed professor of astronomy in Göttingen.

1810 – Founding of Berlin University during the Napoleonic period following the educational reform principles of Wilhelm von Humboldt.

1813 – Riemann's father, Friedrich Bernhard, interrupts his studies in Göttingen to take part in the wars of liberation that culminate with Napoleon's defeat at the Battle of Leipzig.

1826 – Bernhard Riemann is born on 17 September in Breselenz as the second child of Friedrich Bernhard and Charlotte Riemann née Ebell.

1833 – Family moves to vicarage in Quickborn; Riemann and his five siblings are educated at home by their father.

1837 – Physicist Wilhelm Weber and six of his colleagues – the "Göttingen Seven" – are dismissed from their professorships for refusing to abandon their oaths to the liberal constitution, which King Ernst August had annulled.

1840 – Riemann lives with his grandmother, Luise Ebell, in Hanover while attending the city's Lyceum.

1842 – Following the death of his grandmother, Riemann transfers to the Johanneum Gymnasium in Lüneburg.

1843 – Death of Riemann's mother.

1845 – Riemann graduates from the Lüneburg Gymnasium with excellent grades.

1846 – He enrolls as a student of theology at Göttingen University, but already plans to switch to mathematics.

1847 – Riemann begins two years of study in Berlin, taking courses offered by Dirichlet, Jacobi, and Eisenstein among others.

1848 – Political demonstrations in Berlin are suppressed by the Prussian army leading to over 200 deaths.

1849 – A delegation from the Frankfurt Parliament arrives in Berlin, but King Friedrich Wilhelm IV refuses to accept its proffered imperial crown. Riemann returns to Göttingen, where Wilhelm Weber has now been reappointed as professor of physics.

1850 – Riemann joins the newly formed Seminar for Mathematics and Physics; Richard Dedekind begins studies in Göttingen.

1851 – Riemann receives his doctorate with a dissertation on the foundations of complex analysis, introducing Riemann surfaces for the first time.

1852 – Dirichlet visits Göttingen and gives Riemann helpful assistance for his postdoctoral dissertation on trigonometric series.

1854 – Riemann's postdoctoral dissertation is accepted for Habilitation; his lecture on "Hypotheses underlying the Foundations of Geometry" will later be regarded as his most famous single work. Both works remained unpublished during Riemann's lifetime. Dedekind habilitates immediately after Riemann.

1855 – Death of C.F. Gauss, Riemann's father Friedrich Bernhard, and his sister Clara at age 24. Gustav Dirichlet is appointed as Gauss's successor.

1855 to 1859 – Rebecka Dirichlet brings Berlin salon culture to Göttingen, where the Dirichlet home becomes a lively meeting place for academics and musicians. As a talented pianist, Dedekind soon becomes a favorite of the family.

1857 – Death of Riemann's brother Wilhelm and his sister Marie. Riemann publishes his famous paper on Abelian functions and is appointed associate professor.

1858 – Riemann's two remaining sisters, Ida and Helene, come to live with him in Göttingen. Three Italian mathematicians – Betti, Brioschi, and Casorati – visit Göttingen. Dedekind leaves to accept a professorship at the Zurich Polytechnic. Death of Rebecka Dirichlet.

1859 – Gustav Dirichlet succumbs six months after his wife's death. Weierstrass turns down Dirichlet's chair and Riemann becomes his successor. He and Dedekind visit Weierstrass and Kronecker in Berlin. As a newly appointed corresponding member of the Berlin Academy, Riemann offers his short paper on the properties of the zeta function, including the later famous Riemann hypothesis.

1860 – Riemann spends several weeks in Paris, meeting many distinguished mathematicians there, including Charles Hermite. This is likely when he learned about the prize problem on heat conduction posed by the Paris Academy.

1861 – Riemann submits an entry for the Paris Academy's contest, but his solution is deemed insufficiently complete to be awarded the prize. He offers lecture courses on partial differential equations, gravity and electromagnetism, and complex analysis (Abelian and elliptic functions). His student Karl Hattendorff prepares extrapolated texts from all three courses.

1862 – in July Riemann marries Elise Koch, a friend of his older sister Ida. One month later he contracts pleurisy and decides to spend the winter in Italy to recover. They depart from Marseilles in November and arrive in Messina, where they are warmly received by the family of the German Consul Julius Ewald Jaeger.

1863 – Riemann and his wife stay in Pisa, where he again meets Betti. He is offered a professorship in Pisa, but declines due to his health condition. After returning to Göttingen in June, they undertake a second trip to Pisa in August, together with Riemann's sister Helene, who becomes ill. Birth of their daughter Ida in December.

1864 – Pisa experiences a harsh cold winter. Helene's condition worsens and she dies the following summer. The Riemanns decide to remain in Italy through the winter.

1865 – in Pisa Riemann's condition grows worse. He spends time with Enrico Betti and his colleagues; his former student Friedrich Prym visits for a few weeks. In October they return to Göttingen.

1866 – in mid-June, shortly before the outbreak of war between Hanover and Prussia, they depart again for Italy. Riemann's condition quickly worsens and he dies on 20 July in the village of Selasca on Lago Maggiore. Ernst Schering delivers a eulogy before the Göttingen Scientific Society (first published in 1909).

1867 – Dedekind recovers Riemann's thesis and lecture from his Habilitation in 1854. Riemann's dissertation from 1851 is also republished in Göttingen. Schering again memorializes Riemann in the *Göttinger Nachrichten*; he sends this text and his earlier eulogy to Hermann von Helmholtz.

1868 – The Göttingen Scientific Society publishes Riemann's postdoctoral thesis and his Habilitation lecture. Soon afterward, Helmholtz reacts to Riemann's lecture.

1870 – Elise Riemann approves the plan to publish Riemann's Collected Works under the editorship of Alfred Clebsch.

1872 – After making substantial progress on this project, Clebsch suddenly dies in November.

1876 – The first edition of Riemann's *Werke* is published under the editorship of Heinrich Weber, assisted by Richard Dedekind.

# Part I
# Riemann's Life

# Chapter 1
# Riemann's Formative Years, 1826–1847

## 1.1 Home Life in Quickborn

Georg Friedrich Bernhard Riemann was born on September 17, 1826, in Breselenz, a remote village in the Kingdom of Hanover not far from the town of Dannenberg. This scenic region, which lies in the meadowland west of the Elbe River, has changed little even to this day. Bernhard was the second of the six children of Friedrich Bernhard Riemann and his wife Charlotte, née Ebell, whose father, Georg August Ebell, had held the title of Court Counselor in the city of Hanover. Friedrich Bernhard hailed from Boitzenburg on the Elbe. He had studied theology in Göttingen, but although a man of God he apparently had no steadfast pacifist leanings. Otherwise it would be hard to account for the pride he took in serving as a lieutenant during the Wars of Liberation of 1813. Modern German history essentially began with the defeat of Napoleon's armies, thereby ending French dominance on the continent of Europe. In the wake of the conference of Vienna, the former state of Hanover was restored in 1814 and the British monarch who ruled over it, George III, now took the title King of Hanover (which had previously been an Electoral Principality in the Holy Roman Empire).

The elder Riemann had begun studying theology in 1811, so he was surely one of the many who dropped out in order to take up arms. His son later recalled these circumstances in a conversation with the Berlin astronomer Johann Franz Encke, who seemed to have remembered Riemann's father from the years 1811 to 1813 when Encke was studying under Gauss in Göttingen. Back then, Encke had taken a break from staring at the nighttime sky, hoping to make a small contribution toward Napoleon's defeat. Military life held many attractions for young men in those days, a tradition dating back to medieval times, if not earlier. Manly virtues have long been identified with bravery on the battlefield, especially when thinking of those who lost their lives fighting for some noble cause. For those survivors who lived to pay homage to their fallen comrades in arms, nothing could have seemed more dishonorable than to doubt that their cause was just. To believe otherwise would seem a mockery of the



D. E. Rowe, *Bernhard Riemann: His Life and Wondrous Mathematical Legacy*,
https://doi.org/10.1007/978-3-032-25457-3_2

human condition. Yet, by the end of the century, when modern technology had dramatically changed the character of warfare, many had come to question the notion that death on the battlefield was ennobling. Such was not the case during Bernhard Riemann's lifetime, however. He, like nearly everyone else, considered traditional military virtues to be practically sacrosanct. Though never explicitly articulated, his views accorded with the Biblical phrase: "the Lord giveth and the Lord taketh away," as illustrated by the story told in Job 1:21.

Little is known about the genealogy of the Riemann family, other than that Bernhard's grandfather, Johann Christian, had been a wine merchant and member of the council of commerce in Boitzenburg. He married a woman from Hamburg in 1773 and died of gout some twenty-five years later at age 54. Nothing is known as to whether they had other offspring or about their financial situation after the birth of Friedrich Bernhard in 1789. We are better informed regarding Riemann's mother, who came from a distinguished family with strong connections to the Lutheran church. Her paternal grandfather, Georg Wilhelm Ebell, had been the Abbot of the Cistercian Monastery of Loccum, which became a Protestant institution in 1555 as a result of the Treaty of Augsburg. His son, Georg August, eventually became a government official in Hanover. He was well-educated and known for his book exposing the dangers of lead poisoning. In 1779, he married Luise Magdalene Schlemm, who was some twenty years younger than he. Ebell and his wife had several children, a number of whom died at a young age, as was then common. Their daughter Charlotte was born in 1786; a second daughter, Julie, was two years older. She later played a significant role in Bernhard Riemann's life after his mother's death.

During the 1760s, Charlotte Ebell's father had studied law in Göttingen, during which time he met and befriended the physicist and aphorist Georg Christoph Lichtenberg. It was during a trip to England in 1770 that Lichtenberg met King George III, an amateur astronomer who took him to his newly built Royal Astronomical Observatory. Lichtenberg returned to Göttingen with a recommendation from the King that he be appointed associate professor, and so began one of the more colorful careers in the university's history. Lichtenberg remained friends with Ebell and other Hanoverian officials in his circle up until his death in 1799. Georg August Ebell had many connections with the city of Bremen as well and served for some twenty years as postmaster there. His grandson, Wilhelm Riemann, Bernhard's only brother, would later hold a position in Bremen's postal service.

When Bernhard Riemann's parents married on 11 June 1824, his father was 35 and his mother already 38.[1] The ceremony took place in the village of Hoya, where Charlotte's parents had been living in retirement. This must have been a bittersweet occasion, because Georg August Ebell had died only two months before the marriage took place. Both he and his wife were originally from Hanover, which is where Charlotte's mother later returned as a widow. This familial connection on the mother's side later became important, as Bernhard began attending the Gymnasium in Hanover while living in the home of his maternal

[1] This according to https://www.genetalogie.de/gg/alriemann.pdf).

grandmother. Her death in 1842 was what necessitated finding another school where he could continue his secondary education.

Soon after marrying, Friedrich Bernhard became pastor of the Lutheran church in Breselenz. One year later, Charlotte gave birth to their daughter Ida on 1 March 1825. She and her younger brother Bernhard were still very young when the family moved to the nearby village of Quickborn, some three hours northeast by foot, though still west of the Elbe. In Quickborn Pastor Riemann proved to be the dominant figure in the lives of his children, but especially so for his oldest son. Indeed, Bernhard's interest in learning was first awakened by his father, who tutored him until 1840, when the teenager left home to begin his schooling in Hanover.

The atmosphere in the Riemann home was pious, though high-minded, in line with the standards for middle-class families of that day. The children were deeply attached to each other as well as to their parents, who devoted themselves to their upbringing. The church's vicarage thus became Bernhard Riemann's beloved home, to which he gladly returned up until his father's death in 1855. His mother died already in 1843, when he was still a teenager. By then he was living in Lüneburg, where he attended the local Gymnasium. The oldest of the six children, Ida, would be the only sibling to survive her brother Bernhard. Moreover, what little we know about his early life came mainly from her, as she imparted that information to Richard Dedekind for his biographical essay [Dedekind 1876a].[2] There he described how as a five-year-old boy Riemann grew up deeply fascinated by history. He loved to hear about legendary stories from antiquity, but even more the story of Poland's tragic fate. Ida recalled how he would over and again ask his father to tell this story.

Following Dedekind's account, this phase in Bernhard's intellectual development went by quickly once his talent for arithmetic broke through.[3] Henceforth he derived the greatest pleasure from devising difficult problems and then challenging his siblings to solve them. By the age of ten, the children were taught arithmetic and geometry by a man named Schulz, who quickly realized that the second child of Friedrich Bernhard and Charlotte Riemann was a true *Wunderkind* when it came to problem solving. An anecdote from when he was eleven or twelve fits this picture perfectly. Like other children in their village, Bernhard had learned how to make handicrafts out of cardboard or other materials. One of the traditional activities for youngsters at Christmastime was to produce *Bastelobjekte* as decorative presents to give to their parents. One day a group gathered around to show off what they had made. When Bernhard's turn came, they could not believe their eyes: he had constructed a perpetual calendar that actually worked! [Riemann 1990, 851].

---

[2] Beyond the sources cited in Dedekind's biography, I have also drawn on assorted documents in the Riemann family papers held at the Staatsbibliothek Berlin, see [Neuenschwander 1988, 103].

[3] Dedekind wrote nothing about Riemann's lifelong interest in history and biography, a trait that surely had roots in the stories his father told when he was a child.

During the first decades of the nineteenth century primary-level education gradually became compulsory throughout the German states. Naturally, schooling took many forms and remained highly rudimentary in the many small villages sprinkled throughout rural areas. The churches often ran these elementary schools, whereas trade schools served to train young men in fields previously dominated by craft guilds. Co-educational institutions only emerged in the twentieth century, whereas schools for girls went no further than the tenth grade. Families that could afford to offer their sons and daughters some cultural education typically hired home tutors, a role Pastor Riemann apparently assumed himself. None of his four daughters would marry, however, which eventually meant that the two sons had to provide for them. Since almost nothing is known about the financial situation of the Riemann family, one can only infer from Bernhard's letters that providing for his education put them all under considerable strain. To judge from the few documents that have survived, he was often short of money, whether due to a sparsity of resources, mismanagement, or both.

For boys, pursuit of higher learning began by attending a Gymnasium, the secondary schools that prepared young men to pursue professional careers in medicine, law, theology, or as employees of a state government. The curriculum at these Hanoverian schools was dominated by classical languages: Latin and Greek. Some students – Riemann's case was by no means exceptional – studied Hebrew as well, a virtual necessity for those who wished to study theology. In Riemann's time, admission to a university was only possible by first attending a classical Gymnasium, though graduates were not always required to take a leaving examination, the forerunner of today's *Abitur*. In Prussian schools, this suddenly changed in 1812: those who wished to graduate thereafter had to complete six written exams, heavily weighted to test their knowledge of ancient languages and literature. They wrote essays in German, Latin, and French, translated texts in these languages into Greek, and wrote a paper in mathematics. Pupils who, like Riemann, wished to study theology and philology, also had to translate an Old Testament passage from Hebrew into Latin. A major incentive for students to achieve a high grade on these examinations was the chance this offered to be awarded a financial scholarship.

## 1.2 Struggles in School

Shortly before Easter 1840, Bernhard was confirmed by his father at the age of thirteen and one-half. He was then sent off to live with his grandmother in Hanover, where he entered the city's Lyceum. His father had surely done an excellent job in preparing him for this challenge, but he must have also realized that his son's natural shyness would make this a difficult transition. Many of the other boys had already been attending this school for three years or more, whereas Bernhard began in the fourth grade, the *Untertertia*; he was promoted to the *Obertertia* the following year. One counted the classes in classical Gymnasien backwards, so to speak, starting with what today is called grade five, then the *Sexta*. After this came *Quinta* and *Quarta*, followed by *Untertertia*. Thus, the last six years carried the prefix lower- and upper- for *Tertia*, *Sekunda*, and *Prima*.

It was not uncommon for young boys to skip the first three years, as Riemann's family decided he should do, but this did not mean he would automatically be placed in *Untertertia*, since placement was not based on age, but rather intellectual maturity. Thus, on 3 April he informed his mother that the school examiner gave him permission to enter *Untertertia*, whereas another boy the family knew was placed in *Quarta*. In a letter to his parents from 3 May 1840, he wrote about the eight different subjects he was required to take. These were Latin and Greek (6 hours each), German, Arithmetic, and Religion (4 hours), Geography, History, and Natural History (2 hours). Bernhard complained that he was finding Greek difficult and wondered whether he needed private lessons. He was also unhappy about having no lessons in French or real mathematics, just reckoning. He also wished there were more emphasis on history.

Throughout this first year, Riemann had to overcome feelings of loneliness, as he longed for nothing more than to be back home with his family in Quickborn. He knew, though, that the journey was both long and expensive, so all he could do was write regularly and ask to receive news from his parents and siblings. He regularly reported to his father about his grades in the individual subjects, which were more than satisfactory. Moreover, Bernhard always received praise for his diligence and obedient behavior. He got off to a bad start with his mathematics teacher, however, because he made the mistake of correcting him once in class. Luckily, this same teacher got over the incident and later showed his appreciation for the precocious newcomer [Schering 1867, 306]. He may have realized that this young charge was anything but a cocky showoff type, except on those occasions when he impressed others with his speedy grasp of a mathematical problem and how to solve it.

By the following year, Bernhard had largely adjusted to the school routines. He proudly informed his father about how he had received several high grades for his work in Latin. He must have been pleased that French was offered in the *Obertertia*, as he made an "Eins" in that subject on his Easter mid-term exams. In a letter to his father from February, he had complained about not having enough time to prepare for these exams, but the grades he received (all 1's and 2's except for a 3 in natural history) show that he had acquitted himself well. Oddly enough, he was only given a 2 in mathematics. Probably this reflected his lack of interest in the dry and repetitive subject matter taught. The next year, though, his enthusiasm for the subject burst out for all to see at a new school where his mathematics teacher helped him to realize his natural talent.

Bernhard's maternal grandmother, Luise Ebell, passed away around this time, thereby breaking the family connection with Hanover. His academic performance at its Lyceum was never a problem, but Riemann's leaving certificate also noted that he had been admonished at times for his childlike behavior (*kindisches Benehmen*) [Neuenschwander 1981a, 127]. In any case, the young lad was eager to leave and hoped he could attend a Gymnasium closer to the family home in Quickborn. The Johanneum in Lüneburg was roughly 50 kilometers away, a distance one could hope to cover by foot on a nice day. Not that the teenager could manage this trek easily. His mother wrote him loving letters full of encouragement during that first year in Lüneburg, but she also admonished him to take care not to overexert himself [Dedekind 1876a, 511]. No doubt she witnessed her son's exhaustion when he came home after walking all day. Sadly, she passed away the next year, the first in a long chain of deaths in the family. In the end, only Bernhard's older sister Ida would survive him. Lüneburg was an old Hanseatic city with a population of roughly 10,000 at this time. Its most famous architectural adornment was the St. Johannes Church (Fig. 1.1), constructed over a period of two centuries beginning in 1289. By the time it was completed, the city of Lüneburg had founded the Johanneum as a school for all social groups, at least in principle.

Fig. 1.1: St. Johannes Church in Lüneburg.

The Riemann family surely knew about the Johanneum in its contemporary form, i.e., as a prestigious Gymnasium connected with the long tradition of its northern Gothic-style cathedral. As impressive as the church itself was its baroque organ, first built in the mid-sixteenth century. For over thirty years, up until his death in 1733, Georg Böhm had been the organist and choir director at St. Johannes. One of Böhm's private pupils was a young man by the name of Johann Sebastian Bach, who came to Lüneburg in 1700 to complete his schooling. By the 1840s, however, Bach's compositions were only beginning to gain a wider appreciation,[4] so Bernhard Riemann probably never knew that the composer had spent two years in Lüneburg, where he had once played the organ in St. Johannes, an instrument the young Riemann undoubtedly often heard on Sundays.

Bach had attended the Michaelis Latin school, which closed in 1819. By that time, the Johanneum was still recovering from the period when Lüneburg was under French occupation. Initially, French troops confiscated the school building and converted it into a military hospital, though instruction continued in the homes of the teachers. Eventually, they returned to conduct their classes in the old building, but since this had suffered badly in the meantime, plans were made to build a new school. The building Riemann learned to know first opened in the school year 1828/29. It was located directly across from the north side of the church with the appropriate address: Bei der St. Johanniskirche 21. He arrived on April 1, 1842, and had no trouble passing the examination to enter the *Untersekunda*. Here he would spend the next four years, completing his secondary education and preparing himself to study theology in Göttingen, just as his father had done before him. During those first two years, his brilliance in mathematics became evident, but unfortunately his performance in two other principal subjects, Latin and German, brought him and his teachers to the brink of despair.

When looking back on the many problems Bernhard Riemann faced over the course of his life, one can detect certain character traits already apparent from his formative years at the Johanneum in Lüneburg. These traits suggest that some of the difficulties he later faced were not entirely circumstantial. This, to be sure, had nothing to do with his provincial background or a narrow intellectual outlook; on the contrary, Riemann's interests were always exceptionally broad and far-ranging. In fact, one might wonder whether his insatiable curiosity produced an adverse side effect: for throughout his life, Riemann only rarely focused his immense mental energy on a fixed topic of research. He loved to read, and he had an uncanny ability to absorb abstract ideas as well as technically demanding texts practically in no time. Intellectually restless, he was easily distracted and almost never well organized. He could certainly force himself to concentrate when needed, but his lifestyle had all the hallmarks of an impulsive personality. His family seems to have learned to accept his eccentricities, including the long periods when he voluntarily remained incommunicado. In later years, his letters invariably began with apologies for not having written earlier. As with much else, Riemann was a highly unreliable correspondent.

On the other hand, already as a teenager there was an unhealthy imbalance between the mental and physical sides of Riemann's daily life. Some of this

[4] A major breakthrough occurred in 1829 when Felix Mendelssohn Bartholdy directed a performance of Bach's St. Matthew Passion at the Sing Academy in 1829.

reflected his upbringing and the demands elite schools placed on the privileged few who attended them. But Riemann also suffered from a weak constitution that left his body prone to contracting various illnesses. During his school days, he was never sick for a prolonged period of time; still, he often wrote home about his health issues, complaining about sharp headaches or problems with his digestive system that caused him to miss classes. Later, at the university, he often suffered from severe constipation. Yet, beyond these physical maladies, Riemann suffered from an inability to take decisive action when such was called for, such as when he needed to monitor his living expenses. He was, in a word, one of those gentle souls who could never really manage the little things in life, and these eventually turned into very big matters that left him feeling helpless and paralyzed. Although he never seems to have articulated these feelings, one can hardly follow the events in his troubled personal life without thinking back on his days as a pupil in Lüneburg, where his teachers helped him get over the first small hurdles he encountered there.

Riemann later stayed in touch with two of those teachers, and after his death, their memories of him were preserved because of a fortuitous set of circumstances. Riemann's former student and later colleague, Ernst Schering, happened to have attended the Johanneum in Lüneburg only a short time after Riemann had matriculated there. Owing to this, he knew the two teachers who had helped the brilliant young man get through the curriculum during those difficult years. Schering and Riemann were both members of the Göttingen Scientific Society at the time of the latter's death, and thus it was Schering who delivered the eulogy in a public meeting of the society held on 1 December 1866. In preparation for that occasion, Schering wrote to these two former teachers at the Johanneum, asking them for their recollections of the school's star pupil.

Riemann's mathematics teacher in Lüneburg was the Director of the Johanneum, Constantin Schmalfuss. Already during his first year at the Johanneum, young Bernhard's abilities made a deep impression on him:

> His grasp of mathematical subjects immediately revealed itself to me. Riemann only needed to detect the hint of a mathematical law in order to clarify its consequences and put these into solid form, and indeed in the greatest generality. ... Of course, I did not expect him to merely accompany his classmates while he could fly ahead of them all, but rather I tried to offer him something in every lesson that was appropriate to his abilities, and each time he went beyond what I considered to be his limits and probably mine as well, and regularly produced a wealth of results that I had not expected to such an extent.

Schering asked about Riemann's early interest in the foundations of geometry, about which Schmalfuss commented:

> In his last year he had, among other things, invented the theory of spherical geometry for himself, as it were, as a sideline; for he did not happen to know the [relevant] works at all and had proceeded entirely from transferring plane geometry, as much as possible, to the sphere. I very much regret that I no longer remember how he proved these things and developed the formulas for them. Even back then, he was a mathematician whose ability made the teacher feel inadequate.

Schmalfuss had acquired a sizable library filled with older as well as newer mathematical books, so he had no difficulty keeping his new pupil busy.

> All that I had on Euclidean things, including commentaries ...; everything dealing with the Archimedean literature, Apollonius Pergaeus etc., all this he read, and all of it became his sure possession. Newton's Arithmetica Universalis and Cartesius' Geometry interested him no less as well ....

One can see from the works listed that – for Schmalfuss as for many who taught mathematics at German Gymnasien – the classics occupied an exalted place. Naturally, most of their charges received only small doses of the works Schmalfuss cited. More typical would have been the types of problems found in the collections published by Meier Hirsch, a private teacher in Berlin who had instructed the Humboldt brothers, among others.[5] Riemann's teacher reported that his pupil solved all of the problems but one in the *Meier Hirsch Sammlung*.

As might be expected at an elite school, this environment further stimulated Bernhard's sense of personal intellectual ambition. One of the older students, a fellow named Stisser, graduated a year before Riemann. He then also went to Göttingen to begin his university studies. Bernhard reported in a letter to his father, written 1 February 1845, on how Stisser and one other pupil had fared during the final Abitur examinations:

> There was a written Matura exam eight days ago, during which we had to work at school. I wrote seven folio pages for the Latin essay we were assigned at school and finished the German one. In the written exam, Runge got a 1 in all subjects; Stisser "vorzüglich" (excellent) in mathematics and Hebrew; in the rest a solid 2. So a final 1 is pretty sure for both of them. That Stisser received 'excellent' in mathematics gives me hope for a successful exam, since Stisser himself admits that I have more knowledge than does he. Schmalfuss said to [a colleague] that Stisser was the best mathematician to graduate from school for a long time; but I could do much better ... [Neuenschwander 1981a, 92]

The previous summer Bernhard longed to return to Quickborn and he was granted permission to leave Lüneburg early. As a condition for receiving this special treatment, he promised to send his German and Latin essays (the latter one on the role of the Comitia in the Roman Republic), but he never completed the latter. When he returned to Lüneburg, he at first faced difficulty moving back into his former boardinghouse. He then spoke with Schmalfuss, who accused him of breaking the promise he had made before leaving school early. The headmaster felt he had to take this situation in hand, but wanted to help Riemann find a way out of his predicament. In lieu of the Latin essay he had been assigned, Schmalfuss gave him an easier topic: "The Athenians as liberators of Greece."

Bernhard reported on these unpleasant developments in a letter to his father, written on 8 August 1844. This was by no means the last of the problems he would face in school, but this episode ended happily enough, as he eventually received a positive evaluation for his essays. Bernhard's father was greatly relieved when he learned that this mini-crisis had a happy ending. Afterward, he sent his son these words of encouragement:

[5] On the career of Meier Hirsch, see [Wiescher 2025].

> Thank God that you write to me contentedly, because everything is over, happily overcome, also that S[chmalfuss] has written such an encouraging testimonial for you on your essays. So you see, you can if you want, and you certainly don't lack willpower. So let your willpower become action and be careful not to cause gloomy hours and days for yourself and me again. What did Junghans say about your Latin essays, how did he express himself about what happened, and how is he behaving towards you? One can well expect that the whole matter has become known, but since all went well overall and is now settled, you and I no longer have to fear a report on Michaelmas with a threat of expulsion, we can now look calmly into the future again. Avoid everything that fills you with anxiety about the future and disturbs your inner serenity. Also, remember that if you do well in everything, we might also receive the scholarship.

Pastor Riemann did not let this opportunity go by without conveying in strong language the larger moral lesson he hoped Bernhard had learned from this unpleasant experience:

> If S[chmalfuss] accuses you of not having acted truthfully and of having obtained permission to travel by deception, you certainly feel that he was not wrong and that he treated you mildly regarding this affair. May you never again commit the vice of saying an untruth; even in small things one must be truthful. The small is the cradle of the great. If you are in an embarrassing situation, do not try to wiggle out of it with a lie. You will then put yourself in even greater embarrassment and increase your guilt. So terrible are the consequences of lying. In the last issue of the Kulturblatt, I read that a certain Joubert in southern France, who was being led from prison for execution, addressed several onlookers with the words: "Beware of any lie, gentlemen, it is the cause of my present fate."[6]

Schmalfuss obviously took great delight in feeding Riemann's insatiable appetite for challenging mathematics, but he eventually came to realize that this special treatment had a downside. As school director, he continued to hear that his favorite pupil was falling behind badly in his other courses, particularly in Latin and German, because he regularly failed to turn in written assignments on time. As a result, Schmalfuss spoke with the religion teacher, Gustav H. Seffer, who boarded students in his home. He also spoke with Riemann directly, as he had to inform him that he could no longer continue to lend out books from his library, as Riemann needed to concentrate on his other school subjects. An arrangement was then made with Seffer for Riemann to lodge at his home. He lived there for the next two years, along with a few other boarders, though the landlord agreed to charge a reduced fee for Riemann. Seffer remembered him as

> ...quiet, modest, unassuming and faithful to the house rules. He never gave us any cause for reprimand or complaint about his behavior and in this respect made our duty of supervision very easy. He was very embarrassed in the presence of ladies and never completely lost his self-consciousness in his dealings with with my wife. [Riemann 1990, 850]

Seffer's final remark was surely telling and not a little surprising. After all, Bernhard grew up with his mother and four sisters, so this shyness in the company of females can hardly be attributed to lack of familiarity with the opposite sex.

[6] This translation is based on a transcription of the letter made by Norbert Schappacher.

Yet, this awkwardness would remain an inhibiting factor throughout much of Riemann's life.

What Schmalfuss emphasized most about Riemann's character was the difficulties the young man had when trying to express his thoughts fluently. In his letter to Schering, he assumed the latter was well aware of this, but Schmalfuss wanted to describe nevertheless how this handicap affected Riemann's performance at the Lüneburg Gymnasium.

> ...no expression was sufficient for him that did not encompass everything it was intended to describe, and he was extremely hesitant to recognize as correct an exposition that was not of impeccable precision, something like a general formula encompassing all individual cases. His essays in German as well as in Latin were written under severe logical labor pains that required overcoming constant obstacles, and after subjecting whatever he had just found to sharp thinking, he rejected it, and so he was unable to finish writing a text. He never wrote badly, but he also never wrote quickly or with ease. He possessed sufficient grammatical and linguistic knowledge to understand [classical] writers, but he lacked ease of flow and fluency when translating them. Still, when it came to difficult passages he was the equal of the best pupils; he left nothing unexplained simply because it was difficult. [Riemann 1990, 852–853]

He also remembered hearing that Riemann's fellow students occasionally asked for his help when they had difficulty with a passage in Sophocles, Thucydides, or Plato.

Schmalfuss continued to take a strong interest in Riemann, probably all the more so after realizing he had not foreseen the struggle the young lad was facing in other subjects. He asked Seffer to make a statement at the school's teachers' conference pledging that he would ensure Riemann would promptly deliver all required written work in the future. This he did, of course, but Seffer soon came to recognize the source of Riemann's delinquent behavior: he was a perplexed perfectionist. This explained why he could never complete an essay on time; he was just never satisfied with what he wrote. Sometimes doubts crept in, other times new thoughts came to him that he felt needed to be included. These, though, often undermined the whole structure of his essay, forcing him to start all over again.

Once Seffer realized the nature and severity of Riemann's problem, he took it upon himself to fulfill the pledge he had made. Whenever Riemann had to write about a particular question, Seffer would ask him to explain how he planned to answer it. Once they had discussed Riemann's plan for his essay, the ordeal of writing it up began, often lasting into the wee hours of the night. It sometimes happened that Riemann could still not finish, at which point Seffer told him he had to stop, whether with or without a suitable conclusion.

Schmalfuss remembered an episode from Riemann's school days that later became particularly famous. During his penultimate year in the *Unterprima*, Riemann came to ask him for something to read, though nothing too easy. So he came away with Legendre's *Theory of Numbers*. One week later, he returned it saying, "that's a wonderful book; I know it by heart! das ist ja ein wundervolles Buch; ich weiss es auswendig)"[Schering 1909, 439]. More than a year later, when he took his final oral exams, Schmalfuss asked him about certain results in number theory given by Legendre. Riemann answered everything as if he had just studied Legendre's book in preparation for this very exam.

As experts on the history of number theory know, Legendre was the first to formulate what later came to be known as the prime number theorem. Riemann dealt with it later in one of his most famous papers [Riemann 1859/1876], but when he first came to contemplate this problem will probably never be known. Although one cannot rule out the possibility that the seed was already planted in Lüneburg, this would seem highly unlikely. First, the book Riemann borrowed from Schmalfuss was most likely the first volume of Legendre's *Théorie des nombres* [Legendre 1830], which was published as an expanded version of the earlier editions of 1798 and 1808. Those two works presented Legendre's conjecture for the approximate distribution of the primes (see the discussion in Section 8.2). It should be noted, however, that this topic was omitted from [Legendre 1830], so Riemann would not have encountered it in this book. Secondly, one should not overlook the possibility that this famous story about how Riemann more or less memorized Legendre's *Theory of Numbers*, as reported by Schmalfuss more than twenty years later, might be largely a legend. In telling this, he added the remark that Riemann was particularly drawn to number theory, which certainly cannot be confirmed by his subsequent mathematical pursuits. Were this true, one would have expected that he later studied this or other works on number theory; yet, apart from Gauss's *Disquisitiones Arithmeticae* [Gauss 1801/1966], he apparently never borrowed such books from the Göttingen library [Neuenschwander 2022].

Seffer's recollections of Riemann shed a good deal of light on other sides of his character. Since he was preparing to study theology, Seffer taught him Hebrew and remarked that "he was one of my best students, albeit without showing any particular talent for the language." But then he related how Riemann had helped him in preparing his book of exercises in Hebrew for the secondary schools in Germany and Switzerland:

> I had set myself the task of compiling exercises, preceded by the pertinent grammatical rules, but drawing only on passages from the Old Testament. At the same time, I sought to make this collection, at least as far as possible, form a coherent whole. Of course, this was a quite difficult task at times, and one in which Riemann took a keen interest. He often sat for hours and collected such passages from the Old Testament for me to use in the exercises. So I can certainly say that my elementary Hebrew book owes several of its exercises in large part to the great mathematician Riemann.

Seffer was convinced that Riemann was a pious Lutheran at the time he knew him (he taught religion to the *Primaner*). Later, when Riemann was already teaching as a Privatdozent, he saw him again in Göttingen. During that visit, Riemann spoke

> a lot about a philosophical work that was occupying him at the time. Starting from some mathematical assumptions, he finally (or perhaps fortuitously) came to prove the biblical story of creation and other basic Christian doctrines as correct and necessary. That is precisely why he explained these arguments to me so vividly. I must confess that I couldn't follow him at all, indeed I understood next to nothing about it, but I had to admire the magnificence of his goals. He does not seem to have completed that work, which would have created a real stir.

No trace of this can be found in Riemann's posthumous papers, but it is inconceivable he ever intended to publish such a work. Practically all his more philosophical writings, most of which only came to light in the appendix to his *Werke* [Riemann 1876/1892/1902, 1876: 477–506], were wholly unknown during his lifetime. In matters of religion, Riemann kept his thoughts and beliefs to himself.

One year after Bernhard first moved into Seffer's home, he wrote a long letter to his father on 28 April 1845. His performance in Latin and German must have stabilized by this time, as he was already looking ahead to the next year when he would begin studying in Göttingen. In a key passage, he made plain his personal hopes for the future:

> I now feel more and more desire and inclination towards mathematics, and Schmalfuss also seems to be of the opinion that I should study mathematics; of course, for the sake of the scholarship[7] alone, I would register for the exam here in theology and enroll in that subject in Göttingen, but I have to decide for myself what I should actually do there, because otherwise I won't be able to achieve anything respectable in any subject.

This letter may well have caused Riemann's father real distress. Surely he could not have imagined anyone, let alone his own son, taking up theology as what was then called *Brotstudium*. Bernhard needed to earn a living, of course, but for that he could have learned bookkeeping or some comparable trade. Now the elder Riemann must have recognized that his 19-year-old son had already come to a crossroads in his life. Bernhard's letter wasn't exactly a plea, but the issue was now clear, though he also went on to say:

> ... if you should prefer for whatever reason that I stay with theology, I will now give up mathematics completely and instead concentrate more on languages. The costs should also be considered, because I would probably have to study mathematics a half-year longer, and as a theologian I could probably earn something earlier, at least as a tutor.

Bernhard tried to keep his father abreast of his progress at the Johanneum up until the all-important final examinations (*Maturitätsprüfung*). These were a truly grueling affair. Riemann's took place from Monday to Saturday, February 23–29, with a different subject each day. Apparently the only written parts were in Latin and German, so in all other subjects, some of which were elective, candidates were tested orally. In Riemann's case, he began by writing an essay in Latin, then came Greek, followed by an essay in German, then Hebrew, French, and mathematics. No record or other indication of his performance during that week survives, but we may safely assume he acquitted himself well. As already indicated, Schmalfuss came away flabbergasted by his performance in mathematics.

After passing his exams, Bernhard returned to Quickborn, where he and his father discussed his future plans. He would, indeed, enroll as a theology student, but his father graciously allowed him to take courses outside his official field of study. Some of these they surely discussed, based on the offerings listed in the

---

[7] This presumably refers to a stipend offered by the Lutheran church to outstanding pupils who wish to study theology at the university.

course catalog for the summer semester of 1846, which began officially in late April and ended in mid-September. It was far too soon for either of them to imagine what might come next, though both surely believed that God's wisdom would help point the way. If Bernhard Riemann was already dreaming of a career in mathematics at the German universities, he likely kept that idea to himself. He had already achieved an impressive first goal.

Those who managed to graduate from a classical Gymnasium formed a tiny elite in Germany: at mid-century they represented well under one per cent of the male population in a given year. Yet only a very small number among these would set their eyes on pursuing a university career. Since 1820, only two Göttingen graduates had gone on to habilitate in mathematics: Moritz Abraham Stern and Benjamin Goldschmidt. Both came from Jewish families and took up studies in Göttingen under Carl Friedrich Gauss. During his first semester, Riemann was introduced to both by the ophthalmologist and general medical practitioner Theodor Ruete, who likely knew these colleagues through his friend and collaborator, Benedict Listing, another pupil of Gauss. Riemann's Aunt Julie probably referred him to Ruete, a physician who treated Riemann up until 1852, when he left Göttingen to take up a professorship in Leipzig [Neuenschwander 1981a, 102]. Göttingen's academic community formed a very small world, so that even a shy soul like Bernhard Riemann soon came to know a number of its members.

## 1.3 Initial Studies in Göttingen

Graduates leaving a Gymnasium with the *Abitur* were generally allowed to study any field at a German university. Many students chose to attend more than one institution, as transferring only involved minimal administrative paperwork. Another shared feature of these universities made this process even simpler: there were no course grades or any requirements to take tests or turn in written assignments. This radical form of *Lernfreiheit* (freedom to learn) could easily be abused, of course, and often was; yet for those with a thirst for knowledge, this system gave a highly select group of young men an unusual opportunity to explore new ideas, but also to sink if they didn't learn how to swim quickly. For Riemann, who had dreaded the dreary regimentation of Gymnasium routine, these refreshing university waters must have seemed like an academic paradise.

Riemann enrolled as a student in philology and theology, but he ended up taking only a few courses in those fields. He enrolled in a 5-hour course on logic with Heinrich Ritter, a then well-known scholar who published a history of philosophy in 12 volumes. The Göttingen faculty offered several courses in ancient Greek, but Riemann's choice seems quite remarkable: he took a 5-hour course from the philologist Ernst von Leutsch on two plays by Aristophanes: *The Birds* and *The Frogs*. In mathematics, he chose Moritz Abraham Stern's 4-hour course on the numerical solution of algebraic equations. He might have chosen to take Gauss's regular course on practical geodesy, except that it conflicted with the course on terrestrial magnetism, which he preferred to take instead. It was taught by

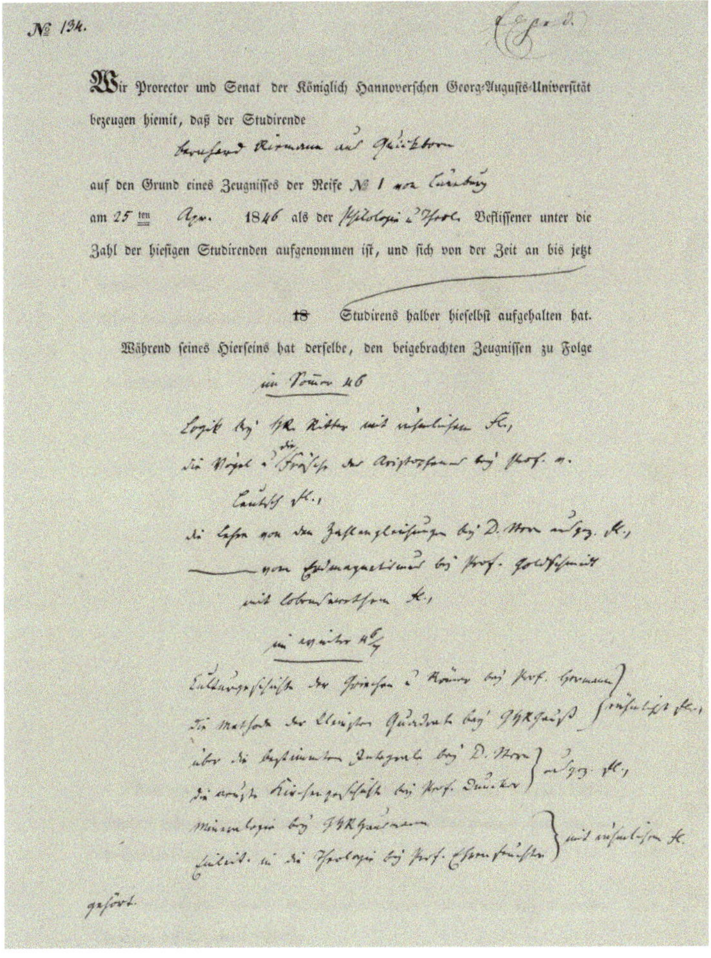

№ 134.

Wir Prorector und Senat der Königlich Hannoverschen Georg-Augusts-Universität bezeugen hiemit, daß der Studirende

Bernhard Riemann aus Quickborn

auf den Grund eines Zeugnisses der Reife № 1 von Lüneburg

am 25 ten Apr. 1846 als der Philologie u. Theol. Beflissener unter die Zahl der hiesigen Studirenden aufgenommen ist, und sich von der Zeit an bis jetzt

18 Studirens halber hieselbst aufgehalten hat.

Während seines Hierseins hat derselbe, den beigebrachten Zeugnissen zu Folge

Fig. 1.2: Riemann's *Abgangszeugnis* (Leaving Certificate) after his first two semesters in Göttingen.

Gauss's former assistant Benjamin Goldschmidt, from whom Riemann probably first learned about potential theory. During the period from 1843 to 1849, when Wilhelm Weber was no longer in Göttingen, research on geomagnetism was largely in Goldschmidt's hands. It was likely on the latter's advice that, during his first semester, Riemann borrowed Gauss's 1833 study "Intensitas vis magneticae terrestris ad mensuram absolutam revocata" (The Intensity of the Earth's Magnetic Force reduced to Absolute Measurement) [Gauss 1833/1995]. For an account of how this famous work likely influenced Riemann, see Section 10.1. These four courses were the only ones recorded in Riemann's leaving certificate (Fig. 1.2), though he likely sat in on others without enrolling.

When borrowing books from the university library, students were required to obtain written permission from a faculty member, and during Riemann's first years in Göttingen he invariably turned to Goldschmidt for that purpose. This inconvenience apparently did little to discourage Riemann's trips to the library; over the years he lived in Göttingen, he checked out around 500 different books and journals [Neuenschwander 2022, 42]. Thanks to Erwin Neuenschwander's diligent research on Riemann's temporary acquisitions from the Göttingen library, a quite vivid picture of his intellectual development emerges.

Not surprisingly, the works of Gauss were foremost on his mind. Already during the first week of the semester, he checked out Gauss's *Disquisitiones Arithmeticae* [Gauss 1801/1966], a pattern he repeated when he returned to Göttingen for the winter semester. He also borrowed the sixth volume of *Commentationes Societatis Regiae Scientiarum Gottingensis recentiores* from the regular series of studies published by the Göttingen Royal Society. Gauss published nearly all his more substantial Latin works in this series, including three articles that appeared in this sixth volume. Perhaps Riemann knew that Volume 6 contained his first study on biquadratic residues [Gauss 1828a] as well as his classic paper on the differential geometry of surfaces [Gauss 1828b]? The latter paper contained the theory of Gaussian curvature, which he derived for surfaces embedded in Euclidean 3-space. In 1854, in his Habilitation lecture (see 3.2), Riemann would indicate how Gauss's notion could be extended to manifolds of arbitrary dimension. What came to be called Riemannian curvature was later understood to be a tensor quantity independent of any embedding space [Spivak 1979].

Not until late July did Riemann borrow two more works from the library. One of these was Legendre's *Traité des fonctions elliptiques* [Legendre 1825/26]. This latter topic was not taught in Göttingen, as Riemann undoubtedly knew, so perhaps he was already hoping to learn more about it under Carl Gustav Jacob Jacobi in Berlin. The other work was Gauss's famous essay on his method of absolute measurement for terrestrial magnetism; Goldschmidt very likely recommended this to him (see 10.1). Gauss's text began with these remarks:

> For the complete determination of the Earth's magnetic force at a given location, three elements are necessary: the deviation (declination) or angle between the planes in which it acts and the meridian plane; the inclination of the direction of the horizontal plane; and third, the strength (intensity). The declination, which is to be considered as the most important element in all applications to navigation and geodesy, has engaged astronomers and physicists from the beginning, and for a century they have also given their constant attention to the inclination. In contrast with these, the third element, the intensity of the terrestrial magnetic force, which is surely just as worthy as a scientific subject, has remained neglected until more recently. Humboldt rendered the service, among so many others, of having been the first to direct attention to this subject, and in his travels he assembled a large array of determinations concerning the relative magnetic strength. These resulted in showing a continuous increase in strength from the magnetic equator toward the pole.[Gauss 1833/1995, 1995: 1]

During the winter semester of 1846/47, Riemann continued to take courses in mathematics and the natural sciences mixed with a few outside these fields. Toward the beginning of that semester, he wrote his father to inform him about

Fig. 1.3: Moritz Abraham Stern (1807–1894).

the six courses in which he planned to enroll. He still had not reached a final decision, though, in part because the scheduling remained unclear. As usual, he touched on financial matters, noting that the costs for these would come to 3 Louis d'or[8] and 3 Taler, or the equivalent of 18 Taler total. Bernhard's preliminary course choices were these:

Karl Friedrich Hermann: Politics and Culture of Ancient Greece and Rome;

[8] The name for these gold coins derived from the Kings of France, as they first appeared during the reign of Louis XIII. Thus, despite the demise of the Kingdom of Westphalia more than three decades earlier, Hanoverians still kept these French gold coins in currency.

Friedrich Ehrenfeuchter: Introduction to Theology;
Hermann Lotze: Logic and Encyclopedia of the Systems of Philosophy;
Johann Georg Reiche: New Testament Exegesis;
M.A. Stern: Definite Integrals;
C.F. Gauss: Method of Least Squares with Applications to Astronomy, Higher Geodesy, and Natural Sciences.

The two theology courses on this list were public, which meant these did not require permission from the instructor to attend. In most cases, public lecture courses were taught by full professors, whereas private lecturers usually offered *privatissime* on more specialized topics. Bernhard also mentioned two other courses: B. Goldschmidt: Probability Theory; and August Griesbach: Introduction to Natural History. He seemed quite interested in Goldschmidt's course on probability theory, scheduled to meet 5 hours a week at 2. Riemann may have even promised the instructor he planned to attend so that this course could take place (with 2 students!). Probably Goldschmidt canceled it since Riemann's leaving certificate (Fig. 1.2) contains nothing about such a course.

Ultimately, he decided to take Stern's (Fig. 1.3) course on definite integrals as well as Gauss's on the method of least squares.[9] He probably took Hausmann's course on crystallography instead, though this was recorded as mineralogy in Riemann's certificate. This course replaced Lotze's and instead of Reiche's, Riemann chose to attend the lectures given by the theologian Ludwig Duncker on the history of the Christian church in the modern era. In all of these courses, the instructors acknowledged the student's positive performance, but Goldschmidt even more. He wrote that Riemann performed "with the most praiseworthy diligence" (mit lobenswertstem Fleiß).

Whether or not Riemann enjoyed any significant personal contacts with Gauss, he probably left him with a positive impression. Over the course of this semester, he borrowed three more volumes of the *Commentationes* containing articles by Gauss on topics in astronomy and number theory. Beyond his continued studies of *Disquisitiones Arithmeticae*, Riemann also began reading Gauss's classic *Theoria motus corporum coelestium in sectionibus conicis solem ambientum* (Theory of the motion of the heavenly bodies moving about the sun in conic sections) [Gauss 1809/1857]. Along with much else, this famous work presented the method of least squares, which was most likely the reason Riemann wished to borrow it. Gauss cited Legendre's work from three years later, but claimed priority of discovery, since he had used it already in the 1790s.[10]

---

[9] For a vivid account of Gauss's teaching style in that course, see [Dedekind 1901]. In a letter to his father from 5 November 1846, Riemann noted that he still had not finalized the official list of six courses he would take. He had hoped to take Friedrich Hausmann's minerology course, but this conflicted with Gauss's lectures [Neuenschwander 1981a, 93].

[10] Gauss also argued that this was the best possible method: "But of all these principles ours is the most simple; by the others we shall be led into the most complicated calculations. On the other hand, our principle, which we have made use of since the year 1795, has lately been published by Legendre in the work *Nouvelles méthodes pour la determination des orbites des comètes*, Paris 1806, where several other properties of this principle have been explained, which, for the sake of brevity, we here omit" [Gauss 1809/1857, 1857, 270].

This comment provoked what has since been called the most famous priority dispute in the history of statistics, though Stephen Stigler may have published the last word on it [Stigler 1981].

One naturally wonders whether Riemann's preternatural abilities in analysis gained anything substantial from the courses he took during his first year in Göttingen. In this respect, the key figure was Stern, who many years later fondly remembered how the young Riemann "sang like a canary" [Klein 1926, 249]. Moritz Abraham Stern was one of the last important exponents of the "combinatorial school," which he promoted in his *Lehrbuch der algebraischen Analysis* [Stern 1860]. This opened with a sweeping attack on Cauchy's infinitesimal analysis.[11] For a brief time, Riemann was drawn to this approach to analysis, a tradition that was not remembered kindly by Germany's leading mathematicians after 1850. During his second semester, Riemann wrote an "Essay on a General Theory of Differentiation and Integration," which he completed on 11 January 1847.[12] This dealt with a new method for computing derivatives and integrals based on an arbitrary real exponent, thereby generalizing the usual power series representations for a function (see [Laugwitz 1996/1999, 49–50].) Although he was only 20 when he wrote this, one can already see Riemann's remarkable mastery of the methods he describes. Stern thought highly of the paper and even suggested that he offer it to August Leopold Crelle for publication in his journal. Yet by the time Riemann met Crelle in Berlin (Section 1.4 below), the whole subject had lost its appeal for him.

Why he lost interest becomes less mysterious once we realize that just two weeks after completing this paper, Riemann had borrowed Cauchy's *Cours d'analyse* from the University library. Alongside this classic, he also studied Abbé Moigno's *Leçons de calcul différentiel et de calcul intégral d'apres Cauchy* (1840). Riemann also borrowed the first of Cauchy's four-volume study *Exercices d'analyse et de physique mathématique* as well as the 1844 issue of the *Comptes Rendus*, which contained numerous papers and reports by Cauchy [Neuenschwander 2022, 42]. In all likelihood, his hunt for Cauchy's pioneering studies in complex analysis only began in earnest once Riemann arrived in Berlin. Thus, Riemann's introduction to analysis through Stern proved to be only a momentary passing phase.

Even before he began reading Cauchy's works, Riemann continued to study elliptic functions with burning interest. Not only did he again borrow the two volumes of Legendre's *Traité des fonctions elliptiques*, he supplemented this with C.G.J. Jacobi's exceedingly difficult *Fundamenta nova theoriae functionum ellipticarum* [Jacobi 1829]. For "lighter reading," he also took out the three volumes of Euler's *Institutiones calculi integralis* (1768), which Legendre drew on

[11] For a comparison of the approaches of Stern and Cauchy, see [Laugwitz 1996/1999, 48–50, 53–55].

[12] Heinrich Weber eventually published this text as [Riemann 1876/1892/1902, 331–344] with a note suggesting that Riemann had soon abandoned the methods employed in this paper. Detlef Laugwitz remarked that those methods break down when applied to trigonometric functions, as Riemann may have learned when studying under Dirichlet [Laugwitz 1996/1999, 51].

heavily in his study of elliptic functions. Riemann probably owned Euler's other two textbooks on analysis: *Introductio in analysin infinitorum* (1748) and *Institutiones calculi differentialis* (1755), works he may have already studied as a pupil at the Lüneburg Gymnsium. In any event, he never had occasion to borrow those books from the library. Although Riemann never published directly on the theory of elliptic functions, this topic played a central role in his approach to Abelian functions, for which Jacobi's classic study served as a model (see Section 7.1). By this time, though, he was running back and forth from the library almost constantly; he checked out books no fewer than eleven times over the course of the semester.

Among other works, he borrowed Volume 1 of Alexander von Humboldt's *Kosmos*, which had just been published in 1845. As an instant bestseller, Riemann was eager to hold it in his hands; whether he actually read it, though, remains an open question. In any event, Humboldt's influence would make itself felt in other ways, as Riemann's interest in geomagnetism became a long-term passion; indeed, this field would remain a constant source of inspiration in the years to come. Toward the end of the semester, Riemann took out the 1840 volume of *Transactions of the Royal Society of Edinburgh*, no doubt to study an article by James David Forbes reporting on measurements of terrestrial magnetism made throughout Europe, which aimed to determine the relationship between magnetic intensity and elevation.

Although nothing seems to be known about Riemann's personal relations with Benjamin Goldschmidt, one can easily imagine that the latter's course awakened the young man's fascination with the Gauss-Weber enterprise, about which no one knew more than Goldschmidt himself. Today memory of that famous collaboration between Gauss and Weber has largely fixated on the electrical telegraph they invented and utilized for a number of years, beginning in 1833. For city dwellers, it became an integral part of their surroundings for over a decade. Riemann never saw it, however, as a year before his arrival the line came crashing down during a lightning storm. What he learned about instead concerned the actual findings Gauss and Weber published during their most fruitful years of collaboration (see Section 9.2).

By the end of his first trial year in Göttingen, Riemann's determination to study mathematics was stronger than ever. His father apparently now acquiesced to Bernhard's new plan, which was to continue his studies in Berlin. He likely knew that many of the courses taught there were on a far higher level than the standard offerings in Göttingen [Schubring 1992]. What he otherwise knew about its university one can hardly say, but he was poised for adventure when he began his studies there after Easter of 1847. From its founding by King Friedrich Wilhelm III in 1810, Berlin not only emerged as Prussia's leading university, but also as one of the foremost centers of the neo-humanist movement which for several decades came to dominate German education. Riemann's experiences at the Gymnasien in Hanover and Lüneburg conformed fully with those same educational ideals. This had great import for mathematics teaching in Germany, which contrasted sharply with the emphasis on science and engineering in post-revolutionary France.

## 1.4 Riemann in Berlin

After arriving in the Prussian capital, it took Riemann some time to get settled. He was still deeply interested in a wide range of subjects, though eventually he focused on mathematics and physics. During his first semester in Berlin, Riemann enrolled in five courses (as shown in Fig. 1.4). Two of these were taught by Dirichlet, one on pure number theory, the other on applications of analysis to the theory of numbers, though these were likely taught as a single course. In a letter to his father from 23 July 1847, Bernhard called this his "principal subject" and "the field of mathematics to which Gauss owes all his fame" [Neuenschwander 1981a, 95]. Having spent a good part of his time in Göttingen reading Gauss's *Disquisitiones Arithmeticae*, he could now cite the opinion of Dirichlet in attesting to the significance of Gauss's book for modern number theory.[13] Riemann wanted his father to know that he was making great efforts to learn it and hoped to be successful. No doubt he did master the terrain Dirichlet covered, but this nevertheless would prove to be Riemann's last intensive encounter with number theory.

Although Riemann left no account of Dirichlet's lecture style, others thankfully did, one being the itinerant English geometer, Thomas Archer Hirst, who wrote:

> Dirichlet cannot be surpassed for richness of material and clear insight into it: as a speaker he has no advantages [he had a somewhat harsh voice] – there is nothing like fluency about him, and yet a clear eye and understanding make it dispensable: without an effort you would not notice his hesitating speech. What is peculiar in him, he never sees his audience – when he does not use the blackboard, at which time his back is turned to us, he sits at the high desk facing us, puts his spectacles on his forehead, leans his head on his hands, and keeps his eyes, when not covered by his hands, mostly shut. He uses no notes, but inside his hands he sees an imaginary calculation, and reads it out to us – that we understand it as well as if we too saw it. [Gardner/Wilson 1993, 622]

Riemann also took a course on optics, taught by the meteorologist Heinrich Dove four days a week from 11 to 12; Riemann found his lectures very interesting. They seem to have left a lasting impression as well, considering the importance Riemann would later place on an integrated mathematical theory accounting for light, gravity, and magnetism. Immediately after his course with Dove, he often sat in on Leopold Ranke's lectures on modern history, which were highly popular and attracted many auditors, including students like Riemann who were not enrolled. After his lunch break, he then went to hear Dirichlet lecture on number theory from 2 to 3. Riemann's fourth course, on spherical astronomy, met from 3 to 4. This was taught by Franz Encke, whose lecture style Riemann found quite dry and boring. On the other hand, he gained quite a lot from Encke's practical demonstrations, which he offered one night a week at the astronomical observatory. Riemann's fifth course on the theory of elliptic

[13] As Goldstein and Schappacher show in [Goldstein et al. 2007], *Disquisitiones Arithmeticae* established a paradigm for algebraic number theory in the nineteenth century.

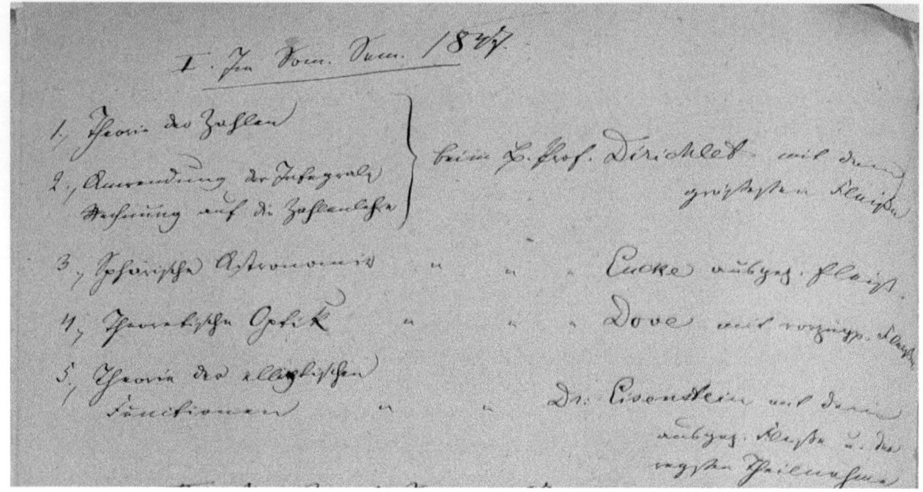

Fig. 1.4: From Riemann's *Abgangszeugnis* after attending Berlin University, 1847–1849.

functions was attended by a small group of students. Several were, no doubt, curious to hear the first university lectures given by the newly appointed *Privatdozent* Gotthold Eisenstein, a rising superstar only three years older than Riemann himself. At first, Riemann often spoke with him after class or at an arranged time, but he later distanced himself from Eisenstein for reasons that will be considered later (see Section 1.5 below).

Riemann was surely disappointed by the fact that C.G.J. Jacobi announced no courses during Riemann's first year in Berlin. As a member of the Berlin Academy, Jacobi had the right to teach, but no obligation to do so. When Riemann returned to Berlin for the 1847/48 winter semester, however, he experienced a pleasant surprise. He related this to his father in a letter from 29 November:

> When I arrived here, I was delighted to learn that Jacobi, who had not listed any lectures in the catalog, had changed his mind and intended to give lectures on mechanics; I would have stayed here for an extra semester, if necessary, just to hear him, so nothing could have pleased me more than this. [Neuenschwander 1981a, 95–96].

Jacobi (Fig. 1.5) had been recovering from a long illness, which had previously prevented him from teaching. Starting in October 1847, he taught analytical mechanics for the third and last time. Riemann followed Jacobi's lectures with enthusiasm, though as one of its 17 attendees the instructor may not have paid him much attention (Jacobi accorded him the standard accolade "outstanding diligence" in rating his performance). By his second year in Berlin, when he attended Jacobi's courses on higher algebra and elliptic functions [Jacobi 1881–91,

Fig. 1.5: Carl Gustav Jacob Jacobi (1804–1851).

7: 411–412], Riemann surely had more direct contact with the era's great algorist.[14]

Jacobi always lectured freely without a prepared text, but he made a practice of collecting copies of the extrapolations made by his auditors.[15] For the course on analytical mechanics that Riemann attended, such an *Ausarbeitung* was prepared by another auditor, Wilhelm Scheibner, later professor of mathematics in Leipzig.[16] During the previous semester, Jacobi had attended a meeting of the Berlin Physical Society held on 23 July 1847. It was on this occasion that Hermann von Helmholtz delivered his controversial lecture "Ueber die Erhaltung der Kraft" (On the Conservation of Energy), which Riemann later read.[17] Jacobi not only attended Helmholtz's lecture, he was also one of the few who reacted

[14] On Jacobi's influence on Riemann, see [de Paz/Ferreirós 2020].

[15] Since many German professors at this time lectured extemporaneously, rather than relying on a written text, in such cases the only records available come from notes taken by students. They were expected not merely to transcribe what was said, but also to elaborate on the professor's remarks by preparing an *Ausarbeitung* of the lecture. By the latter part of the century, some mathematicians, including Weierstrass and Klein, exerted their influence over this practice by appointing students who prepared authorized elaborations for designated lecture courses.

[16] [Jacobi 1996]; for an overview, see Helmut Pulte's preface. [Scheibner 1860] was one of the first explications of the results in Riemann's paper on the zeta function (8.2).

[17] He borrowed it from the University Library on April 12th, 1858 [Neuenschwander 2022, 57].

to it positively [Caneva 2021, 154–155]. The mathematician Leo Koenigsberger, who wrote extensive biographies of both Helmholtz [Koenigsberger 1902] and Jacobi [Koenigsberger 1904], was in an excellent position to report on these circumstances. Moreover, there is reason to believe that this had a real influence on the new direction Jacobi took in his course on analytical mechanics, which broke quite sharply from the highly mathematical tradition that characterized Lagrange's *Mécanique analytique.*

Only a week before he attended Helmholtz's lecture, Jacobi had himself delivered an unpublished lecture on the principle of least action, a topic he also treated in his course.[18] The principle of least action long carried the aura of a teleological principle, which naturally appealed to religiously inclined minds. Whether or not Jacobi ascribed to such views, Riemann definitely did. He explicitly linked causality with purpose, a natural enough position for a religious thinker. What matters even more, though, is that Riemann, Jacobi, and Helmholtz were all in search of physical principles that extended beyond the horizon of classical mechanics.

While in Berlin, Riemann also took the opportunity to meet August Leopold Crelle, editor of *Journal für die reine und angewandte Mathematik.* In a letter to his father, he informed him that this meeting went well:

> I paid my respects to Crelle; he was quite friendly towards me. Incidentally, he doesn't seem to be particularly luminary. I didn't give him the work that Stern deemed suitable for publication in his journal because it seemed to me to have too little practical use;[19] but I hope to be able to replace it now with something better. [Neuenschwander 1981a, 95]

The fact that Riemann no longer identified with his earlier work tells us a great deal about his intellectual ambition. What other twenty-one-year-old student would have passed up such a golden opportunity to publish a paper in Germany's leading mathematical journal? Still, Riemann was his own harshest critic, which at least partly explains why his mathematical works only came to fruition slowly and after long periods of brooding. He only wanted to publish his work when he had something significant to say. It took him another decade before he finally submitted a paper to Crelle's distinguished journal, by which time its founder was no longer alive (see Section 3.5).

Dedekind speculated in [Dedekind 1876a, 512] – apparently on the basis of remarks Riemann had once made – that the latter had already found the essential ingredients for his definition of a complex analytic function by the au-

---

[18] A number of distinguished mathematicians and physicists wrote about its colorful history over the years. One knows from these accounts how Euler and Lagrange restored the principle's dignity, following Maupertuis' misguided and largely unintelligible claims, which took some version of this principle as the key to the universe. For an historical survey up to Kant, see [Schramm 1985]; on the resurgence of the principle of least action toward the end of the nineteenth century and afterward, see [Stöltzner 2002].

[19] The paper alluded to here was the manuscript mentioned in 1.3 from January 1847, which Stern had commended.

tumn semester break of 1847.[20] If so, then he would probably have encountered Cauchy's works on complex functions during his first semester in Berlin. This may seem plausible enough, given that Riemann had studied Cauchy's *Cours d'analyse* and other works on real analysis in Göttingen. Still, taking into consideration the heavy course load Riemann carried during his first semester in Berlin, plus the fact that none of those five courses had a direct relevance for complex functions, this conjecture begins to look rather dubious. Dedekind apparently leaped to this conclusion based on Riemann's remark that he and Eisenstein had fundamentally disagreed about the role of complex numbers in analysis.

This comment, however, might very well pertain to an argument they had during the second semester, when Riemann attended the sequel to Eisenstein's course on elliptic functions (see 1.5). During that term, Riemann was taking Jacobi's course on analytical mechanics and Dirichlet's on partial differential equations, but just these three courses. Moreover, his enthusiasm for Eisenstein's course apparently diminished very markedly over the course of the two semesters: Eisenstein praised his "outstanding diligence and enthusiastic participation" after the first semester, whereas he merely wrote "present" for his performance during the second part. During Riemann's second semester in Berlin, he had more free time to think about those parts of higher mathematics that appealed most to him. Prior to this time, his exposure to complex functions came mainly through various works of Gauss, which mainly offered hints but nothing that pointed directly to a general theory. Dirichlet's lectures would later serve as the model for Riemann's single most influential course, which treated partial differential equations in direct connection with their applications to mathematical physics [Riemann/Hattendorff 1869]. There were many connections between Riemann's later work and what he learned from Dirichlet, but these ties are difficult to reconstruct given the paucity of sources available.

Dirichlet was the first professor in Germany to offer regular courses on number theory; moreover, he virtually created analytic number theory. Over the course of his 27-year career in Berlin he taught this subject twenty-three times out of a total of eighty-eight lecture courses. Dirichlet also taught a standard course in potential theory and published four papers on the topic between 1846 and 1852 [Merzbach 2018, 186–190]; these built on Gauss's classical study [Gauss 1840a]. He offered this subject no fewer than eight times over the course of his career, including two courses in Göttingen that served as models for Riemann.[21] Overall, Dirichlet's lecture courses fell into three main areas: number theory, foundations of real analysis, and mathematical physics [Biermann 1959, 34–39]. He continued this general teaching pattern during his far briefer career in Göttingen, where Dedekind took up the first area of research and Riemann the third. It should be noted, however, that Riemann was exposed to all three areas

[20] In Riemann's time, the summer semester began before Easter and typically ran from mid-March until late August. The winter semester began after St. Michael's Day on September 29. It usually ran from late October until early February.

[21] As Uta Merzbach pointed out, however, this broader influence has largely been overlooked because of the immediate controversies surrounding "Dirichlet's principle" [Merzbach 2018, 250].

of mathematics during his two years as a student in Berlin. In fact, Dirichlet's earlier work on representations of real functions by Fourier series served as the background for Riemann's postdoctoral thesis (see Section 6.1).

In Berlin, Riemann continued to build on his earlier studies, including potential theory in the tradition of Gauss and Dirichlet. At the same time, he interacted off and on with the young and immensely talented Gotthold Eisenstein, who had just become a junior member of the Berlin faculty. Writing to his father on 23 July 1847, Riemann complained that it took him 35 minutes to walk to Eisenstein's apartment, where he lectured three times a week for two hours on elliptic functions. Riemann reported further that

> ...for quite a while I saw him regularly, as we undertook daily walks together, and since he possesses a good deal of talent I believe this was very useful in furthering my mathematical education; since then I have withdrawn from him somewhat for reasons I will explain orally. [Neuenschwander 1981a, 94]

Riemann's ambivalent attitude toward Eisenstein may have reflected a clash between their personalities, but more likely he regarded him as highly ambitious and prone to engage in controversy. In any event, he soon learned that Eisenstein was embroiled in a priority dispute with his older colleague Jacobi, a truly formidable opponent.

## 1.5 Jacobi vs. Eisenstein

Ferdinand Gotthold Max Eisenstein was born in Berlin on April 16, 1823 [Biermann 1980]. Although only three years older than Riemann, by the time Bernhard met him he already enjoyed a glowing reputation for his work on algebra and number theory. Both of his parents were Jewish, but they converted to Protestantism before their son's birth. Like many others of Jewish background, he was keenly aware of the precarious situation of the German Jews, some of whom became leading figures in liberal politics. Up until 1842, Eisenstein attended Gymnasien in Berlin, but he also sat in on lecture courses taught by Dirichlet at the university. In the meantime, he was already reading works by Euler, Lagrange, and Gauss. It was the astronomer Franz Encke who first drew Gauss's attention to Eisenstein. Encke knew him fairly well, because his sons had attended the same Gymnasium as this brilliant young mathematician.

Eisenstein's mathematical ascent was so swift and dramatic that he even greatly impressed the usually quite reticent C.F. Gauss. During the single year of 1844, Eisenstein managed to fill volumes 27 and 28 of Crelle's Journal with a total of twenty-five papers, far more than Riemann would publish over the course of his entire lifetime. These dealt with topics on the cutting edge of pure mathematics: quadratic and cubic forms, reciprocity theorems for quadratic, cubic, and biquadratic residues, plus cyclotomy and related results on elliptic and Abelian transcendentals [Schappacher 1998]. Eisenstein sent some of these papers to Gauss, who praised them highly and looked forward to meeting their author. Thus, in June 1844, Eisenstein set off to see Gauss in Göttingen, where he stayed for two weeks. During that visit he also met and befriended Moritz Abraham Stern, probably one of the few friends he ever made.

On July 7, 1844, Gauss wrote to the astronomer Heinrich Christian Schumacher:

> I recently had the opportunity to personally meet a young mathematician who came here with a recommendation from Humboldt and who possesses a most outstanding talent. His name is Eisenstein, and he appears to be of Jewish descent. [Gauss 1863–1933, 4: 265]

German Jews were by no means uncommon in the rarefied world of higher mathematics, but they had virtually no chance for professional advancement unless they converted to Christianity. Stern, in fact, was the first to ever break that barrier, when he gained an *Ordinariat* (full professorship) in Göttingen in 1859, thirty years after habilitating there under Gauss [Rowe 1986].

Riemann was one of six students who attended Eisenstein's first course, which dealt with elliptic functions. At first they spoke often with one another, but these meetings did not go on for very long, as Riemann indicated in the letter to his father cited above. After the semester break, Riemann bumped into Eisenstein on the street, a chance encounter that led to a friendly conversation [Neuenschwander 1981a, 96]. Eisenstein invited him to attend gratis the sequel course he was offering, and Bernhard agreed to do so. He later reported to his father that this was a good decision: Eisenstein's lecture style had improved and he was now better organized. Still, they never became close, which probably led Eisenstein to wonder why. At some point Riemann learned that Jacobi had publicly attacked Eisenstein in Crelle's Journal. This controversy turned particularly awkward for Riemann, since he also learned that Gauss had taken Eisenstein's side in this dispute. As a young newcomer in Berlin, Riemann understandably had no desire to get entangled in this highly visible conflict, which concerned priority rights in the field of reciprocity laws in number theory. This was an area of research pioneered by Legendre and Gauss, which meant the two combatants were playing for very high stakes.

Jacobi had earlier supported Eisenstein, but he now insinuated that the latter had plagiarized his work on cubic and biquadratic reciprocity. Since these results were never published, Jacobi tried to argue that Eisenstein nevertheless must have known them. To make this case, Jacobi republished his earlier paper from 1837 on cyclotomy as applied to number theory, which included the following passage:

> My main application of cyclotomic division was for the theory of cubic and biquadratic residues, by means of which one can prove with great simplicity and ease the beautiful theorem of Gauss in his second study on biquadratic residues, a theorem whose proof (still unknown, but probably derived in a completely different way) he called a *mysterium maxime recondite*. The reciprocity theorem for cubic residues has the greatest simplicity, since its proof follows almost with a single stroke from the known formulas for cyclotomy. The proofs of these theorems were presented without difficulty to my auditors in lectures from the past winter [of 1837]. [Jacobi 1846, 171–172]

Jacobi then attached a footnote to this text, asserting that copies of those same lectures had circulated widely since 1837, so that both Dirichlet and Kummer had known about these proofs for years. In spite of these circumstances, Jacobi remarked, Eisenstein published these same results in the 27th and 28th volumes of Crelle, presenting them as if they were his own intellectual property.[22]

Jacobi's charges amounted to a remarkable volte-face, as only a short time earlier he had helped Eisenstein receive an honorary doctorate from Breslau University. Writing to his friend Stern on 20 April 1846, Eisenstein assured him that

> ...the whole trouble arose because, when I learned of [Jacobi's] work on cyclotomy, I did not immediately and publicly acknowledge him as its originator, while I frequently have done this in the case of Gauss. That I omitted to do so in this instance was merely the fault of my naive innocence. [Hurwitz/Rudio 1895, 181–182]

Jacobi chose to reprint his earlier paper as a convenient ploy in order to insinuate that his rival must have known about Jacobi's prior work, though he chose to ignore it. Clearly, Eisenstein had to answer this charge, and he decided to so in kind, though not by republishing a single paper: his reply came in the form of an entire book containing various works he had recently published in Crelle [Eisenstein 1847]. In all likelihood, Eisenstein also wrote to Gauss, requesting that he write a preface for this collection; somewhat remarkably, the latter obliged.

Gauss was never one to engage in a public dispute, but on this occasion he must have felt compelled to throw his weight behind Eisenstein. He began his preface as a paean to the beauty and mysteries of higher arithmetic, calling on two famous witnesses from long before Eisenstein's time, namely Euler and Lagrange. Gauss then ended with these words:

> The present essays contain so much that is excellent and noble that their author is ensured an honorable place next to his predecessors, whose works are hereby worthily continued. This compilation obliges all to give thanks that these studies are now more easily accessible. [Eisenstein 1847, iv]

Surely no one was in a better position to judge these matters than Gauss, whose words of praise for Eisenstein were clearly also intended to rebuke Jacobi for publicly attacking a gifted young mathematician.

Eisenstein showed this preface to Riemann, who also read the note Eisenstein added to counter Jacobi's claims. We know this because on 29 November 1847, Riemann wrote to his father, saying that he had visited Eisenstein shortly beforehand and found him busy with his book; probably he was then working on the proofs. Riemann certainly viewed Eisenstein more like a competitor rather

[22] As Eisenstein noted in a letter to M.A. Stern [Hurwitz/Rudio 1895, 182], Gauss was the first to announce a proof of these results, whereas he (Eisenstein) was the first to publish a proof. Jacobi's claim, based on a proof he presented in his lectures, detracted in no way, according to Eisenstein, from Gauss's priority. This is but one example of the confusion that reigned throughout the nineteenth century, when no clear standards existed for adjudicating a priority dispute. Thus, a mathematician of high authority, like Gauss, could be credited for a new result, even when his "proof" was never presented publicly.

than as a friend, and he apparently distrusted him. Having noted Eisenstein's tactics in attempting to rebut Jacobi's remarks, Riemann had doubts this reply would have the desired effect. Eisenstein began by quoting Jacobi's words verbatim, and then commenting that this accusation:

> ...might raise the suspicion that my proofs were perhaps nothing more than copies or elaborations from those lecture courses. To this I must object by remarking that at the time I wrote and published those proofs, I was completely cut off from oral and literary communications (except for the works of Gauss and some volumes of Crelle's Journal), so that the investigations of Prof. Jacobi were completely unknown to me. Up until the present day, I have not found the opportunity to study any of the mentioned lecture notebooks in detail. [Eisenstein 1847, 333–334]

Eisenstein then threw some sand in Jacobi's face, by noting that in his very first paper in Crelle he had taken pains to explain how his work had arisen. In case Jacobi had missed those remarks, he now reiterated that his own work derived from the researches of "the creator and founder of the entire theory" – namely Gauss. As a closing touch, Eisenstein added his personal conviction that the lecture notebooks Jacobi claimed had circulated so widely after 1837 were, in fact, nowhere to be found in Berlin at the time Eisenstein composed his papers (*ibid.*).

In the letter to his father, Riemann cited Gauss's praise from the preface, but he also mentioned Eisenstein's note, about which he wrote: "he tries to wash himself as clean as possible from his dealings with Jacobi, but it seems to me he hasn't completely succeeded with that" [Neuenschwander 1981a, 96]. He related, furthermore, that Eisenstein's book ended with some open problems, about which Riemann mentioned the following:

> The main problem is the one that I worked on constantly during the holidays and which I brought close to an end. When I told him about some of my results without describing the path I took in finding these, he appeared to have great interest in learning about this matter. After my previous experiences with him, though, I thought it advisable not to pursue this line further with his accompaniment. He returned my visit the same evening and stayed with me for two hours. On leaving, he invited me to come visit him for tea, but I have not responded to his invitation. (*ibid.*)

Adding to the mystery surrounding this passage is the problem of discerning what Riemann meant by the "main problem" (*Hauptaufgabe*).[23] This probably referred to the following challenge: given a function satisfied by two different differential equations, to find a method for obtaining the *simplest* differential equation that has this function as a solution.

This appears especially intriguing since this problem may have been related to Riemann's earliest attempt to frame a theory of complex analytic functions

[23] Eisenstein listed essentially only three problems [Eisenstein 1847, 335–336]. The first dealt with the extension of Cayley's results on algebraic invariants, a problem Eisenstein had worked on himself. The second concerned certain division properties of elliptic integrals relating to the field of Gaussian numbers, a problem that fit Eisenstein's research agenda in number theory, though this was a field Riemann largely ignored. The third problem would thus appear to be the only one close to Riemann's general analytical interests.

based on the so-called Cauchy-Riemann differential equations. In any case, this would seem to accord with Richard Dedekind' mentioned above:

> Riemann later recounted that [he and Eisenstein] had also discussed the introduction of complex quantities into the theory of functions, but had completely differing opinions on the underlying principles; Eisenstein remained focused on formal calculations, while Riemann himself saw the essential definition of a function of a complex variable in terms of its partial differential equation. It is likely that he first thoroughly developed these ideas, which would prove decisive for his entire subsequent career, during the autumn holidays of 1847. [Dedekind 1876a, 512]

For a number of reasons, including mutual mistrust, Eisenstein's influence on Riemann was far more limited than in the cases of Dirichlet and Jacobi. After taking several courses with both men, he was surely struck by the contrast between their respective approaches to analysis. Jacobi was perhaps the greatest algorist since Euler, whereas Dirichlet's work reflected the principle of "less is more." Riemann, although a technician of the first rank, placed himself decidedly in Dirichlet's camp. As emphasized by Detlef Laugwitz, Riemann's conceptual approach to complex analysis marked a major shift that pointed to a new trend in higher mathematics [Laugwitz 1996/1999]. This takes us ahead of our immediate story, however, which toward the end of Riemann's second semester involved a dramatic and wholly unanticipated turn of events that profoundly affected life in Berlin.

# Chapter 2
# Years of Adventure, 1848–1853

Riemann's two-year stay in Berlin coincided with a period dominated by tumultuous political events. The year 1848 famously saw a chain reaction of revolutionary disruptions throughout much of Europe, including several German states. These uprisings effectively marked the end of the restoration era, the period that began in 1815 when the Congress of Vienna reinstated the monarchies Napoleon had toppled so swiftly. This system had kept established elites in power for over thirty years before unraveling, a process Riemann witnessed at close range in Berlin. Within Prussia, the events of 1848–1849 brought to a head the inherent conflict between the army as the guarantor of the monarchy and its purported neutral role in politics [Craig 1955, 82–135]. As for the larger context of German history, what transpired during those two years spelled the beginning of the end for the liberal movement. The ensuing period of reaction, dominated by Bismarck's politics of "blood and iron," culminated with the founding of a Prussian-dominated German empire in 1871.

## 2.1 Winds of Revolution

In February 1848, revolutionaries in Paris forced Louis Philippe, the so-called Citizen King, to abdicate. For those who lived in remote countryside villages, like Quickborn, news traveled slowly, so Riemann tried to keep his family abreast of events beyond those they could read about in regional newspapers. His father sent him a letter dated February 24th: the day Parisian revolutionaries drove Louis Philippe from the throne. Before going into exile, the King desperately tried to salvage power for the House of Orléans, but his opponents were determined to break with the past in order to found a Second Republic.

It took Bernhard two weeks to reply, a delay he tried to explain. His letter, written on 12 March, a Monday, indicated that all lectures would end that week due to the political turbulence. As it happened, his father's letter arrived just after the dramatic news from Paris had reached Berlin. This set off considerable



D. E. Rowe, *Bernhard Riemann: His Life and Wondrous Mathematical Legacy*,
https://doi.org/10.1007/978-3-032-25457-3_3

commotion in Berlin, especially among those who had long been calling for reform. Commenting about this, Riemann wrote that the 24th of February was a day whose consequences might well transform half of Europe:

> You can easily imagine the excitement here when the first telegraphic news arrived from France. Just as I received your letter, the news came that the Republic had been proclaimed and Louis Philippe's castles had been devastated; therefore, I had no peace of mind to answer your letter immediately, as I had planned. The letter to Ida also didn't arrive; I still wanted to write to her post festum, but the holidays are now so close that I can soon tell her verbally everything I now had to write. I also won't write about political news, like those occurring every day now, because I don't know how far your newspapers reach; but perhaps I'll still be able to tell you quite a lot during the holidays. [Neuenschwander 1981a, 98–99]

One might well wonder what Riemann meant by "telegraphic news," since Siemens and Halske would only begin constructing installations for electrical telegraphy one year later. Once this technology came on board, earlier optical systems for transmitting messages became obsolete. One such system, however, was not forgotten because it gave its name to the Telegraphenberg just outside Potsdam. In 1832, King Friedrich Wilhelm III ordered the construction of an optical telegraph line in order to transmit administrative and military messages to and from Berlin. The entire telegraph line comprised 62 stations, each furnished with a signal mast equipped with six cable-operated arms. Adjacent stations used telescopic sightings to read coded messages and then forward these to the next station. This system spanned nearly 550 kilometers from Berlin to Koblenz; at those two localities, as well as a third dispatch department located in Cologne, the messages were coded and decoded. Riemann thus knew about this means of conveying telegraphic news; that the citizens of Berlin knew so soon about the events in France also implies that the Prussian government decided to release this information before the arrival by courier of more detailed news reports from Paris.

On March 13, one day after Riemann wrote home, revolutionary violence broke out in Vienna, leading to clashes between the military and demonstrators. To restore calm, the government accepted Prince Metternich's resignation. News of his departure reached Berlin by the 15th, setting the stage for the events that followed. Tensions in Berlin had already been running high, as just two days earlier soldiers at the Brandenburg Gate clashed with demonstrators who were returning to the city. The scene turned ugly, leading to the death of one activist from a stab wound. The events in Vienna encouraged Berlin's demonstrators to push for immediate reforms, while for some the time seemed ripe to topple the monarchy in the name of a people's revolution. But who were the German people and who spoke for them? By mid-century one could hear many competing voices, some representing powerful new economic interests, others spoke in the name of the urban proletariat, and still others wanted only some new form of the status quo.

For the liberal bookshop owner Julius Springer, the issue of censorship stood at the top of his list of desirable reforms. In fact, in 1842, when he opened his shop just minutes away from the Royal Palace in Berlin, Springer was primarily

Fig. 2.1: *Der deutsche Michel* (Gullible Fritz), printed in 1842.

issuing political pamphlets and caricatures. Best known among the latter was *Der deutsche Michel* (Gullible Fritz), a lithograph by R. Sabatky (Fig. 2.1). In the background, guardsmen are commanded to stand at attention so as not to awaken the sleeping giant, while the Pope screams in rage, holding St. Peter's key. On the left, Prince Metternich draws Michel's blood, which magically turns into money. On the right, the British bulldog grabs a pouch of money from a beer stein, while a Frenchman saws away on his left arm, symbolizing the territory west of the Rhine. At the same time, the Russian czar wraps a bib around Michel showing 39 small numbered squares representing the states in the German Confederation. Note especially the padlock on Michel's mouth should he ever awaken. Caricatures fell under the censorship laws, which were rigid enough to discourage Springer from continuing to promote this form of political satire. In fact, during the reactionary period of the 1850s, he gradually shifted his publishing program away from political topics, though not completely. His sons would later expand the company into the fields of medicine and engineering, whereas his grandson, Ferdinand Jr., made Verlag Julius Springer into a publishing powerhouse for mathematics and physics during the Weimar era.

On the 17th, a Friday, news circulated throughout the city that a mass protest would take place before the Royal Palace on Saturday afternoon. That morning, Friedrich Wilhelm IV already feared the worst. He conferred with State Minister

Bodelschwingh about a plan to flee the city and take up refuge in nearby Potsdam. As crowds began to assemble just outside the palace, the King realized this course of action would be too risky. Shortly after noon, the throng reached a peak of perhaps ten thousand assorted demonstrators and onlookers, some of whom began calling out for the monarch to appear.

Fig. 2.2: A caricature of King Friedrich Wilhelm IV backed by Crown Prince Wilhelm as they barricade the palace door from a group of petitioners. The caption reads: "No scrap of paper will come between me and my people!".

Jacobi's patron, King Friedrich Wilhelm IV, was a throwback to the romantic ideal of a Christian monarch who saw himself as a kind and just sovereign. He tried to convince his people of his benevolence by various acts of political charity, only to become disappointed and dismayed that his critics kept returning with new demands (see the caricature in Fig. 2.2). On 11 April 1847, he brought about the United Prussian Parliament to represent the citizens in all regions of the state, but he regarded this institution as nothing more than a body representing the traditional estates. His opening speech openly rejected the notion of a constitutional monarchy based on shared powers. Instead of such liberal ideas, he spoke as a monarch who wanted to rule through the faith and love of his people and not by some artificially constructed constitution.

Around 1:30 p.m., the King stepped out onto the balcony to greet a wildly cheering crowd; others fell silent and waited anxiously to hear his message to the people. He was joined by Minister Bodelschwingh, who began reading the King's response to the people's demands. Probably few could hear him, but a text was distributed afterward listing a series of concessions, each beginning with the phrase: "The King wants ...." Recognizing the inevitability of a constitutional monarchy in Prussia and throughout the other German states, Friedrich Wilhelm IV tried to place himself ahead of these anticipated events. He set forth his views for a future government, pretending not only to support the liberal movement but to stand at its forefront. He thus proclaimed his desire to have a German national flag, while assuring his subjects that Prussia would take the lead in pursuing the national movement. He also committed himself to supporting freedom of the press, a state parliament, and a constitution for all German states based on the most liberal principles.

These were all fine words and sentiments, expressed in language thoroughly suitable for an autocrat. Those who could follow what was said would have realized that the King was not responding to the people's demands: he was telling the crowd what *he* supposedly wanted. Bodelschwingh knew, of course, that Friedrich Wilhelm IV based his authority on the divine right of kings; if his subjects happened to receive special rights or privileges from His Majesty, these had to be understood as *freely* given. In fact, as events later showed, this whole balcony scene was an absurd ruse, since what the King really wanted was to escape the fate of his counterparts in Paris and Vienna. His goal was to hang on to power by appeasing the liberals, which required playing along for the time being.

After the government's response was read, the crowd pushed forward trying to get closer to the scene of action. This worried the soldiers stationed in the courtyard, who feared they were facing a mob intent on storming the palace. Parts of the crowd began shouting loudly for the soldiers to withdraw, which was the one key demand the King had failed to address. He had called in the army to maintain order, but the reformers demanded that his soldiers be replaced by armed civilians (a *Bürgerwehr*). In response to this proposal, Friedrich Wilhelm appointed a new local army commander, who took immediate steps to secure the armory and other vital outposts.[1] As the situation at the palace grew more tense, the King apparently became angry and flustered. He gave orders that the soldiers were to disperse the crowd, a task more easily said than done. They were to move forward on horseback with sheathed swords into a hostile crowd that was urging their departure from the city. Chaos soon followed and some of the dragoons drew their sabers against the protesters, but then two shots rang out. The crowd panicked and fled into the streets. Within only a few hours, Berliners across the city blocked the streets with whatever they could find. On several barricades, they hoisted a black-red-gold flag, the traditional symbol for German unity and freedom. What followed can only be described as a bloodbath.

---

[1] This commander was General Karl Ludwig von Prittwitz, who gave a detailed personal account of the events of those days, later published in [Heinrich 1985].

## 2.2 Berliners at the Barricades

One of the more horrendous battles took place just down the street from Julius Springer's bookshop at the *Köllnisches Rathaus* (old City Hall). The novelist, Theodor Fontane, an eyewitness to the events of that day, later reported what he learned from one of its few survivors. He and a group of others were sharpshooters who defended the barricade from inside that building. They had succeeded in repulsing two charges on the barricade, when

> "immediately afterward, the battalion advanced for the third time, but more for show. While we again greeted its approach from our window, certain of seeing it retreat once more, we suddenly heard the heavy footsteps of grenadiers on the stairs leading up to us. They had entered the city hall ... from behind and to the side. Each of us knew we were lost. In a senseless urge for escape, everyone hid behind the large black tiled stove, while an inner voice called to me: 'Anywhere but there.' That saved me. I confronted the officer who was entering at the head of his men, received a saber blow over the head, and collapsed half-unconscious, but immediately afterward I heard shot after shot, for everyone who had hidden behind the stove, rifle in hand, was now dead ...."
>
> In the manner recounted here, most of the victims perished on March 18th, especially in the corner houses of Friedrichstrasse; the defenders retreated from stairway to stairway to the attics, hid behind the chimneys, were brought out, and massacred. [Fontane 1898, 339]

Riemann undoubtedly heard about what happened to Gotthold Eisenstein after he got swept up in this violence. Eisenstein testified about the mistreatment he received, a statement published already in 1848:

> During the night of the 18th to the 19th, I had taken refuge in a house when soldiers broke in to seize those who had fired from there. ... Although I assured them that I was a peaceful teacher at the local university, I was taken away without further ado. When we reached the corner of Leipziger Strasse, a soldier approached us with a drawn saber and, taking a wide swing, struck me with all his strength on the back, so hard that I fell into the middle of a group of soldiers, overcome by the force of the blow and the pain. Here I was met with violent blows from rifle butts on all sides, so that I immediately sank down and, thinking that they wanted to kill me, stretched out my hands in supplication. Nevertheless, they continued to hit me with the rifle butts; there were continual blows to my head, back, sides, and in short, every part of my body. Finally, I crawled, as it were, towards a place where several other prisoners were already located. [Biermann 1988, 74–75]

This barbaric treatment continued when Eisenstein and the others were marched off to the prison in Spandau. Already in frail health, he was barely able to recover from the physical abuse he suffered during those two days. Yet what bothered him especially was the mentality of his tormentors:

> All this torture was not characterized by agitation or anger, but rather by a devilish sneer, which often erupted in malicious jokes. ... I stand in amazed wonderment: how was this barbarism possible in the 19th century, in the midst of such an enlightened nation, on the part of such a disciplined and, until then, highly regarded military? [Biermann 1988, 75]

Over two hundred civilians, referred to as the "March Martyrs," lost their lives during the fighting at the Berlin barricades, the heaviest losses suffered during the course of the Revolution. The royal troops suffered comparatively fewer losses, with less than 50 dead. One of the victims was Gotthold Heine, brother of the mathematician Eduard Heine, who happened to be Gustav Dirichlet's brother-in-law. The twenty-nine-year-old historian died after being shot in the head in a local restaurant. In addition, over 600 others were wounded or, as in the case of Eisenstein, taken prisoner.

Like every literate person living in Berlin, Riemann read about these events, which were covered in detail by local journalists. Summarizing these reports for his brother Wilhelm, he wrote that the newspaper accounts indicated the demonstrators had successfully warded off the attacks on the street barricades as of Saturday night, March 18. This seemed to suggest that, militarily, the fighting had ended in a standoff, but in fact only one barricade actually withstood the attacks on that day. Several others, however, remained outside the zone of the soldiers' attacks. Riemann reported further that, according to these unnamed sources, the chaos had reached the point that had the fighting continued on Sunday, the people's anger would have likely led to a mob storming of the King's palace.

## 2.3 Politics of the Prussian Crown

Friedrich Wilhelm IV realized he had to make amends for this disastrous turn of events, if he hoped to remain in power. At the same time, he knew that Berlin was not Paris. He counted on his subjects' servile mentality, not to mention the strong aversion its bourgeoisie felt toward the rabble. It later emerged that nearly all those who died fighting at the barricades came from the lower rungs of Berlin society. As for those in the middle classes, their appeals aimed at democratic reforms that the King's father had already promised before his death in 1840.

Already on the afternoon of March 19, Friedrich Wilhelm began making amends by paying homage to those killed in the fighting. The bodies of 150 protesters, decorated with flowers and branches, were carried into the courtyard on carts while the King watched this procession from the balcony. Their upper bodies were exposed to show the wounds inflicted by the guns and bayonets of his soldiers. In response, he removed his headwear in a gesture of humility. That same day, he issued a proclamation, reaffirming his commitment to support a constitution that would uphold German unity and freedom. Two days later, on March 21, he rode through the city wearing a black-red-gold armband. Before him, a soldier in civilian dress carried the flag symbolizing German unity.

No one had foreseen such a tragedy, but the question now loomed: who would take responsibility for it? One figure who came under suspicion for his role in this fiasco was Crown Prince Wilhelm, who fled his palace on Unter den Linden on the evening of March 19, after which it was declared national property. Civilian guards thenceforth stood watch over the building to protect it from vandalism and looting. One of these happened to be Gustav Dirichlet. Uta Merzbach recalled a humorous episode from those days:

> While pacing up and down with his weapon near a gunpowder-holding structure, puffing on one of his usual cigars, Dirichlet was stopped by one of the guards, who pointed out to the professor that smoking was not permitted in the vicinity of firearms and gunpowder. The guard happened to be one of his students. Dirichlet gave up his cigar. [Merzbach 2018, 162]

A letter that Riemann wrote to his brother on 12 April 1848 provides a glimpse into the activities of the citizens' guard at the Royal Palace, where he and a large contingent of students from Halle kept watch over the Swiss Chamber, the antechamber of the royal family. They were on duty from 9 a.m. Friday morning on March 24 until 1 p.m. the next day.

> Toward evening, the King himself inspected the watch, asked the names of each of us, and spoke a few words with everyone. Also we had the opportunity in the company of a few noblemen, e.g., Prince Lichnowsky, to have a few words with the Crown Prince. We were treated to the royal kitchen, where we could write letters or play chess and other games. Of the 120 students from Halle who were there, a few even wore their fraternity uniforms.

Although Riemann probably heard nothing about what was planned for the next day, the King and Prince Wilhelm were preparing to depart in the morning for Potsdam. The timing for this trip was a sensitive matter, as many in Berlin, including those who were students, held the Crown Prince largely responsible for the massacres of the previous week. Rather than risk facing his accusers, he would later leave Potsdam and go into exile in London.

In the meantime, the National Assembly in Frankfurt began working on a constitution that would serve as the legal framework for a modern German state. Those charged with this task, the assembly's elected delegates, convened in Frankfurt's St. Paul's Church. In the spring of 1849, when Riemann was about to leave Berlin, thirty-two of those delegates arrived to present the Prussian court the results of their deliberations. King Friedrich Wilhelm IV summoned this delegation the following day, on April 3rd, but he only allowed them to enter the royal palace by way of the delivery entrance, thereby signaling the frigid reception that would follow. Since Prussia's formal response to the proposed National Constitution was crucial, the Kings of Hanover, Saxony, and Bavaria awaited Friedrich Wilhelm's decision before issuing their own decrees.

They did not need to wait long, as on 21 April the Prussian King formally rejected both the constitution as well as the imperial crown that went along with it. Although he had earlier signaled his willingness to accept leadership of a united Germany, Friedrich Wilhelm would not countenance any plan that would bind the Prussian crown to a "people's constitution." The other German monarchs now followed suit, recognizing that the political scales had been tipped against the liberals. There would be no peaceful solution to the constitutional crisis unless the liberal movement abandoned its goals, which meant acknowledging that the National Assembly's attempt to create a unified political framework for a German state had failed. This set the stage for the swift reaction that led to the collapse of the revolution.

We have no way of knowing how Riemann viewed this historic turn of events; he had already left Berlin in early April 1849 soon after the arrival of the Frankfurt delegation. Not much is known either about his final semester in Berlin, though a letter to his father underscores a continued interest in the works of Gauss, some of which, oddly enough, he had not been able to obtain in Göttingen. On 30 March, thus toward the end of his stay, he wrote:

> Easter is now just around the corner and I still haven't gotten around to writing a letter. You know only too well how I usually feel about it, and I therefore hope that you at least weren't worried about me. Thank God that things went pretty much as I wanted on the trip as well as here. ... When I arrived, the bayonets were staring at me, as the platform was occupied by soldiers and a non-commissioned officer went from coupé to coupé to ask for credentials. Anyone who could not provide sufficient identification was transported to the city bailiff, but I was allowed through with my student ID. ...

Already as a first-year student in Göttingen, Riemann had studied the pioneering works of Legendre and Jacobi on elliptic integrals and their associated functions. This was also the principal topic in the courses he took with Eisenstein. Not until this final semester, though, did he have the opportunity to hear Jacobi himself lecture on this topic. Jacobi's course gave him a solid grounding in the theory of elliptic functions, including the latter's approach via theta functions. He mentioned it briefly in the same letter to his father:

> I came at a very opportune time for the Collegia of Dirichlet and Jacobi; the latter had just begun a new part of his course on the theory of elliptic functions. He once again derived [everything] in a very clear and simple way. It was thus easy for me to grasp the context, and I have been attending regularly over the last four weeks. Unfortunately, I haven't been able to get a copy of these lectures yet, but I have been promised them for the next semester.[2] Dirichlet also arranged for my use of the Library, which led to none of the difficulties I had feared. Usually I arrive in the reading room around nine in the morning; I have been reading two papers by Gauss which I had been unable to obtain before elsewhere. Another work of Gauss, which won the prize in Copenhagen [Gauss 1825a], I looked for in vain in the data-books of the Royal Library, but now I have finally obtained it through Dr. Galle of the Observatory.[3] I am now ready to start studying it. [Neuenschwander 1981a, 99–100]

It comes as quite a surprise that Riemann had been unable to get his hands on [Gauss 1825a] during his first year in Göttingen. Of course as a first-year student in Göttingen, he would never have dared ask Gauss for his help. When he wanted to borrow a monograph from the university library, he always went to Benjamin Goldschmidt to obtain written authorization. One can only guess whether Riemann already knew what he was likely to find in Gauss's paper, but in any case it provided him with key support for a new theory of analytic functions of a complex variable, the topic of his dissertation. Translated

[2] Jacobi's lectures often circulated in written copies made by students who sold these to their fellow students.

[3] Johann Gottfried Galle had been Franz Encke's assistant since 1835. He is remembered today as the first astronomer to sight and identify the planet Neptune based on information he received from Urbain Le Verrier on the morning of 23 September 1846. Le Verrier had studied the perturbations of Uranus from which he derived the position of a yet unseen celestial body. Galle found it the very same night in the corresponding section of sky.

into Riemann's conceptual framework, Gauss's study showed that an analytic function could be described as a conformal mapping (see Chapter 5).

Riemann was already making plans for his lodging in Göttingen. He asked his father to send some money to his brother Wilhelm, since he intended to visit him in nearby Osterode before returning to Göttingen.[4] Once there, he planned to take Friedrich Wöhler's laboratory course in chemistry, perhaps the most innovative science offering at that time. For this course alone, he would need some extra funds, as the fee to attend was 2 Louis d'or.

Remarkably, Riemann's two-year stay in Berlin came just as Eisenstein had exploded onto the scene and when Jacobi stood at the height of his fame. Both, in different ways, bore scars from the political reaction that followed the events of 1848, when German liberals had high hopes for reforming their monarchies. Those who pushed for a full break, as had happened in France, constituted only a small minority, though one should not overlook that 1848 was the year in which Marx and Engels wrote their famous *Communist Manifesto*. Jacobi's involvement with politics amounted to no more than a speech he delivered at one of Berlin's constitutional clubs [Ahrens 1907]. Although little could be recalled afterward about the position he took on that occasion, some claimed he had maintained that the word "Republik" never gave him goose bumps.

No doubt the abdication of the King, to whom Jacobi owed his livelihood, was the last thing on his mind. Still, some sound bites can stick, especially in a heady atmosphere and when spoken with eloquence. Some of his rivals in that club surely knew that Professor Jacobi was a baptized Jew and dependent on royal largesse. Rumors eventually spread about a great controversy that had broken out over this speech, and reports about his alleged former political activities found their way into the Berlin press. Jacobi scrambled to save his reputation, but the King decided that the matter was too serious to be ignored and suspended part of his salary. He stayed in Berlin, but his family had to move to Gotha, where the cost of living was more affordable. Jacobi hung on, trying to rehabilitate himself, but then in February 1851 he contracted smallpox. He quickly succumbed to it and died at the age of 46.

Eisenstein never fully recovered from the brutal treatment he received in March 1848. His meager salary was afterward reduced, adding to his longstanding feelings of misery and isolation. He felt despised by his relatives in Berlin, who could never comprehend how a brilliant young man would choose a life of self-inflicted poverty. At the same time, his estrangement from Jacobi left him feeling bitter toward the entire academic establishment in Berlin. Eisenstein's dramatized self-pity only came to light many years later, when Alfred Stern,[5] the son of M.A. Stern, oversaw the publication of Eisenstein's letters to his father. Riemann was mentioned therein only once in a letter written shortly

[4] The town of Osterode lies 40 km northeast of Göttingen on the edge of the Harz Mountains. Wilhelm likely took a temporary position there or perhaps was doing a practicum to prepare for regular employment in the postal service.

[5] Alfred Stern became a prominent historian, best known for his 10-volume study *Geschichte Europas seit den Verträgen von 1815 bis zum Frankfurter Frieden von 1871*, written over the course of more than thirty years.

after his arrival in Berlin. When he met Eisenstein, Riemann presented a card of introduction written by Stern [Hurwitz/Rudio 1895, 184].

In a letter from November 1848, Eisenstein congratulated Stern on his promotion to full professor, which he characterized as "an attainment of our period of freedom, which will now soon come to an end,"

> because it seems to me that the events in Vienna and the [citizens'] disarmament taking place here puts the final stone in place for the suppression of the movement that began in France. I write this now as Berlin is under a state of siege. [Hurwitz/Rudio 1895, 192]

Immediately after Jacobi's death, Gauss wrote to Humboldt on 28 February 1851, pleading for Eisenstein, whom he later nominated for a corresponding membership in the Göttingen Scientific Society. In 1852, Eisenstein was elected to the Berlin Academy as Jacobi's successor. That same year, however, he died from tuberculosis at age 29.

In the meantime, Friedrich Wilhelm IV had assumed his former political colors. He revoked the 1848 Prussian Constitution, replacing it with a revised version that the Prussian Assembly ultimately adopted on 5 December, 1848. This document incorporated some of the fundamental rights demanded by liberals, but it also anchored Prussia's three-class voting system, which left workers and the poor without effective representation. Perhaps most important of all, it left the armed forces in complete control of the King, thereby making the military a state within a state. When the liberals later tried to block funding for reform of the army, President Otto von Bismarck provoked a constitutional crisis by bypassing the legislature, claiming it had no authority over the military budget.

As Riemann was readying to leave Berlin, he still had his eye on political developments that might yet reshape Hanover along with the other German states. A key figure among the liberals in Hanover was Johann Carl Bertram Stüve, a member of the Assembly of Estates who in 1837 lodged an official protest against King Ernst August when he annulled the constitution. That stroke of the pen had, indeed, set off protests throughout the land, the most famous of which were those of the "Göttingen Seven." In the wake of the March 1848 Revolution, the King tried to mollify the liberals by appointing Stüve as Interior Minister in the Hanoverian government. This led to a series of immediate reforms, as Stüve abolished censorship, eliminated certain class privileges, and separated the judiciary from the government administration. These reforms remained in place even after October 1850, when the so-called March ministers were forced to resign. Riemann had been following these and other political events, as reported in Berlin newspapers, and on 30 March 1849, he queried his father about the local situation in Dannenberg:

> Have you perhaps recently heard how Stüve's dispute with the 2nd Chamber is being addressed in Dannenberg, and what do the gentlemen say about the election of the King of Prussia as hereditary German Emperor? [Neuenschwander 1981a, 101]

This was written, in fact, only weeks before Friedrich Wilhelm IV categorically rejected that crown.

## 2.4 Resuming Studies in Göttingen

Throughout Riemann's lifetime travel was an arduous affair, especially over longer distances. The state of Hanover only began to lay down tracks for passenger trains in the 1840s, and these mainly served to connect the city of Hanover with other urban centers in the state. The main hub for train connections was actually in the town of Lehrte, approximately 20 kilometers east of Hanover. Beginning in 1846, Gauss's oldest son, Joseph, played an instrumental role in planning the railroad lines that eventually connected the capital city with other parts of the country. The line running southward from Göttingen to Kassel only opened in 1854. Not long thereafter, Gauss, who otherwise virtually never left town, went on an excursion with his daughter Therese to see what it looked like. That little trip by horse-drawn carriage turned into a near-nightmare experience. Just as they approached the tracks, a train went by, spooking the horses. This caused their carriage to flip over, seriously injuring the driver, yet rather miraculously the elderly Gauss and Therese emerged unscathed [Sartorius 1856/2012, 72].

In early April, Riemann was back in Göttingen and happy to report to his brother Wilhelm that his journey had gone well. Unfortunately, however, things had not been going well for him since his arrival because of mental and physical health problems. This is how he described his state at that time:

> I have felt very unwell since my arrival; I am terribly exhausted, every mental exertion is very difficult for me; I fear I have ruined myself ... through excessive work. ...

In the meantime, he paid their Aunt Julie a visit, and she advised her nephew to make an appointment with Dr. Ruete.

> I went to see him that afternoon – Aunt J. seemed to have already spoken to him about me – and he examined me for quite some time about my condition. He then prescribed a powder and advised that I take long walks and avoid studying mathematics or philosophy. He suggested instead lighter reading and history, but also that I go out in company. I have indeed taken long walks, but unfortunately I am unable to work; I haven't had the opportunity [to socialize] yet. Some of my acquaintances are no longer here, others I have neglected too much, and I therefore feel very much alone.

As usual, Bernhard was anxiously awaiting money from his father, but now he feared that his aunt might have written him about his son's present condition:

> I still hope that I will soon feel better and be able to continue my studies here. Since my illness seems to be located in my abdomen, it was probably exacerbated by the poor diet I maintained during the journey. Aunt J. will probably have already written to father about me, and according to her comments when I was with her, she may have advised that I spend the summer in Quickborn. Should this letter have already passed through your hands, please forward it to me as soon as possible. Even if, in the worst case scenario, I cannot continue my studies as I had planned and cannot work as much, it is still better if I remain here. [Neuenschwander 1981a, 101–102]

Before he sent this letter off, Bernhard was relieved to receive a message from Wilhelm that reassured him greatly; it seemed their aunt had not written to Quickborn after all. Moreover, Bernhard could also report back that over the last 24 hours he was feeling somewhat better and was no longer so melancholic and exhausted. During the course of that summer semester, he also learned that Wilhelm Weber had been reappointed to his chair in physics. Weber now offered what would become his standard course on experimental physics, together with a private practicum open to qualified students who obtained his permission. From this time forth, Weber's work and personality exerted a lasting influence on Riemann. Which courses he actually took that summer seems not to be known, but beyond Weber's lectures it seems likely Riemann would have been eager to take Goldschmidt's course on the theory and praxis of magnetic observations. This met at 8 o'clock in the morning, immediately before Sartorius von Waltershausen taught a 4-hour course on crystallography.

Riemann had already enjoyed good relations with Benjamin Goldschmidt during his first semester in Göttingen, so re-establishing contact with him no doubt helped provide the guidance he still needed. From this time forth, his interest in geomagnetic research, including the fine points of the instruments developed by Gauss and Weber never abated, so he would hardly have passed up this chance to deepen his knowledge of the subject. He passed up taking Friedrich Wöhler's 6-hour chemistry course with lab exercises, but took this later. Instead, he chose to take Sartorius's course and checked out three books on crystallography at the beginning of the semester [Neuenschwander 2022, 42].

Some of the other books Riemann was reading at this time reveal the breadth of his early interests. He would later delve deeply into the writings of Immanuel Kant, whereas during this first return semester he began with his *Prolegomena* (1783). Riemann also borrowed books and journals on astronomy, physics, and mathematics. He continued studying Gauss's works in the *Commentationes* and the *Göttingische Gelehrte Anzeigen*, but also studied books by Lagrange on algebra and analysis. He checked out several volumes of Liouville's *Journal* from the 1830s, among other publications. Beyond these scientific interests, Riemann read about cultural and political issues from the early 1840s in the periodical *Deutsche Jahrbücher für Wissenschaft und Kunst.* As a quite different indication of his intellectual interests, Riemann also borrowed a well-known survey of the history of Göttingen University covering the period 1820 to 1837 [Oesterley 1838]. The author was the recently deceased Georg Heinrich Oesterley, who came from a distinguished family in the town. Riemann was especially fascinated by the period when Gauss and Weber collaborated so fruitfully in their research on terrestrial magnetism. He may have even hoped to pursue a career in this fast-breaking area of research.

In the coming years, Riemann's relationships with Wilhelm Weber and Wolfgang Sartorius von Waltershausen would take on growing significance. Both were on very close terms with Gauss, who otherwise retained a comfortable distance in his relations with other members of the faculty, except for these two and his son-in-law, Heinrich Ewald. Riemann probably saw these three and a throng of other dignitaries enter the University Aula, where they attended a special event that took place there on the 16th of July, 1849. As a fourth-year student, he would have taken a seat toward the back, since all the seats up front were

Fig. 2.3: Bernhard Riemann, ca. 1855.

reserved for prominent professors, officials, and guests. They came to participate in a public meeting of the Göttingen Royal Society of Sciences celebrating the the fiftieth anniversary of an important academic event. Fifty years earlier to the day, Carl Friedrich Gauss took his doctoral degree in Helmstedt with a Latin dissertation presenting a new proof of the fundamental theorem of algebra (see Section 10.2). To commemorate this achievement, Gauss returned to this theme for the fourth time by submitting an essentially revised version of the original proof with commentary on it [Gauss 1863–1933, 3: 73–102].[6] Some of his general reflections in that publication left an impression on young Bernhard Riemann, as discussed in Section 10.5.

[6] All four proofs were reprinted in the third volume of his *Werke*, which appeared soon after Riemann's death in 1866 [Gauss 1863–1933, 3: 3–122].

Recognizing that Gauss's creative years were essentially over, Riemann only occasionally sought to draw him into a scientific exchange, evidently never with any real success.[7] This accords with what Sartorius later wrote about how Gauss's life changed after he completed this final work:

> From then on, Gauss seemed to rest on his laurels. He often explained to his closer friends that he did not want to get carried away with his scientific work and that his working hours were noticeably shorter compared with previous years. [Sartorius 1856/2012, 69]

Gauss also complained about the burden of having to teach in his seventies, though he continued to offer the same regular courses, in which he presented the material conscientiously. Who could or would have asked for anything more? Certainly not Riemann, who adopted his usual deferential behavior toward an esteemed professor. Their relations remained cordial, yet formal, in accordance with the standards of the time. Riemann seems to have viewed Gauss as a proud, but remote figure; possibly he also saw him as a model for the kind of scientific career he dreamed of one day attaining.

Riemann never observed any such Olympian remoteness on the part of Wilhelm Weber, a gregarious bachelor remembered for his outbursts of nervous laughter. Weber came from a very large family: it was even twice the size of Riemann's own. His father had been a theology professor, whereas two of Weber's brothers became scientists in Leipzig, where he himself taught until his position in Göttingen was restored in 1849. Riemann was on friendly terms with nearly everyone in this extended Weber family, several of whom regularly came to Göttingen for lengthy visits. At some point, he even fell in love with one of Wilhelm Weber's younger nieces, Laura Weber, but this turned into a traumatic experience that may have contributed to Riemann's emotional instability [Neuenschwander 1981a].

Laura's father was Ernst Heinrich Weber, a Leipzig physiologist who was nine years older than Wilhelm. This elder Weber is particularly remembered for the Weber-Fechner Law in pyschophysics, which he found together with his Leipzig colleague Gustav Fechner. Whether or not Riemann ever met Fechner remains unclear, but he definitely studied some of Fechner's highly speculative writings, which made a strong impression on him [Riemann 1876/1892/1902, 1876: 482–486]. Perhaps this was only a passing phase, but the young Riemann was hardly averse to bold ideas. Today a philosopher might imagine that Fechner was struggling with a contemporary version of the mind-body problem, though a closer account would stress the importance of empiricism and atomism as these relate to the divide separating materialism and vitalism.[8] Yet Fechner went so far as to argue that plants have souls, which live on as stable parts of the biosphere, an idea that for a time appealed to Riemann.

[7] Jacobi fared no better in this respect. After attending the Gauss jubilee in 1849 with Dirichlet, he complained about how impossible it was to engage in a scientific discussion with Gauss [Merzbach 2018, 172].

[8] For a sympathetic view of Fechner, see [Heidelberger 2004]; see also the older study [Laßwitz 1896], written by the author of an authoritative history of atomism [Laßwitz 1890].

Fig. 2.4: Richard Dedekind (1831–1916).

Richard Dedekind, (Fig. 2.4), who was five years younger than Riemann, arrived in Göttingen in 1850 from the Collegium Carolinum in Brunswick [Scharlau 1981]. As a native of that city, Dedekind had surely heard many stories about the great Gauss, who attended that same institution more than a half century earlier. During his first year in Göttingen, he naturally took Gauss's regularly offered course on the method of least squares, an invention Gauss had devised for handling data from astronomical observations. Dedekind recalled it fondly in [Dedekind 1901], but he also later admitted that during that era Göttingen could not hold a candle to the mathematics courses regularly taught in Berlin.[9] After only two years, he had exhausted everything the Göttingen faculty had to offer. Nevertheless, he stayed, later lamenting that much of what he needed for a deeper study of mathematics he had to learn on his own:

> There were no lectures on modern geometry or higher number theory, nor on cyclotomic equations and higher algebra, elliptic functions, and mathematical physics, all of which were brilliantly taught in Berlin by Steiner, Jacobi, and Dirichlet. In the years after I

[9] For an overview of how the Berlin-Göttingen rivalry unfolded, see [Rowe 2018, 3–68].

> took my doctorate (1852–54) I met with the greatest difficulties trying to fill these gaps somewhat. [Lorey 1916, 82]

Dedekind's appreciation for Riemann's mathematical talent was second to none, and beginning in the mid-1850s he became Riemann's single most important source of intellectual support. Nevertheless, as we shall see, Dedekind often felt frustrated by his friend's withdrawn and morbid outlook. By 1857, this had reached a point that Riemann was on the brink of suffering a debilitating nervous breakdown (see Section 3.6). As the passage above indicated, Dedekind's earlier years as a student in Göttingen were filled with other sorts of frustrations. He and Riemann surely interacted occasionally during the early 1850s, but virtually nothing is known about the nature of these encounters. Still, the world in which they then moved was so small that we can easily imagine some shared experiences which shaped their early thinking. One particularly relevant topic of mutual interest concerned the foundations of real analysis.

As noted in Section 1.3, Riemann's initial exposure to this subject came from M.A. Stern, an advocate of the methods favored by the so-called combinatorial school. Stern was generally regarded as a somewhat dry lecturer, whose courses nevertheless provided beginning students with a solid foundation. He also espoused views regarding the foundations of infinitesimal calculus and its history that appear strikingly similar to Riemann's own. We know this in Stern's case thanks to a notebook Richard Dedekind kept when he took Stern's course on definite integrals in the summer semester of 1850. In it, Dedekind recorded these brief remarks about the respective roles of Newton and Leibniz:

> Newton was its inventor. Leibniz introduced the concept of the infinitely small into its principles, which cannot actually be justified. Newton regarded the ratio of two infinitesimally small quantities as the fundamental conception, and Euler adopted this as well. Lagrange's methods merely hid the difficulties of the differential calculus without resolving them. For that reason, one has returned to Newton's original standpoint, if not quite generally. [Dugac 1976, 154]

These remarks should be set against comments Riemann made during the early 1850s about Newton's approach:

> The method Newton applied in grounding the infinitesimal calculus was the method of limits. This is recognized by the best mathematicians of this century as the only method that delivers secure results. Instead of employing a continuous passage that takes the values from one position to another ... it first locates a finite number of intermediate positions and then allows this number to grow, thereby making the distance between two neighboring places to decrease accordingly. [Riemann 1876/1892/1902, 1892: 525]

In effect, Riemann's argument suggested that mathematicians should avoid trying to tackle properties of continuous magnitudes synthetically. Newton spoke of fluid quantities that vary with time, of course, but his method came down to analyzing these by means of finite entities that were at each stage tractable. We find this same idea in Stern's course, though expressed more clearly: he emphasized that the notion of continuous magnitudes that change over time is not amenable to further mathematical analysis.

> ...In mathematics, all our investigations start from the world of bodies, from which one first arrives at the concept of a continuum or a continuous magnitude. We can very well conceive of something fixed and complete, but not continuously changing, such as the passage from part of a continuum to the next or the passage from rest to motion. Mathematics does not analyze the concept of the continuum, but rather leaves its justification as a problem for philosophy, while accepting this notion as something that exists for us conceptually. [Dugac 1976, 154]

Dedekind also spoke highly about Stern's course, which appealed to his general philosophical inclinations regarding the foundations of mathematics. On the other hand, Dedekind later grew dissatisfied with the overall state of real analysis during the 1850s, which led him to propose an entirely new arithmetical theory of the continuum in [Dedekind 1872]. The conception behind this famous booklet had already occurred to him when he began his teaching career in 1858. His daring conception of what came to be known as "Dedekind cuts" pointed in the direction of a modernist trend that would dominate mathematical research for nearly a century. Riemann had far fewer logical scruples, nor did he find the challenge of securing the foundations of an established theory an attractive task. Nevertheless, he was certainly aware of the need to rigorize the existence theorems he based on Dirichlet's principle. Many who found Riemann's approach to complex analysis dissatisfying, focused their criticism on one simple fact: that a sequence of decreasing positive real numbers need not attain a minimum value.

In the meantime, Riemann took the opportunity in Göttingen to deepen his growing interests in natural philosophy. In November 1850, as a member of the pedagogical seminar, he composed an essay entitled "On the Scope, Organization, and Method of Science Teaching in Secondary Schools." The following excerpt reflects how Riemann was striving to find a unified conception of nature:

> ...a completely self-contained and comprehensive mathematical theory can be formulated, which progresses from the elementary laws applicable to individual points to the processes that unfold in the actual continuous plenum of space, without distinguishing whether these pertain to gravity, electricity, magnetism, or the equilibrium of heat. [Dedekind 1876a, 545]

This passage, vague though it may be, suggests how Riemann's views were inspired by the work of Gauss and Weber on applications of potential theory. Such ideas and methods played a major role in his own work throughout the 1850s.[10]

The pedagogical seminar was under the direction of Karl Friedrich Hermann, professor of classics and archeology. Students majoring in mathematics and physics could also take the mathematical-physical seminar, which was headed by two professors in each field. This institution was established in 1850 on the initiative of Moritz Abraham Stern, following a model introduced a short time earlier at the Prussian university in Halle [Folkerts 2024]. Its main purpose aimed to improve the educational standards for future teachers, especially those qualified to teach higher-level classes at the Gymnasien. Since those without other employment possibilities inevitably chose to become teachers at secondary schools,

---

[10] During this semester, Riemann checked our several issues of Karl Mager, Hrsg., *Pädagogische Revue. Centralorgan für Pädagogik, Didaktik und Culturpolitik.*

students usually sought to pass the state examinations required to attain such positions, which were civil service appointments. Riemann, however, apparently never made plans to do so, as members of the philosophical faculty in Göttingen learned early on that he hoped to join them as a private lecturer (*Privatdozent*).

Dedekind wrote about how Riemann's interests in physical experiments distracted him from writing his dissertation, which he only submitted in November 1851 [Dedekind 1876a, 545]. In fact, one might view this the other way around, as throughout the early 1850s his two dissertations only interrupted his ongoing work in theoretical and experimental physics. As a student of Weber, he naturally enrolled in the physics section of the mathematical-physical seminar, and he was paid 20 Taler for teaching exercises to the younger members [Folkerts 2024, 107]. He also delivered a lecture on 22 May 1851, in this seminar on the properties of the bifilar magnetometer invented by Gauss. He based his presentation on Gauss's paper [Gauss 1840b], which explained the theory behind this instrument (see Section 9.2). Riemann had surely first learned about it from Goldschmidt, either in his introductory course on geomagnetism or when he took his course on the experimental basis of terrestrial magnetism. Only three months before Riemann presented his talk, forty-three-year-old Benjamin Goldschmidt was found dead in his apartment at the astronomical observatory. As a multi-talented scientist, Goldschmidt had served for a number of years as Gauss's mainstay of support at the observatory, which now raised the question of what could be done to replace him. Time was not on Riemann's side.

## 2.5 Promotion to Dr. Phil.

Part of the mystery surrounding Riemann's mathematics involves the debts he owed to other mathematicians, whose works he only rarely cited. At times, he may have consciously suppressed such information, but in those days authors were not expected to acknowledge their sources diligently. In Riemann's case, we know he was extremely well-read; moreover, he had immersed himself in the works and ideas of Gauss from the time he first entered the university, a point emphasized in [Neuenschwander 1981b, 89–90]. He occasionally interacted with Gauss and sometimes even imagined himself competing with him, particularly in mathematical studies of electricity and magnetism. Nevertheless, despite their close physical proximity, Gauss's actual influence on Riemann was rather akin to an action-at-a-distance force. Nor should we imagine that Gauss acted as Riemann's mentor in the modern sense of that term; mentoring was essentially a twentieth-century innovation. A close examination of his promotion procedure will help clarify the differences between now and then.

On November 14, 1851, Riemann applied to the Dean of the Philosophical Faculty to be admitted to take the final oral exams for his doctorate. Besides his dissertation [Riemann 1851/1876], he also submitted a formal letter of application and his curriculum vita, both written in Latin as stipulated by the faculty. The presiding Dean to whom Riemann presented his application was the oriental philologist Heinrich Ewald, Gauss's son-in-law. Ewald had also been one of the "Göttingen Seven," so he lost his Göttingen professorship in 1837, but like Wilhelm Weber he was later allowed to return.

The day after he received Riemann's documents, Ewald wrote to inform a committee of full professors on the faculty:

> I have to put before my distinguished colleagues the treatise of a new candidate for our doctorate, Mr. B. Riemann from Breselenz; and entreat Privy Counselor Gauss for an opinion on the latter and, if it proves to be satisfactory, for an appropriate indication of the day and the hour when the oral examination could be held. The candidate wants to be examined in mathematics and physics. The Latin in the request and the vita is clumsy and scarcely endurable: however, outside the philological sciences, one can hardly expect at present anything better, even from those who like this candidate are prepared to enter on a career at the university. [Remmert 1993, 45]

Riemann's opinion of Professor Ewald as an administrator was scarcely any better (see below).

On 24 November, Bernhard wrote his younger brother Wilhelm about where matters stood. In the meantime, Gauss had read the dissertation and reported on it, though the contents of such reports were for internal use alone. Riemann undoubtedly never saw it, but he did meet with Gauss a short time before the latter sent his report to the faculty. He thus could inform Wilhelm about this conversation, during which Gauss made some mysterious remarks:

> When I was with Gauss on Saturday, he had not yet read my treatise, but told me that he had been preparing a paper for years (and was working on it right now), the subject of which was the same, or at least partly the same, as the one I had dealt with. He had indeed already indicated in his doctoral dissertation 52 years ago[11] that he intended to write about this subject. I have not yet heard his opinion of my work, . . . however, I must conclude from the comments of others, which I could not take to be mere pleasantries, that he was satisfied with my work. [Neuenschwander 1981a, 104]

In his biographical sketch of Riemann, Dedekind cited the passage above, indicating that Gauss was preparing a paper on a related topic [Dedekind 1876a, 545]. Gauss often made such vague remarks, whereas Riemann often worried about what Gauss might still publish, but of course never did. If, as Riemann wrote, Gauss had not yet read his dissertation, surely this was only a harmless remark, even though Riemann clearly took it seriously.

Riemann's letter makes clear that Gauss took no part in advising him on dissertation work, nor would such mentoring have conformed with the norms of that day. As Dedekind noted, Riemann discussed his work with Gauss *after* it had already been submitted, thus immediately before taking his orals, for which Gauss acted as the examiner in mathematics and Weber in physics. It would appear likely that Riemann mentioned this same discussion with Gauss when he and Enrico Betti spoke about how one could use cross cuts of a surface to derive its connectivity, as Riemann had done in his dissertation. Betti apparently

[11] Riemann surely attended the special public meeting of the Royal Society held on 16 July, 1849 (see Section 2.4). This is when Gauss spoke about his dissertation, in which he gave his first proof of the fundamental theorem of algebra.

misconstrued what he heard when he later wrote to Placido Tardy that Riemann first learned this method from Gauss [Weil 1979].

The above letter was written two days after he met with Gauss, by which time Riemann had evidently spoken with faculty members who either saw or knew that Gauss would file a positive report. That news surely relieved a good deal of the pressure before he took his oral exams. Gauss's brief report read as follows:

> The paper submitted by Mr. Riemann bears conclusive evidence of the profound and penetrating studies of the author in the area to which the topic dealt with belongs, of a diligent, genuinely mathematical spirit of research, and of a laudable and productive independence. The presentation is circumspect and concise, in part even elegant: yet the majority of readers might well wish in some parts a still greater transparency of arrangement. The whole is a worthy and valuable work, not only meeting the requisite standards which are commonly expected from doctoral dissertations, but surpassing them by far. [Remmert 1993, 47]

Gauss was often stingy with his praise, so these few words spoke volumes, even if he said nothing whatsoever about the content of the text.[12] This gave a new definition of analytic functions and described their fundamental properties, while deriving their existence by means of what he later called Dirichlet's principle. It introduced Riemann surfaces for multi-valued functions, and ended by giving a proof of the Riemann mapping theorem. In short, it was an astoundingly original piece of work in a virtually new field, complex analysis, to which both Gauss and Cauchy had made fundamental contributions. Even though Riemann published his doctoral dissertation, a standard requirement, very few people read it at that time. Mathematicians only became aware of it after the appearance of his monumental study on Abelian functions [Riemann 1857b], in which he referred directly to [Riemann 1851/1876]. As Gauss anticipated, those who tried to read it found the text difficult and rather obscure, partly due to its novelty. Nevertheless, Riemann's dissertation holds a very special place among his many varied works. In it, he already laid the foundations for his single most fruitful contribution to mathematical knowledge: the theory of complex-valued functions, culminating with his revolutionary theory of Abelian functions.

What now followed should help illuminate how the ritual of obtaining a doctorate in Riemann's day was strikingly different from what we imagine today. The philosophical faculty as a whole was responsible for conferring a doctoral degree, and those professors who took part in the procedure received payment from the state (in this case through the Ministry in Hanover) for their services. As the Dean had requested, Gauss suggested possible times for the oral, and these met with the approval of the other signatories: the 91-year-old classical philologist Christian Wilhelm Mitscherlich, the minerologist Friedrich Hausmann, the philosopher Heinrich Ritter, the classical archeologist Karl Friedrich Hermann, the historian Georg Waitz, Karl Hoeck, a classical philologist and librarian, and Wilhelm Weber. This group of nine represented half of the full faculty, and its membership was apparently based on seniority alone. One notes also the large

[12] For a discussion of the contents of Riemann's dissertation, see Section 5.2.

number of classical philologists, in accordance with their long dominant role in the philosophical faculty.

The oral examinations were now to be held on December 3, which Riemann assumed would be the end of his ordeal. He had applied for dispensation from the custom of defending theses in a public dispute, but this plea was refused. On 13 December, he vented his anger over this in a letter to his brother:

> Unfortunately, I have had to deal with a dean (Professor Ewald) for my doctorate who is a terrible pedant and troublemaker; he kept me waiting for more than five weeks after I handed in my manuscript, whereas the process should take barely a fortnight. My exam finally took place on Wednesday, eight days ago. Now he is demanding a public disputation, even though this has not taken place in the philosophical faculty for ten years. ... I have agreed to do so, in order to avoid further delays, and today I finally received word that he has scheduled the disputation for Dec. 16. [Neuenschwander 1981a, 104–105]

Riemann's anger was perhaps understandable, but its focus was misplaced. Ewald had by no means acted alone; he simply wanted to uphold the statutory requirements. Moreover, Gauss had independently reached the same conclusion. Only Weber had voted in favor of granting Riemann dispensation from the formal dispute, but for an odd reason. He knew Riemann was a shy loner and thought he might find it difficult to persuade a fellow student to play the role of his opponent in a formal dispute. Weber thought that such a circumstance might place the faculty in the awkward predicament of having to appoint such an opponent. Gauss, however, was unmoved: "Without a disputation, a doctor is no real doctor." He also noted that Riemann wanted to become a private lecturer (*Privatdozent*), which meant he would have to meet the requirements for Habilitation. These stated that candidates who had not previously passed a formal dispute in Latin needed to do so as part of that procedure. This being the case, Gauss saw no point in waiving this requirement, and his colleagues concurred [Remmert 1993, 47–48].

Riemann probably never knew that Gauss, and not Ewald, had played a leading role in quashing his petition, nor will anyone likely discover the name(s) of Riemann's opponent(s) in the disputation that took place on 16 December. Perhaps none were needed after all, though what happened that day remains quite unclear. Since no record of what transpired during Riemann's disputation survives, all that can be said with assurance is that the candidate survived this ordeal. Otherwise, this remains yet another mystery in Riemann's life (for more details, see Section 5.1). Dean Ewald's final report to the Ministry was as succinct as humanly possible:

> Bernhard Riemann propter eximian eruditoricum in scientii mathematicis et physicisi commentatione typis excusa, examine, disputatione publica approbatam (in rough translation: Bernhard Riemann, because of extraordinary learning in the mathematical and physical sciences as discharged in the printed treatise, was examined and approved in public disputation).

As a final remark, we might recall that Riemann claimed to his brother that a formal Latin dispute had not taken place in the Göttingen philosophical faculty for a good ten years. Whether or not that was true, we might guess that Riemann's was one of the last. It also appears very doubtful that he spoke for long in Latin, and this would no doubt have soothed Professor Ewald's nerves. Alas, we don't know whether Riemann actually faced a formal opponent who took the other side in the argument. Given the technical nature of the propositions he defended, this would have surely led to a contrived situation, and Weber's argument may have prevailed in the end, albeit in the form of this compromise solution. A thesis defense, as we have seen, was not part of the requirements for the doctorate, and yet Riemann's disputation probably had much in common with what later became the standard procedure at German universities and elsewhere. Having cleared this first hurdle, he could now prepare himself for the next: *Habilitation.* In Göttingen, those who held the Dr. Phil. were expected to wait a minimum of two years before applying.

## 2.6 Preparing to Habilitate

Pursuing an academic career at a German university in Riemann's day still had much in common with applying to enter a medieval craft guild. Instead of guild masters, one faced the professors in a faculty, who were the guardians of the traditions they held dear. A student thus began in the role of an apprentice (*Lehrling*), eventually becoming a journeyman-doctor (*Wandergeselle*). Usually after studying at two or more universities, a doctor could apply to habilitate somewhere. To reach this first rung of the academic ladder, the candidate presented the philosophical faculty with a postdoctoral thesis, the *Habilitionsschrift.* This requirement corresponded to the production of a "masterpiece" (*Meisterwerk*) when applying to become a full member of a guild. In addition, the faculty usually expected the candidate to deliver a lecture on a suitable topic that they approved in advance. If the thesis was deemed acceptable, the lecture (*Habilitationsvortrag*) was then rarely ever an obstacle. Once approved, the candidate was granted the *Venia legendi*, which gave him the right (and duty) to offer lecture courses as a *Privatdozent* and to collect course fees from students while the university paid him nothing. In Riemann's case, these formal hurdles proved to be particularly eventful.

Little is known about Riemann's activities during the first several months after he took his doctorate. He surely felt gratified to learn, even if only indirectly, that Gauss had lauded the originality of his work, and in a letter to his father he painted an optimistic future:

> With my dissertation now completed, I believe I have significantly improved my prospects; I also hope that I will learn to write more fluently and quickly over time, especially when I seek more contact and also have the opportunity to give lectures. I am, therefore, now in good spirits. [Dedekind 1876a, 545]

Riemann knew, of course, that his family could not afford to support his venture for a long time. Nevertheless, by the time he submitted his doctoral dissertation, he had received assurances – from whom we do not know – that he could confidently prepare to habilitate in Göttingen. Given the limited number of professorships at the German universities, few young men had the ability and ambition to seek such a position. Riemann, too, undoubtedly thought about the usual option for those who took doctorates in mathematics, namely to teach at a Gymnasium. Whether they completed that degree or not, the crucial step was to prepare for and pass the state examination for teachers. Nearly all candidates took this exam, but Riemann apparently chose not to do so, focusing instead on the pursuit of a university career.

In this connection, Dedekind noted that Riemann apologized to his family for not having exerted himself in an effort to gain Goldschmidt's former position at the observatory, though at this stage of his career, when he still lacked a postdoctoral credential, he would not have been formally qualified. Later, as we shall see in Section 3.6, he did try to gain a comparable post at the observatory, but without success. Curiously, Dedekind added a parenthetical remark made by Wilhelm Weber relating to the earlier situation in 1851/52:

> According to a communication from W. Weber, Gauss himself did not want Riemann to take on this position; although he did not doubt his theoretical and practical aptitude for it, he already had such a high opinion of Riemann's scientific importance that he feared Riemann might be too distracted from his actual field of work by the time-consuming and partly subordinate official duties associated with this position. [Dedekind 1876a, 546]

Leaving aside the fact that Weber was recalling what Gauss allegedly said more than twenty years earlier, nothing suggests that Gauss undertook any significant steps to promote Riemann's career.

Nor did the topic of Riemann's postdoctoral thesis owe anything to the work of Gauss. In fact, its theme was closely related to an earlier study published by Dirichlet on the representation of real-valued functions by trigonometric series [Dirichlet 1829]. When Riemann first learned about this remains unclear, but Dirichlet took up this problem long before in connection with the theory of heat as represented by Fourier series. Dirichlet had the opportunity to learn about that theory firsthand when he studied under Fourier in Paris during the 1820s. His paper from 1829 appeared in Volume 4 of Crelle's Journal, a widely accessible publication, but we have no clues as to when and how Riemann first read it. One possible scenario suggests itself, however, namely that this topic came up when Riemann was studying in Berlin, at which time Dirichlet may have given him an offprint of this paper. If so, he may have also informed Riemann of a subsequent paper on the same topic published by the astronomer Friedrich Wilhelm Bessel in *Astronomische Nachrichten* [Bessel 1839].

Whether or not such a conversation actually took place, we can be sure that Riemann was already familiar with Bessel's paper even before he had completed his doctorate. In fact, during the winter semester of 1851/52 and the summer semester following, library records indicate that Riemann only borrowed a single work: Volumes 15 and 16 of *Astronomische Nachrichten*, checked out on 11 November 1851, and 5 May 1852. In [Riemann 1868/1876], he would argue that Bessel's presentation was misleading since it failed to improve on the results given in [Dirichlet 1829]. By the fall of 1852, Riemann was putting the last touches on his *Habilschrift*, which began with a lengthy historical account starting with the famous eighteenth-century debates over the vibrating string problem. Riemann had the good fortune that just at this time Dirichlet arrived to visit with Gauss and other colleagues in Göttingen.

In September 1852, Riemann learned from Gauss that this distinguished visitor had arrived, an event Bernhard later reported to his brother:

> Tuesday afternoon I visited Gauss, who told me that Prof. Dirichlet was here from Berlin. I attended many of his lectures in Berlin and consider him, after Gauss, to be the greatest living mathematician; Gauss also thinks very highly of him. [Neuenschwander 1981a, 106–107]

Dirichlet was staying at the Crown Hotel near the middle of town, where Riemann went to look for him the next day. When he arrived, Dirichlet had just finished eating, but he joined him for a cup of coffee and some pleasant conversation. They then went to Dirichlet's room and began discussing mathematics. Before long, they were joined by the physicist Benedict Listing and the geologist Wolfgang Sartorius von Waltershausen, who came to fetch Dirichlet for a walk up the Hainberg. Since it was raining, they decided to wait another hour before leaving. They then hiked up to the Kehr, a traditional resting spot, only returning that evening around eight o'clock. Riemann gratefully accepted an invitation from Sartorius to dine with him and Dirichlet at the Krone on the following day. They were joined by Listing and another physics professor, Heinrich Wilhelm Dove, whose lectures Riemann had attended in Berlin. As he reported to his brother, "the session lasted until 9 in the evening and was, as you can imagine, extremely instructive and interesting for me."

A few days later, on 30 September, Riemann wrote Wilhelm once again at length about his recent experiences:

> On Friday morning, in order to continue my report, Dirichlet came to see me in my room. I had asked him for advice on my work and he gave me the necessary notes so completely that it made things much easier for me. I otherwise would have had to search for some things in the library for a long time. D[irichlet] was generally extremely kind, told me what he had been working on over the last few years, went through my dissertation with me; and so I hope that he will not forget me later and will give me his sympathy. He is still here now and I had the opportunity to see him twice later for a longer period of time. [Neuenschwander 1981a, 107]

Soon after Dirichlet's visit, Riemann checked out several works from the University Library, including Fourier's *Théorie de la Chaleur* (1822), Volume 1 of d'Alembert's *Opuscules Mathématique* (1761), and a volume containing an early paper by Fourier from 1807 [Neuenschwander 2022, 45]. These were among the works Riemann later cited in the historical introduction to his thesis. This even included mention of an unpublished paper by the elderly J.L. Lagrange, attacking Fourier's claim that an arbitrary function can always be represented by an infinite trigonometric series. Riemann added in a footnote that he owed this information to Dirichlet, who informed him that this paper had been deposited in the archive of the Paris Academy [Riemann 1868/1876, 233].

Riemann's report to his brother about Dirichlet's visit continued as follows:

> Weber had just traveled to the naturalists' meeting in Wiesbaden the day before my arrival, and he was expected back any day at the end of last week. After I had been to his house a few times in vain, I met him on Monday in the company of Listing and Dirichlet. After the first greetings I was invited to join them on a trip to the Upper Hagen. After a good breakfast, the four of us, W[eber], L[isting], D[irichlet] and myself, set off at about 12 o'clock on foot and a good hour later the rest of the party caught up with us coming from Göttingen with the omnibus. This group consisted of Weber's two nieces, who are now with him (Sophie and Anna) and Professor Ewald together with his two parents, and finance councilor [Heinrich August] Schleiermacher from Darmstadt. We drank coffee near the top of the mountain under the green beeches; we only left at sunset and returned home in the most beautiful moonlight and to the singing of our beautiful companions.
>
> Afterward I saw Dirichlet again, yesterday at noon, where Weber had invited me to dine. In addition to him, Schleiermacher and four local professors were also there. You can see that I haven't stayed home much so far, but I worked all the more diligently in the morning. The philologists' meeting, which you have probably read about in the newspapers and which has been going on since the beginning of this week, is now in full motion here; but I have not let it tempt me from my room. [Neuenschwander 1981a, 107–108]

Over the course of this winter semester of 1852/53, Riemann continued to study the historical background that preceded Dirichlet's work on representations of functions by means of trigonometric series. In February 1853, he checked out several volumes of Crelle's Journal and, once again, Volume 16 of *Astronomische Nachrichten* containing Bessel's attempt to improve on Dirichlet's results [Bessel 1839]. Something led him to Volume 12 of the series *Commentarii academiae scientiarum Petropolitanae*, which contains Euler's startling calculations related to the zeta-function [Euler 1750]. This may well be an indication that Riemann had already begun to contemplate the distribution of prime numbers many years before he wrote about this in [Riemann 1859/1876].

It was also around this time that he began to explore the historical background that led to the famous priority clash between Newton and Leibniz over the calculus [Hall 1980]. On February 25, he borrowed three volumes of *Acta Eruditorum* – though oddly not the one containing Leibniz's famous paper from 1684 [Leibniz 1920] – as well as Newton's *Principia* and the first four volumes of Samuel Horsley's edition of Newton's *Opera*. A few days later, Riemann checked out James Gregory's *Vera Circuli et Hyperbolae Quadratura*, which presents his use of convergent series to approximate the areas of the circle and hyperbola. Riemann pursued the study of works by Leibniz and Newton in later years, during which he engaged deeply with Newton's ideas about methods of inquiry in natural philosophy.

## 2.7 Physics and Philosophy

During the early months of 1853, Riemann's interests suddenly shifted away from mathematics to a variety of questions pertaining to physics and philosophy.

This shift was apparently sparked by his reading during the previous semester, especially his encounter with Newton's *Principia* [Newton 1713/1934]. Pinpointing the evolution of Riemann's thinking is nearly always a hopeless task, but we are fortunate in this case that he noted how the ideas for one of his more interesting texts came to him on 1 March 1853, thus two weeks before the winter semester ended. Only a few days earlier, on February 25, he had checked out a copy of the *Principia* [Neuenschwander 2022, 46]. After reading parts of it, he proposed a remarkable theory, which he called "New Mathematical Principles of Natural Philosophy" [Riemann 1876/1892/1902, 1892: 528–532].

> [Its] purpose is to penetrate into the interior of nature, going beyond the foundations of astronomy and physics as laid down by Galileo and Newton. ... The foundation for the general laws of motion of ponderable bodies presented at the beginning of Newton's Principia lies in the internal state of those bodies. Let us attempt to form an analogy between these states and our own inner mode of perception. New image masses constantly arise in us and very rapidly disappear again from our consciousness. We observe a constant activity of our mind. Every mental act is based upon something enduring, which is manifest (through memory) on certain occasions, without exerting a lasting influence on the phenomena. Thus (with every act of thinking) something enduring continually enters our mind, which does not, however, exert a lasting influence upon the world of phenomena. Every mental act, therefore, is based upon something enduring, which enters our mind with the act, but at the same moment completely disappears from the world of phenomena. [Riemann 1876/1892/1902, 1892: 528–529]

Riemann's analogy thus sought a Newtonian theory that could describe the inner-workings of mental processes, a novel idea that may have come to him out of the blue. During this period when Riemann was deeply engaged with philosophical as well as physical speculations, he often cited the work of the philosopher and psychologist Johann Friedrich Herbart as an important source supporting these views. Here, however, the speculative writings of Gustav Fechner may have sparked these daring thoughts.[13] Whatever the case may have been, the consequences Riemann drew from these speculations were both startling and seemingly fantastic:

> Guided by this fact, I put forward the hypothesis that there is a kind of space-filling substance which continually flows into ponderable atoms and there disappears from the world of phenomena (the corporeal world). ...
>
> The reason the substance disappears at this particular point can be attributed to the thought matter which was formed in the immediately preceding period: the ponderable bodies are accordingly the place where the world of mind engages the corporeal world.[14] [Riemann 1876/1892/1902, 1892: 529]

It would appear that Riemann explored these ideas for some time, but the mathematics only led him to an impasse.[15]

---

[13] Norton Wise considered Riemann's physical views in relation to those of both Wilhelm Weber and Gustav Fechner; see [Wise 1981, 288–295].

[14] Riemann added a footnote here, noting the relevance of Herbart's psychological theory, according to which the individual mental images have substantiality rather than the soul itself. This reference to Herbart may have been written somewhat later.

[15] See the discussion in [Bottazzini & Tazzioli 1995, 12–15].

The above passages were likely written in March 1853, around which time Riemann may well have written *all* of the unpublished texts Dedekind and Heinrich Weber chose to include in [Riemann 1876/1892/1902, 477–506], the Appendix to the *Werke*.[16] One should note that for the second edition Weber changed the original ordering to reflect the more likely chronology, though what really should have been made clear was the role the editors played in reconstructing this supplementary section. Their objective was surely to make it somewhat more readable, but in doing so they added a tripartite structure to which they gave appropriate headings. Since the original sources no longer exist, a careful reader should place red flags at every clear break in the Appendix, keeping in mind that the ordering of the various pieces reflects the editors' attempts to establish some reasonable coherence.

My conjecture that the fragments in the Appendix were written on or after March 1853 is supported by two sources. First, Dedekind wrote that "the beginning of 1853 saw an almost exclusive preoccupation with natural philosophy; [Riemann's] new ideas took on a firm form to which he always returned, despite all interruptions" [Dedekind 1876a, 515]. Secondly, Neuenschwander's list of works Riemann borrowed from the Göttingen University Library indicates that before the summer of 1853 he had only checked out three philosophical books, namely those by Kant, Hegel, and Trendelenburg. These evidently left no lasting impression, whereas reading Newton clearly did.

The same applied to the works of Herbart, who had studied and taught in Göttingen around the turn of the century, returning later as Professor of Philosophy in 1834, where he taught up until his death in 1841.[17] Riemann may have already encountered some of Herbart's writings when he attended the pedagogical seminar. Herbart was, in fact, a well-known authority on didactics in the tradition of his mentor, Johann Heinrich Pestalozzi. If this were the case, then it would indicate that the fragment "On Psychology and Metaphysics" was likely written during the period 1850–1852, thus before Riemann read Newton. The flavor of this text can be seen from this passage:

> With every simple act of thought, something permanent, something substantial enters our soul. Although this substance appears to us as a unity, it seems to contain (insofar as it is the expression of a spatially and temporally extended unity) an inner multiplicity (*Mannigfaltigkeit*); I therefore call it mental mass (*Geistesmasse*) – all thinking is thus the formation of new mental masses. ...
>
> The mental masses that form merge, combine or complicate to a certain degree, sometimes among themselves, sometimes with older masses. The nature and strength of these connections depends on conditions which Herbart only partially recognized and on which I shall elaborate below. This is mainly based on the inner relations between the mental masses. [Riemann 1876/1892/1902, 1892: 509]

---

16 As indicated below, the dating of these texts largely hinges on when Riemann became familiar with Herbart's writings, since these strongly influenced his views on mental processes.

17 For an analysis of his influence on Riemann's conceptualization of mathematical ideas, see [Scholz 1982a].

Riemann's general notion of a manifold[18] is brought into play here, though the context is framed by German Idealism. The reciprocal relationship between mind and matter takes place through their relative motions. Corresponding to the form of mental mass is an equivalent form of material motion, and conversely. Thus, the mental masses that take form in our cerebral-spinal system coincide with the physical or electro-chemical processes that accompany them. This conception takes us a step closer to Gustav Fechner's psychophysics, but Riemann next makes a giant leap. "Let us now apply these laws of mental processes," he writes:

> which lead to the explanation of our own inner perceptions, in order to seek an explanation for the purposefulness perceived on earth, i.e., to explain our existence and historical development. In order to explain our soul life (*Seelenleben*), we must assume that the mental masses generated in our nervous processes continue to exist as parts of our soul, that their inner connections remain unchanged, and that they are only subject to change insofar as they enter into connection with other mental masses. A direct consequence of these explanatory principles is that the souls of organic beings, i.e., the compact mental masses created during their lifetimes, continue to exist even after death. [Riemann 1876/1892/1902, 1892: 511]

While the dating of this text remains unclear, what we do know is that when Riemann returned to Göttingen for the summer semester of 1853, he took out works by Herbart on psychology [Neuenschwander 2022, 46]. Moreover, he continued to read these and other studies by Herbart over the next several semesters. Like Newton, Herbart sought to ground his philosophy on a logically self-consistent set of first principles.

Riemann by no means slavishly followed Newton's precepts, even if he appreciated the direction of his thought. Already before he composed his own now famous reflections on the hypotheses underlying the foundations of geometry, he noted the conceptual shift in the role of hypotheses that led to a contemporary understanding quite different from the position Newton took in his day. For, unlike Newton, "we are now accustomed to understand by a hypothesis everything added by thought to the phenomena." This applied to Newton's three laws as well:

> The distinction Newton makes between laws of motion, or axioms and hypotheses, does not seem tenable to me. The law of inertia is the hypothesis: If a material point were present alone in the world and moved in space with a definite velocity, then it would continue to maintain this constant velocity. [Riemann 1876/1892/1902, 1892: 525]

During Riemann's lifetime, Newtonian mechanics still dominated natural philosophy, just as inverse-square laws had proliferated into many other fields. In 1846, Wilhelm Weber introduced his new inverse-square law for electrodynamics. Not long thereafter, Riemann began his investigations on gravity and light, for which he took pains to differentiate between two main hypotheses. The first dealt with the corresponding laws of motion, whereas the second addressed their cause. The former he designated as a mathematical hypothesis as opposed to the second, which was metaphysical in character. Regarding the latter, he did

[18] On Riemann's concept of manifolds, see [Scholz 1980, 24–99].

not seek to account for this in terms of action at a distance, that is by means of a theory that only took account of the changes in distance between two particles of matter.

Riemann further noted that those who had adopted this view did so because they believed that Newton's law of universal gravitation was incapable of further elucidation, a view that Newton himself by no means held. To support that claim, Riemann gave a verbatim quotation from Newton's letter to the Master of Trinity College, Rev. Richard Bentley, written in 1692:

> That gravity should be innate, inherent and essential to matter so that one body may act upon another at a distance through a vacuum without the mediation of any thing else by and through which their action or force may be conveyed from one to another is to me so great an absurdity that I believe no man who has in philosophical matters any competent faculty of thinking can ever fall into it. [Newton 1756, 25–26]

Riemann, of course, knew Newton's General Scholium from 1713. In his papers, he cited a passage from it in the original Latin [Riemann 1876/1892/1902, 1892: 525]:

> Et haec de Deo; de quo utique ex Phaenomenis disserere, ad Philosophiam Experimentalem pertinet.
>
> (And thus much concerning God; to discourse of whom from the appearances of things, does certainly belong to Natural Philosophy.)

"Hypotheses non fingo" was one of Sir Isaac Newton's most famous phrases, and it remained so during Riemann's lifetime. It comes from this General Scholium, which Newton attached to the second edition of his *Principia.* There he wrote further:

> I have not as yet been able to discover the cause for these properties of gravity from phenomena, and I do not feign hypotheses. For whatever is not deduced from the phenomena must be called a hypothesis; and hypotheses, whether metaphysical or physical, or based on occult qualities, or mechanical, have no place in experimental philosophy. In this philosophy particular propositions are inferred from the phenomena, and afterward rendered general by induction.[19]

Newton was here answering his critics, who had long maintained that Newton's universal law of gravitation was inconsistent with the principles of mechanical philosophy, since it appealed to an occult force that mysteriously acted over huge distances with no lapse in time. In this famous passage, Newton sought to evade this charge by claiming that hypotheses play no role in investigations based on experimental philosophy. He was thus advancing a methodological principle that sought to draw a sharp line between empirical knowledge and metaphysics. But this raised an obvious question: where do hypotheses come from and how do they arise? Questions such as these long fascinated Riemann, but whatever answers he found he mainly took these with him to his grave.

Riemann was not only a formidably talented mathematician, but also a deep thinker steeped in natural philosophy and, like Newton, theology. He surely

[19] Anne Whitman's translation from the third edition (1999).

noted what Newton wrote to Bentley at the outset of his reply: "When I wrote my treatise about our system, I had my eye upon such principles as might work with considering men, for the belief of a Deity, and nothing can rejoice me more than to find it useful for that purpose" [Newton 1756, 1]. Moreover, Riemann was an avid reader of history and biography, including *The Life of Sir Isaac Newton* [Brewster 1831], which he read with enthusiasm. Newtonianism had long passed its zenith by Riemann's time, but the quest to uncover the laws of nature for microscopic as well as macroscopic phenomena still reflected ideas and principles that had their roots in Newton's works. It is thus my contention that to understand Riemann's world one must look backward, resisting the temptation to see him as a thinker who anticipated what would come long afterward.

The inspiration for Riemann's approach to physical phenomena came largely from potential theory and the theory of elasticity, a subject he taught twice during his short career. One can gauge the extent of his interest in elasticity by the fact that he continually studied Gabriel Lamé's book [Lamé 1852], beginning in the summer semester of 1855. In this brief passage, Riemann described the central idea behind his aether physics:

> Thus the effect of universal gravitation on a ponderable atom can be expressed through (and thought of as dependent upon) the pressure of this space-filling substance in the immediate neighborhood of the atom. It necessarily follows from our hypothesis that the space-filling substance must propagate as the vibrations we perceive as light and heat. [Riemann 1876/1892/1902, 1892: 529]

Riemann combined this physical conception with speculative psychological ideas. For these, he mainly drew on the works of Herbart and his disciple, Moritz Drobisch, the former long dead, whereas the latter was very much alive and active in Leipzig. In this passage, he described his debt to Herbartian ideas:

> Herbart furnished the proof that concepts which allow us to comprehend the world – those whose origin we can trace neither in history nor in our own development, because they are delivered to us unnoticed through our language – can be derived from this source, insofar as they are more than mere forms combining simple sense images; and therefore these concepts need not be derived from some special constitution of the human mind which precedes all experience (such as Kant's categories). This proof of their origin in our ability to comprehend that which is given to us by sense perception, is important for us, because it is only in this way that their meaning can be determined in a manner satisfactory for science .... [Riemann 1876/1892/1902, 1892: 522]

Thus, by the summer semester of 1853, Riemann had begun to delve deeply into various philosophical works, but especially Herbart's writings. As the months went by before he completed his Habilitation, he continued to develop the mathematics for his aether theory, while at the same time undertaking new research with Weber on electrical phenomena. Shortly after submitting his post-doctoral thesis, he went home for the Christmas holiday. From Quickborn, he wrote to his brother Wilhelm on 28 December 1853, with recent news, including the progress he had made on his new theory:

> My other investigation into the connection between electricity, galvanism, light, and gravity I resumed immediately after completing my habilitation thesis, and I have progressed to the point where I can publish it in this form without hesitation. However, it has become increasingly clear to me that Gauss has also been working on this for several years and has confidentially shared the matter with some friends, including Weber. I can probably write this without it being interpreted as presumptuous – I hope that it is not too late for me now and that it will be acknowledged that I discovered these things entirely independently. [Riemann 1876/1892/1902, 1892: 547]

The paper described here would never appear during Riemann's lifetime, but it was likely the text of "Gravitation and Light" [Riemann 1876/1892/1902, 1892: 532–538] (see Chapter 9). Why Riemann ultimately chose not to publish this work, to which he clearly attached great significance, is simply unknown.

We are also ill-informed as to why Riemann imagined that Gauss was pursuing a similar theory. This was by no means the only time he expressed his worries that Gauss would suddenly publish a new study preempting his own work. In the meantime, he was busy cooperating with Weber on experiments on electrical discharges in Leyden jars. He explained the circumstances that led him to this work in a continuation of the above letter from 26 June, 1854 to his brother:

> During the Easter holidays, [Rudolf] Kohlrausch, a son of the school inspector and cousin and brother-in-law of Schmalfuss, who is now a professor in Marburg, visited Weber for two weeks to conduct an experimental investigation into electricity with him. Weber had done the preliminary work for one part of this investigation, and Kohlrausch had done the other, and had conceived and constructed the apparatus. I took part in their experiments and on this occasion met Kohlrausch. Some time before, Kohlrausch had made and published very precise measurements on a previously unexplored phenomenon (the electrical residue in the Leyden jar), and I had found the explanation for it through my general investigations into the relationship between electricity, light, and magnetism. I then spoke with Kohlrausch about it, and this prompted me to work out the theory of this phenomenon and send it to him. [Riemann 1876/1892/1902, 1892: 548]

Riemann's manuscript received a positive response from Kohlrausch, who offered to send it on to Johann Christian Poggendorff, editor of *Annalen der Physik und Chemie*. Kohlrausch also invited Riemann to visit him in Marburg during the fall break in order to pursue these things further. As he reported to his brother, he was more than a little delighted by this turn of events:

> This matter is important to me because it is the first time I have been able to apply my work to a previously unknown phenomenon, and I hope that the publication of this work will help ensure a favorable reception for my larger work. Here in Quickborn, I will probably have to concentrate partly on printing this work – since the proof sheets will likely be sent to me – and also partly on preparing my lectures for next semester. (*ibid.*)

Riemann later met with Kohlrausch again in Göttingen; after another exchange of letters, however, he decided to forgo the publication of his essay on the residue in the Leyden jar, probably because he did not wish to make the modifications requested by the editor.[20] Instead, Riemann's essay on the

[20] It was published posthumously in [Riemann 1876/1892/1902, 345–356].

theory of Nobili's color rings appeared in *Annalen der Physik und Chemie* [Riemann 1876/1892/1902, 54–61]. About this work he wrote to his sister Ida: "This subject is important because it allows very precise measurements to be made, and the laws governing electricity can be tested very precisely" [Dedekind 1876a, 550].[21]

In retrospect, one might wonder why Riemann attached so much significance to these physical investigations, which never evoked much contemporary interest. He could have easily gone straight from his dissertation to an exploration of its applications for solving open problems in complex analysis, itself a virtually newborn field of research. Yet the fact that he did not only underscores that Riemann was a bold and *independent* thinker who went his own way. His intellectual debt to Gauss will be taken up in detail in Chapter 10, but one should resist the idea that Riemann was Gauss's student even when the intellectual affinity between their works was strongest. During these years of high adventure, Riemann went chasing after unexplained physical phenomena in the hopes of finding a deeper explanation of their causes, while seeking to bolster his belief in the underlying unity of the forces of nature. In the face of such enticing thoughts, mathematics could evidently afford to wait. Not until 1856, roughly five years after he had outlined what would later be called the theory of Riemann surfaces, did he again pick up his pen to produce two major papers that would set the mathematical world astir.

[21] For further discussion of Riemann's interests in physics, see Chapter 9.

# Chapter 3
# Years of Struggle, 1854–1858

Throughout 1853, Riemann spent much of his time pursuing physical ideas alongside Wilhelm Weber and his colleagues. That year, Weber made him an assistant in his work group, which meant preparing exercises for students enrolled in the mathematics-physics seminar. One standard activity they engaged in took place at the magnetic observatory, where four times each year measurements had to be taken in intervals ranging from five to fifteen minutes and over a period of 48 hours (or later 24 hours). Students had already participated in this work back in 1835, but with the founding of the mathematics/physics seminar Weber had a steady supply of potential recruits.

As an expert on the properties of the magnetometer, Riemann almost surely helped instruct novice students in the use of this device. Gauss had long before devised a precise protocol for this purpose, which was followed at all the observatories in the network: by Riemann's day, there were over fifty. Besides Weber himself, several other staff and faculty members took part in recording the data, from which the parameters for magnetic fluctuation had to be calculated. During the years 1837 to 1849, these calculations were mainly carried out by Goldschmidt and Ulrich, the latter substituting for Weber. Little is known about what Riemann actually taught in these seminar sessions, but they at least provided him the opportunity to step into the role he would soon assume, that of a *Privatdozent*. To obtain the right to offer lectures under his own name, he needed to acquire the *venia legendi*, which meant the Göttingen faculty had to approve his application for Habilitation.

## 3.1 Riemann's Habilitation

The Dean of the Philosophical Faculty at this time was the philologist Karl Hoeck. On 5 December 1853, he reported to the subcommittee that handled doctoral and postdoctoral affairs on having received Riemann's application. This was the same group of full professors who had overseen Riemann's promotion



D. E. Rowe, *Bernhard Riemann: His Life and Wondrous Mathematical Legacy*,
https://doi.org/10.1007/978-3-032-25457-3_4

to Dr. Phil. Thus, the faculty must have regarded this procedure as a mere formality in order to reaffirm the candidate's outstanding qualifications. Dean Hoeck asked Gauss to report on the postdoctoral dissertation Riemann had submitted and, if he approved it, to choose one of the three topics the candidate had offered for the final qualifying lecture. For his assessment, Gauss simply wrote: "The present treatise contains so many evidences of fine knowledge, good assessment and self-capable skill that it is completely sufficient for its purpose."[1]

Having submitted a successful postdoctoral dissertation, Riemann received permission to prepare this final act leading to his *Habilitation*, the obligatory trial lecture on a topic chosen by the faculty from the candidate's list of three. In his account of this, Dedekind cited what Riemann wrote to his younger brother Wilhelm on 28 December 1853:

> I handed in my Habilitation thesis at the beginning of December and had to propose three topics for a trial lecture, one of which the faculty would then choose. I had finished the first two and hoped that one of them would be accepted, but Gauss chose the third, so now I'm in a bit of a pickle again, as I still have to write it up.[2] [Dedekind 1876a, 547]

Riemann's three topics can be found in Akt. 25 of his scientific estate::

> 1) History of the question on the representation of a function by a trigonometric series.
> 2) On the solution of two second-degree equations in two variables.
> 3) On the hypotheses underlying the foundations of geometry. [Riemann 1876/1892/1902, 1902: 112]

Now that he had submitted his Habilitation thesis, Riemann was hoping this final formality would go quickly, but instead he found himself waiting for several months to learn when he could hold this lecture. As the winter dragged on, he put off working on it, since his new physical ideas had captured his mind completely. In the meantime, Riemann's health once again took a turn for the worse, as he related to his brother:

> ... I resumed my research on the relationship between the fundamental laws of physics and became so engrossed in it that when the topic was assigned to me for my trial lecture at the colloquium, I couldn't immediately put this work down. I soon fell ill, partly as a result of too much brooding, and partly as a result of spending so much time indoors in the bad weather; my old ailment [severe constipation] returned with great stubbornness, and I made no progress with my work. Only after several weeks, when the weather improved and I began to socialize more, did my health begin to recover. [Riemann 1876/1892/1902, 1892: 547–548]

---

[1] Die vorliegende Abhandlung enthält so viele Beweise von feinen Kenntnissen, guter Beurtheilung und selbstfähigen Geschicklichkeit, dass sie für ihren Zweck vollkommen ausreicht (Dekanatsakten von 1854, SUB Göttingen).

[2] Dedekind wrote in this connection that choosing the third topic rather than the first was contrary to the usual custom. Detlef Laugwitz questioned the plausibility of this claim, noting that Riemann had already dealt with the first topic in his postdoctoral thesis and that the second topic appeared far too elementary. Perhaps Riemann was not really disappointed that Gauss chose the third? [Laugwitz 1996/1999, 219].

Riemann's father was deeply concerned about his son's future, but the bond between them was strong, anchored by their religious faith. Some two months after Bernhard had submitted his thesis, thus in early February 1854, they exchanged letters, including this reply from the son:

> As I write this, I received your endearing letter; this reminds me to write to you about myself. How much I feel that it was wrong of me not to answer your letter immediately; and that I have worried you at this time by my behavior. I hope you will forgive me. Immediately after submitting my Habilitation thesis I continued my investigations into the connection between the laws of nature and became so engrossed that I could not get away from them. Prolonged preoccupation with this has also probably been detrimental to my health, at least immediately after the New Year my usual discomfort returned so drastically that I could only find relief by the strongest means.[3]
>
> I was in a very bad way, felt unable to work, and tried to recover my health by taking long walks. For eight days now I have been feeling better again; the qualifying lecture I am to give at the Colloquium is half finished, and your letter and the thought of you should be an incentive for me not to let anything distract me from this work again. When I have finished, I will write to you straight away; unfortunately, I will then have to ask you for money again; my coffers are now completely empty. I will then put everything in order so that I can teach here in the next semester and hope to see you again in Quickborn. [Neuenschwander 1981a, 109–110]

Health issues and financial concerns put Riemann under considerable stress, and these were anything but momentary factors. Such mundane matters far outweighed any worries he had about his forthcoming lecture; he surely had that topic already well under control in his mind.

Six months after Riemann first applied to habilitate, Gauss finally decided on June 9 that he could attend his qualifying lecture. Dean Hoeck immediately took charge of the matter, inviting his colleagues to attend this hastily arranged meeting to be held in his home at 10:30.[4] Five responded that they would be present, so on June 10, 1854, Riemann read his prepared text before six distinguished members of the philosophical faculty. It must have taken him a good two hours to do so, perhaps also answering a few perfunctory questions afterward. In the unlikely event Gauss reacted to Riemann's ideas, his words are lost forever. Riemann recounted what transpired in a letter from 26 June to his brother:

> I had already resigned myself to the inevitable. Then [Gauss] decided suddenly, at my repeated request, "to get the thing off his neck," and so on Friday after Pentecost he scheduled the colloquium for the the following day at half past ten, and by one o'clock on Saturday I was happily finished with it. [Dedekind 1876a, 548]

In keeping with the usual formalities, we might imagine that the candidate left the room for a few minutes so that the professors could confer among themselves, after which the Dekan would have invited him back to the meeting room to receive the good news. Riemann could then pack up his manuscript and prepare for the trip back home to Quickborn. The following day, those who were present,

---

[3] Riemann suffered from chronic constipation.

[4] The information related here and below can be found in the Dekanatsakten der Philosophischen Fakultät, Nr. 137, 1853–54, Universitätsarchiv, SUB Göttingen.

as well as two who were not, signed the official petition sent to the Ministry of Education in Hanover requesting that Riemann receive the *venia legendi* on the basis of the formal requirements he had fulfilled. This final decision rested with the officials in Hanover, who were charged with ensuring that those accorded the right to teach possessed the proper *moral authority* to do so. For although the Carlsbad Decrees from 1819 were no longer in effect, the German states remained keen to keep political troublemakers (i.e., outspoken liberals) out of their universities.

After Riemann read his text, he left Göttingen to begin preparing his first lecture course on partial differential equations as well as a paper on recent experimental findings related to electrical phenomena. Although he never mentioned his Habilitation lecture again, its ideas were clearly related to his physical speculations from this time. These have often been interpreted as an anticipation of later developments in field physics, though they can be more accurately described as a type of ether physics that attempted to unite gravity, light, and magnetism. It was two weeks after completing his Habilitation, on 26 June, that Riemann wrote the letter, cited above, to his brother Wilhelm. It seems noteworthy that this letter contains not a word about the content of his lecture or how it was received, not even a hint. Nor does it appear that Riemann ever again mentioned its truly remarkable ideas in any of his papers or course lectures, although his former student and later colleague Ernst Schering claimed to have spoken with him about this topic [Schering 1909, 377].[5] One should also be wary of attaching too much significance to what Dedekind wrote about Gauss's reaction:

> ...the lecture, which exceeded all his expectations, astonished him greatly, and on his way back from the faculty meeting he spoke to Wilhelm Weber with the highest appreciation and with a rare excitement for him about the depth of the thoughts presented by Riemann. [Dedekind 1876a, 549]

Dedekind added this to his text after speaking with Weber, who then claimed to recall how Gauss had reacted more than twenty years after the event.[6]

## 3.2 On the Hypotheses Underlying Geometry

The hypotheses underlying the foundations of geometry was a famous topic that had challenged mathematicians ever since Euclid codified the subject more than two millennia earlier. Today we are accustomed to reading about old-fashioned Euclidean geometry, which inevitably lost its special status once Einstein successfully predicted that matter produces small curvature effects in

[5] As later events will show, Schering's claims had a self-serving character. Moreover, he resented the work Riemann and Dedekind did for the Gauss edition, the project to which he devoted much of his time and energy.

[6] This conclusion is based on information from a letter Dedekind wrote to Heinrich Weber on 10 November 1875 [Scheel 2015, 79].

space. He went even further in his famous first paper on relativistic cosmology [Einstein 1917]: there he appealed directly to an idea first advanced by Riemann, namely, that the entire universe might form a huge 3-dimensional sphere. Were that the case, then space would carry the structure of a non-Euclidean geometry, an idea very few people, if any, entertained before Riemann suggested this in the year 1854.

Yet, to do justice to both Riemann and Einstein one should frankly admit that their conceptions of space, time, and matter had very little in common. In Riemann's case, his physical theorizing aimed to modify Newtonian mechanics by incorporating an ether that transported both gravity and light. All his work on electricity and magnetism preceded the era of Maxwellian field physics, which eventually led to a major crisis due to its incompatibility with classical mechanics. Einstein was one of several physicists who struggled to solve that problem, and in 1905 he succeeded. Three years later, Hermann Minkowski showed how special relativity could be understood as the invariant theory of the Lorentz group acting on Minkowski spacetime. At that point, the next problem seemed obvious: how to incorporate gravity into this new theory? Yet Einstein saw the situation differently, and his perseverance eventually led to a new spacetime physics (see [Janssen/Renn 2022] for that story).

Riemann was surely a bold thinker, but the present account focuses on the many authors and ideas he knew well, while trying to avoid the temptation of viewing his famous lecture "On the Hypotheses underlying the Foundations of Geometry" [Riemann 1868/1876] from the vantage point of all that came afterward. Looking backward, one is just as likely to cite what Immanuel Kant wrote in relation to this topic. As virtually every well-educated German from Riemann's time knew, Kant grounded his epistemology on the notion that the properties of space are given to us as pure intuitions [Friedman 1992]. For Kant, pure space and pure time anchor all our perceptions of the world. Echoing Newton's notions of absolute space and time, these categories form the foundation stones for Kant's transcendental idealism. Thus, our minds create the spatial framework that structures all our perceptions. Kant's conception of the pure intuition of space placed it outside the natural world given to us empirically. Space in itself is the precondition for our ability to know spatial objects. We cannot know space directly, even though all that we perceive is mediated by spatial structures.

Moreover, Kant's understanding of space entailed Euclid's theory of parallels, which accorded with the geometry underlying Newton's first law of motion, the law of inertia. Thus, an object once set in motion in empty space will continue to move along a straight line at uniform speed, in principle, until the end of time. Accordingly, space contains a system of parallel directions. Any two parallel lines lie in a plane, and all the lines in such a Euclidean plane satisfy the parallel postulate (in Kant's time it was often called Euclid's eleventh axiom).

Riemann said not a word about Kant's ideas, even though he was very familiar with his philosophical views; instead, he mentioned that his own work was partly inspired by the ideas of another philosopher, Johann Friedrich Herbart. By the summer of 1853, Riemann had begun an intensive study of Herbart's writings alongside those of his disciple Moritz Drobisch, works he continued to read over several semesters [Neuenschwander 2022, 46–47]. Already before his Habilitation lecture, Riemann identified himself as a Herbartian, while sketching a "theory of fundamental concepts in mathematics and physics [that can serve] as a basis for the explanation of nature" [Riemann 1876/1892/1902, 1876: 489–493]. For Riemann, the key question concerns how we acquire those fundamental concepts, which he understood as belonging to a long historical process that accompanied human development. Although he expended only a few words on this, the passage cited in 2.7 explains why:

> Herbart has now provided proof that the concepts used to understand the world, whose origins we cannot trace either in history or in our own development because they are passed down to us unnoticed through language, all come from a single source, ... and therefore they do not need to be derived (as opposed to the Kantian categories) from a special constitution of the human mind that precedes all experience. This proof of their origin in the comprehension of given sensory perceptions is important for us because only in this way can their significance for natural science be realized in a convincing manner. [Riemann 1876/1892/1902, 1876: 490]

The main mathematical innovation Riemann introduced in his text – though only by way of a terse description – was his generalization of Gaussian curvature to higher dimensions. As mentioned in Section 1.3, already during his first semester of studies, Riemann borrowed Volume 6 of the *Commentationes*. So he was long familiar with Gauss's famous study of surface theory, *Disquisitiones generales circa superficies curvas* [Gauss 1828b]. In his lecture, Riemann also mentioned two other works by Gauss: one announcing his new results on biquadratic residues based on complex numbers [Gauss 1831], the other containing his more recent reflections on the fundamental theorem of algebra (see Section 10.2). Riemann thus knew that Gauss not only opposed Kant's theory but also that he had refuted one of Kant's arguments in support of it.[7] Of course, Riemann made no mention of this in his extant writings, but he referred directly to Gauss's geometric explanation for complex multiplication at the beginning of his course on function theory [Neuenschwander 1996, 22]. Whether or not he knew anything about Gauss's interest in the new theory of non-Euclidean geometries, as developed independently by the Hungarian János Bolyai and the Russian Nikolai Lobachevsky, remains a matter of speculation (see Section 10.4).

After alluding to the works of Euclid and Legendre, Riemann quickly shifted his attention to an entirely new set of hypotheses for geometry based on the concept of a multiply-extended manifold. His initial remarks highlighted the need to examine the foundations of geometry from this new direction, which had no direct bearing on the far narrower question concerning the status of the parallel postulate. As for the situation at present:

> It is known that geometry assumes as given both the notion of space and the first principles of constructions in space. The definitions for these are merely nominal, whereas the

[7] On Gauss's critique of Kant's position, see [Rowe 2023, 118–125].

> essential determinations appear in the form of axioms. The relations underlying these assumptions remain, however, in darkness; we neither perceive whether and how far their connection is necessary, nor whether they are a priori possible. [Riemann 1868/1876, 272]

For space at large, Riemann claimed that ordinary Euclidean geometry was altogether adequate, though he noted the possibility of a very small positive curvature, which would then give it the structure of a gigantic 3-sphere. Since both cases – flat space or a 3-sphere – assume the universe to be unbounded, Riemann emphasized that the topological hypothesis was far more certain than the narrower hypothesis that assumed the metric relations extended to infinity. Regarding the empirical basis supporting these considerations, he wrote:

> If we suppose that bodies exist independently of position, the curvature is everywhere constant, and it then results from astronomical measurements that it cannot be different from zero; or at any rate its reciprocal must be an area in comparison with which the range of our telescopes may be neglected. [Riemann 1868/1876, 285]

In Section 6.2, Riemann's text will be considered more closely, followed by a consideration of its reception history in 6.3. There it will be shown that Riemann's ideas had a strong impact on Hermann von Helmholtz and Eugenio Beltrami, though only the latter made a direct connection with what came to be known as hyperbolic geometry in 3-space. In this sense, Riemann's now classic lecture occupies an important place in the history of non-Euclidean geometry. Within that context, it was natural to assume that Riemann wrote his essay to clarify the vexing status of Euclid's parallel postulate. Yet many readers have since wondered why his text makes no mention of the property of parallelism.

When Riemann did speak about Euclidean geometry, he focused on the free mobility of rigid bodies, which explains why Helmholtz was later able to connect his own reflections on the foundations of geometry with those of Riemann. Yet Riemann's entire framework was far broader, not only because he treated higher-dimensional differentiable manifolds but also because he took invariant line elements rather than 3-dimensional bodies as the starting point for his analysis. Furthermore, as just noted, he carefully distinguished between topological and metric properties, a viewpoint wholly consistent with later investigations. Mathematicians found it natural to begin with a topological manifold to which one could then attach additional structures. Riemann's text reads today like a blueprint for a new theory of differentiable manifolds (or Riemannian manifolds if we insist that the metric be positive definite). Little wonder that this long-forgotten essay has come to be regarded as the most visionary of all his works.

Which leads us to ask: what significance did Riemann attach to the ideas in his lecture, and why did he never take them up again after delivering it? These are questions that cry out for answers, but in view of his silence these matters will always remain shrouded in mystery. The fourteen-year gap that lies between the composition of this text and its eventual posthumous publication makes it even harder to assess what he might have had in mind. At any rate, given Riemann's desire to avoid scientific controversies – a veritable minefield in German academia – we can at least imagine his reluctance to pronounce his personal views on a question that had long been debated by leading natural philosophers, including Newton, Leibniz, and Kant. Back in 1854, we only know that Riemann read his text before Gauss and other members of the philosophical faculty, but did anyone else in Göttingen actually read it? Nothing suggests that anyone did, although Ernst Schering may have, since he apparently discussed these ideas with Riemann. On the other hand, it seems hard to believe that Riemann saw this essay as anything more than a necessary stepping stone in his career. This was surely not a programmatic text – which his dissertation most definitely was – but rather a speculative essay written to fulfill a formal requirement for Habilitation. Gauss had wanted "to get the thing off his neck," and so did Riemann, though he clearly wanted to leave a positive impression with Gauss as well.

Nor did Riemann ever publish his Habilitation thesis; in fact, he never again worked on this or any other topic in real analysis. Thus, the same questions apply to both manuscripts from 1853/54, although we have hardly any clue as to why he chose not to publish them or even refer to their contents elsewhere. Only after Dedekind received the manuscripts from Riemann's widow did they appear in the *Abhandlungen* of the Göttingen Royal Society as [Riemann 1868/1876] and [Riemann 1868/1876]. These were, thus, occasional works that stood apart from Riemann's main research interests.[8] At the time he prepared these papers, Riemann was busy pursuing his physical researches, and these were anything but a mere distraction for him. That larger program, as he himself made clear, was his principal research goal.

Over the entire two years between taking his doctorate and qualifying for Habilitation, Riemann was deeply involved with physical research projects, some directly linked with Wilhelm Weber's interests. The importance he attached to this work is reflected in a document, in which he described his three main research interests during this phase of his career:[9]

> 1. To introduce the imaginary into the theory of transcendental functions in a manner similar to that which has already been done with great success for algebraic, exponential and circular functions, as well as elliptic and Abelian functions. I have provided the necessary preliminary work for this in my inaugural dissertation (cf. Art. 20).
>
> 2. In connection with this are new methods for the integration of partial differential equations, which I have already successfully applied to several physical problems.
>
> 3. My main research work concerns a new conception of the known laws of nature – expressing these by means of other basic concepts – through which it becomes possible to employ experimental data on the interactions between heat, light, magnetism, and electricity in order to investigate these interrelationships. I was led to this mainly by studying the works of Newton and Euler, as well as those of Herbart. [Riemann 1876/1892/1902, 1892: 507–508]

---

[8] Laugwitz wrote about [Riemann 1868/1876] at great length in [Laugwitz 1996/1999, 181–213], since that paper was an important contribution to integration theory; it also served as a key stepping stone that eventually led to Georg Cantor's theory of infinite point sets, for which see [Dauben 1979, 11–19].

[9] This undated text was published by Heinrich Weber based on a document in Riemann's Nachlass; when and why he wrote it cannot be established with certainty.

Riemann's more expansive remarks with regard to the third topic reflect the great importance he attached to physical theories. As Umberto Bottazzini and Rossana Tazzioli have convincingly argued, Riemann's mathematics was thoroughly permeated by his parallel interests in natural philosophy [Bottazzini & Tazzioli 1995]. It is also significant that he took the opportunity to note more precisely how his views related to those of Herbart:

> As far as the latter is concerned, I was able to follow almost completely Herbart's earliest investigations, the results of which are presented in his doctoral and habilitation theses (from 1802), but had to deviate from the more specific course of his speculations on one essential point. This is due to a difference in relation to his natural philosophy and those propositions of psychology which concern connections with natural philosophy. [Riemann 1876/1892/1902, 1892: 507–508]

One easily sees from this that Riemann's Habilitation thesis fell outside the framework of his main research interests. Nevertheless, it represents an important contribution to a famous theme in the history of mathematics: the representation of real-valued functions by trigonometric series [Riemann 1868/1876]. He had already been exposed to this subject in Berlin under Dirichlet, who gave him additional help in 1852 when he visited Göttingen (see Section 2.6 and [Merzbach 2018, 173]). This mainly concerned relevant earlier studies, which Riemann cited in the historical introduction of his thesis.

Riemann habilitated in the same year as did Gauss's last student, Justus Georg Westphal, a native of the same remote region in Hanover from which Riemann hailed. For a brief time, Westphal seemed destined for a brilliant career. He was awarded a prize for solving a difficult problem posed by Gauss, who praised the author's thorough analytical skills [Günther 1897]. The young man had shown how to approximate the roots of a certain seventh-degree equation by means of a series that converged very quickly. From 1851 up until the death of Gauss in February 1855, Westphal worked as his assistant in the observatory. Since this position was afterward terminated, his prospects for future employment were dim. In a letter to his brother from May 1855, Riemann reported that Westphal was no longer working at the observatory, though he still collected a salary of 200 Taler. For the next several years, Westphal announced a 3-hour course on spherical trigonometry in relation to astronomical research, though how often he actually taught it, no one can say. Riemann only mentioned him on occasion, but not as a potential candidate for a position at the *Sternwarte*, which later became a matter of deep interest and concern to Riemann (see Section 3.6). News of Westphal's death in Lüneburg in 1859 apparently only reached Göttingen through a local newspaper; he was just 35 when he died.

## 3.3 Physical Interests

After a longer stay in Quickborn, Riemann returned to Göttingen in September 1854 to participate in the annual meeting of the Society of German Natural Scientists. Both Weber and Stern had urged him to give a lecture in the Section on Mathematics, Physics, and Astronomy. Riemann's talk gave a synopsis of the work he had done with Kohlrausch and Weber on the phenomenon of electrical residues in a Leyden jar (see Section 2.7). He discussed this with respect to Fechner's dual-stream approach to electricity as well as the older theory advocated by Benjamin Franklin, in which electricity was conceived as a single form of charged matter. Riemann advocated this latter view.

Today this Göttingen conference is remembered because of a controversial lecture delivered by physiologist Rudolf Wagner on "Human Creation and Soul Substance" (*Menschenschöpfung und Seelensubstanz*) [Laugwitz 1996/1999, 279]. The context that led to this polemical speech arose from the ongoing efforts of physiologists to chip away at vitalistic explanations for phenomena that had no clear mechanistic cause. Wagner lashed out against those who took a stringent materialist view in these matters, as he saw these colleagues as advocates for an anti-religious brand of scientific inquiry. Whether or not Riemann actually heard Wagner speak, he surely was aware of the fallout from his talk, which launched the materialism debate (*Materialismusstreit*). The subsequent exchanges between Wagner and his opponents were bitter, but neither side offered arguments of importance for resolving the mind-body problem.

As noted in Section 2.7, Riemann was deeply interested in that problem, which he tackled by proposing to supplement the principles of Newtonian mechanics. Nevertheless, he seems to have never referred to these debates, even though since perhaps 1852 he had begun to develop fairly elaborate ideas related to the fundamental problem itself. His forays into these matters remained an altogether private matter, so far as we know, which means that all references to his proposals took place after the publication of his *Werke* in 1876. Only then did readers begin to appreciate that the young Riemann had attempted to develop a scientific philosophy that accounted for the mental and emotional world in terms of a teleological worldview. Herbart had taken a somewhat similar approach, but Riemann drew on more recent scientific findings related to these issues. An important divide within Riemann's philosophical framework involved the distinction between organic and inorganic matter. The issue came to the fore after 1828, the year in which Riemann's chemistry professor, Friedrich Wöhler, succeeded in synthesizing urea from inorganic compounds. Riemann sought to explain why only four chemical elements and their special compounds arise in life forms. He claimed that their resistance to condensation was the key chemical property they shared [Riemann 1876/1892/1902, 1892: 514].

Riemann's talk was printed in the conference proceedings and later republished in [Riemann 1876/1892/1902, 48–53]. Shortly after delivering it, he described the experience in a letter to his father:

> My lecture was scheduled for Thursday, and since no other lecture was scheduled for this meeting of our section, I elaborated on the subject a little further the evening before in order to fill the usual time of the meetings to some extent. Initially, I had only wanted to briefly state the law I wanted to communicate, but now I applied it to several phenomena and demonstrated its agreement with experience. My lecture was admittedly less fluent in this last part, but I still believe that the overall impression gained from including this; I spoke for approximately one hour and fifteen minutes. The

fact that I spoke publicly at the meeting gave me more courage for my lectures; but at the same time I saw the great difference it makes whether one has long before come to terms with one's ideas or has only developed these immediately before speaking. I hope to be able to think about my lectures more calmly six months from now, so that my stay with you in Quickborn will not be spoiled again the way it was last time. [Dedekind 1876a, 516–517]

Having successfully completed his Habilitation, Riemann could now begin offering lecture courses at Göttingen University. Dedekind had also habilitated, too, so he also began teaching in the winter semester of 1854/55. Unlike, Riemann, however, he offered two standard courses on probability theory and geometry, whereas Riemann taught a single course that dealt with applications of partial differential equations, a topic he had studied in Berlin under Dirichlet [Merzbach 2018, 245–246]. Initially he hoped to attract the requisite number of auditors to make a course, namely two! In a letter to his sister Ida, written on 9 October 1854, when the semester was just beginning, Riemann explained how his group gradually came together:

The first person to announce his pleasing intention to attend my lectures was Mr. Schering, whom you know ....[10] In addition to Schering, Esselbach from Schleswig ...contacted me before the beginning of the semester. He had taken part in the campaign in Schleswig Holstein [which ended in Denmark's favor in 1852] .... The third person who appeared, requesting that I enter my name in his registration book, was a Mr. Bargum from Kiel. Before the campaign in Holstein he was a cadet in the Danish navy; he then served in the Schleswig-Holstein army, followed by three years at the polytechnical school in Hanover. So as you can see, he's a man before whom I must be careful. ... [Neuenschwander 1981a, 110–111]

As these remarks suggest, Riemann shared the widespread admiration for those in the military, independent of the country for which they served.[11] He also clearly enjoyed telling Ida about how his course came together:

So now – tres faciunt collegium[12] – my course was happily complete. Gradually, however, this little group has grown somewhat, and the artillerymen Mr. Quintus and Mr. Gerdes from Hanover have also joined, as well as an astronomer Mr. Pape from Verden, who is already known for several works. To these came Mr. Hans Zinken, called Sommer from Braunschweig,[13] and finally Mr. Oppenheim, who joined after my lectures had begun. There you have the complete list of my auditors. [Neuenschwander 1981a, 111]

Eight was indeed a very respectable number for an advanced mathematics course during these years, when mathematics in Göttingen still lagged far behind

---

[10] This suggests that Ida Riemann must have visited her brother in Göttingen at some point.

[11] Ute Merzbach observed that Dirichlet enjoyed a certain modicum of respect from his students at the Berlin Military Academy because they knew he had studied in Paris and spoke fluent French. German officers-in-training often considered this a sign of high breeding [Merzbach 2018, 58].

[12] This rule actually applied when there were only two students, as the instructor counted as the third!

[13] Hans Zinken regularly attended Dedekind's lectures and became a close friend; both shared a deep love for music as well as mathematics. Ultimately, Dedekind decided to focus on mathematics, whereas Zinken went on to become a composer [Scharlau 1981].

Berlin. Riemann could furthermore rejoice over being allowed to hold his lectures in the auditorium reserved for the senior theology professor Johann Gieseler, author of *Lehrbuch der Kirchengeschichte* in five volumes. Later in the semester, on 18 November, he wrote to tell his father:

> My life here has gradually taken on a fairly regular and monotonous pattern. I've been able to hold my lectures regularly so far; my initial embarrassment has largely subsided, and I've gotten used to thinking more about the audience than about myself, reading their expressions to determine whether I should move forward or whether I need to elaborate further. [Dedekind 1876a, 550]

In truth, Riemann always would have great difficulty reaching his audience in the courses he taught, as Dedekind himself later attested [Dedekind 1876a, 550–551].

For this first course, he wrote out his lectures himself. Karl Hattendorff later published *verbatim* the text Riemann wrote for the opening session in [Riemann/Hattendorff 1869]. As an authentic account in Riemann's words, this deserves repeating, at least in part. After noting that a course on partial differential equations had never before been offered in Göttingen, he continued:

> As is well known, a scientific physics has only existed since the invention of differential calculus. Only since then have we learned to follow the course of continuous natural events by successfully reconstructing the connections between phenomena and abstract concepts. This required two things: first, simple concepts for these constructions, and secondly, a method for deriving, from the simple laws relating to points in time and space, the laws for finite intervals and distances, which alone are accessible to observation (and can be compared with experience). [Riemann/Hattendorff 1869, 1]

The simple concepts that led to mathematical physics were introduced by Galileo and Newton: the former derived the law governing freely falling bodies in a vacuum, the latter the laws governing central forces. Riemann emphasized the limitations of these concepts, asserting that mathematical physicists had yet to devise new ones that went beyond these. The mathematical speculations of Pierre-Simon Laplace, Poisson, and Cauchy, those which go beyond the thread of observations, all endeavor to explain the phenomena in terms of Newtonian principles. In this respect, the conceptual basis for physical theorizing was very thin:

> With regard to the concepts used in the physics of nature, we are today still no further than the standpoint of Newton. No new progress has been made since [his time]; all attempts to penetrate into the interior of nature by going beyond these basic concepts have so far failed. The influence of later philosophical systems, insofar as they pertain to the physical literature, has only had the effect of modifying the original Newtonian theory, while introducing inconsistencies. [Riemann/Hattendorff 1869, 2]

In contrast to the paucity of conceptual progress in physics, Riemann underscored the impressive methodological innovations that followed soon after the invention of the calculus.

> In the first period after the invention of the differential calculus, one could only treat certain abstract cases. For freely falling bodies, their total mass was viewed as forming a point. The heavenly bodies were thought of as mathematical points, and in the theory

> of the pendulum one treated only the mathematical pendulum, i.e., a rigid line turning about a fixed fulcrum and equipped with a mass point. Here one progressed from the infinitely small to the finite simply by considering a single dimension, where this variable quantity was given by the time. (*ibid.*)

Gradually new methods were developed for studying the motion of extended 3-dimensional bodies under the influence of a system of central forces, and this led to the first theories involving partial differential equations. Riemann cited d'Alembert's treatment of the vibrating string problem as the initial breakthrough in this direction. General methods for handling solutions of partial differential equations, however, only emerged following Fourier's revolutionary work on heat conduction. Summing up these developments, he concluded that this second period was nearly as long as the first:

> Newton's *Principia* appeared in 1687 and d'Alembert's solution of the vibrating string problem was published in 1747. Another sixty years later, on 21 December 1807, Fourier submitted his first study of heat conduction to the Paris Academy. [Riemann/Hattendorff 1869, 3]

Since that time, the mathematical basis for all physical theories has depended on calculations derived from partial differential equations. Riemann noted their role in studying sound vibrations in different states of matter – whether liquid, solid, or gas – as in studies of elasticity or of optical phenomena. These studies typically assumed a molecular theory of matter with forces of attraction and repulsion that depend on the distances between particles. The distribution of the molecules and the law governing their motion then lead to certain constants that enter into the partial differential equations. Riemann emphasized, however, that these researches were very far from drawing any definite conclusions regarding either the distributions of the hypothesized molecules or the laws pertaining to them. Nevertheless, here too the mathematical foundations were derived from partial differential equations. In particular, these applied to extended bodies that attract or repel by virtue of an inverse-square law, which was valid for bodies subject to magnetic, electrical, or gravitational forces.

Riemann further noted how Laplace introduced a potential function in 1782, which enabled him to describe the strength and direction of gravitational force at each point for any given distribution of matter. Furthermore, the work of Gauss and Weber showed how these mathematical methods carried over directly to the field of terrestrial magnetism.

> The partial differential equation in which this law is expressed forms the basis for all further deductions. Likewise, the theory of terrestrial magnetism is based on the law according to which the direction and magnitude of the geomagnetic force varies from location to location, a law which is completely independent of the arrangement of the sources of this magnetic force – the magnetic fluids or the Ampère currents – in the Earth's interior. [Riemann/Hattendorff 1869, 3]

As a general summation of the above, Riemann concluded:

> What has emerged in this way via induction as a fact – namely, that partial differential equations constitute the actual foundations of mathematical physics – also follows a priori. True elementary laws for spatial and temporal points can only pertain to the

> infinitely small. Such laws, however, will generally be partial differential equations, and the derivation of laws for extended bodies and time periods requires their integration. Therefore, methods are necessary by which one can derive these finite laws from those that hold for the infinitely small, and indeed, rigorous methods that allow for no error. For only then can one test these laws empirically. (*ibid.*)

These general views regarding the importance of partial differential equations for mathematical physics provided a touchstone for Riemann's physical researches throughout his career.

As planned, Riemann spent the semester break with his family in Quickborn. On his way back to Göttingen, he stopped for a short time in Hanover, where he visited with two of his former Gymnasium teachers. Both were evidently apprised of his present academic position and financial situation. On the day of that visit, 18 April 1855, he wrote to his brother:

> The evening before yesterday I departed from Quickborn and, thank God, left everyone as healthy as possible .... I was here today with Schmalfuss and Seffer and found out from them that I probably will receive an annual remuneration of 200 Taler. At least two officials in the Ministry of Culture ... made the suggestion. I'll probably get 100 Taler on Michaelmas, but I hope to be able to send you the money I borrowed quite soon. I will ask Aunt Julie to take out 100 Taler for me, and I hope she will do that soon. [Neuenschwander 1981a, 111–112]

Dedekind confirmed that Riemann was granted a yearly stipend of 200 Taler, a very modest salary, but enough to live on in Göttingen; a typical laborer earned 3 to 4 Taler a week and some even less than that. As another reference point for comparison, Ernst Eduard Kummer received a salary of 1500 Taler at the time he succeeded Dirichlet in Berlin, an amount his colleagues claimed was insufficient to live on without supplementary teaching at the Military School [Biermann 1988, 292]. Dirichlet's situation had been much the same, another reason why the chair in Göttingen proved attractive to him.

For several years, Bernhard's financial situation would remain a major source of worry for both him and his family. His Aunt Julie Ebell also had deep concerns about her nephew's well-being and mental health. In January 1855, she sent him an invitation to join her for tea. Her message included the remark that it was "not good to be alone with your sad thoughts."[14] Her note was sent to Barfüßerstrasse 18,[15] the home of a mechanic named Ryelshausen, from whom Riemann had rented a room. Today one finds a plaque on this building honoring the mathematician, who lived there from 1854 to 1857. This marker was one of several erected by the city as the result of a petition lodged by Felix Klein [Gierl 2004, 180].

Bernhard had hoped to bring his brother even better news, as two of his supporters had been planning that Riemann be appointed as associate professor,

---

[14] Julie Ebell to Bernhard Riemann, 10 January 1955, Riemann Nachlass, Staatsbibliothek Berlin.

[15] The street name derives the former Franciscan Monastery from the 13th century, one of several orders whose members went barefoot or with sandals; it was closed during the Reformation when Göttingen became Protestant.

but their efforts fell through in the end. Riemann may have known why, but if so, he did not explain what happened:

> For a long time I hoped to be able to surprise you with more pleasant news. I had certain assurances from Sartorius and Weber that Warnstedt[16] had applied for an associate professorship for me. But there was an intrigue against this in Göttingen from another side and therefore nothing will come of it this time. [Neuenschwander 1981a, 111–112]

During the following summer semester of 1855, Riemann and Dedekind both taught standard courses in the curriculum. Dedekind offered the course Gauss had taught for many years on the method of least squares and its applications to astronomy and geodesy, whereas Riemann presented a 4-hour course on definite integrals. Both of them had taken Stern's course on this topic, Riemann already in his first semester (see Section 1.3). As one might suspect, Riemann had no interest in going over ground he had learned four years earlier, and Dedekind was eager to learn whatever his older colleague could teach him. His *Nachschrift* from that lecture course is one of the many now available online from the SUB.[17] Much of the text is highly legible, but in places the writing is sloppy and scattered over the page, a likely sign that Riemann had to correct himself in such places. After Dedekind left Göttingen to take a professorship at the Zurich Polytechnic, Riemann taught the course on least squares in the summer semester of 1860, but never again. In general, all his teaching was confined to topics in higher mathematics, including mathematical physics.

Around this time, Wilhelm inquired about a vacancy at the astronomical observatory. This position arose originally following the death of Benjamin Goldschmidt in February 1851. On the advice of his former student, Christian Gerling, Gauss decided to bring over Wilhelm Klinkerfues from Marburg; later that year, he was appointed as the official observer. Following the death of Gauss four years later, Göttingen was without a competent theoretician in this important field, as Dirichlet had no background in astronomy. Riemann's brother wondered whether Gauss might have mentioned Riemann as a potential candidate for this position.

Bernhard evidently knew nothing more about this, as on 5 May he merely replied:

> ... I believe that the judgments on the papers I submitted for my doctorate and Habilitation, which [Gauss] had to hand in officially, have been sent to Hanover. Be that as it may, I am most pleased about Warnstedt's letter to Weber, especially because I believe it contributes a great deal, or at least some, to reassuring father about my future, ... I fear he has been more worried about it in recent years than I have been myself. The letter will probably arrive [in Göttingen] around now. [Neuenschwander 1981a, 113]

As time went by, the situation at the observatory continued to remain unclear, while Riemann clung to his hopes of gaining an appointment there. Eventually,

[16] Adolf von Warnstedt was one of the two officials mentioned above. He was involved with university affairs in Göttingen and later worked there as general director of the Curatorium.

[17] See Cod. Ms. R. Dedekind 1: 14, SUB Göttingen.

he realized that Ernst Schering had better chances than he of attaining that goal. Back in the spring of 1855, though, he seemed less concerned about what might happen; instead, he merely hoped that the recent signs of support for him would help alleviate their father's worries about his future.

By this time, Friedrich Bernhard Riemann had passed age 65 and was in poor health. As the year wore on, his condition steadily worsened to the point that he became bedridden. Wilhelm came from Bremen, but he probably already arrived too late to be at his father's bedside at his death. He remained to take care of the family's financial arrangements. Shortly before this, the siblings had lost their sister Clara as well.[18] The remaining sisters traveled with Wilhelm to Dannenberg, where they parted company in mourning of their father's death. Returning back to the vicarage in Quickborn, Ida wrote how lonely they all felt.[19] Aunt Julie had come from Göttingen to stay with the remaining sisters for a short time, but they would soon begin making plans to join their brother Wilhelm in Bremen. All of this sad news weighed very heavily on Bernhard, who had been very close with his father. Morose by nature, he would from this time forth face distressing new circumstances, while suffering from long bouts of physical sickness and psychological depression. In the meantime, a major transition had taken place that would have profound consequences for the future of Göttingen mathematics.

## 3.4 Dirichlet Succeeds Gauss

Gauss had already been in precarious health throughout much of 1854, so those who knew him could not have been surprised by the news of his death on 23 February of the following year. Only weeks before, Hanover's King George V had sent a sculptor to Göttingen to design a medallion of Gauss. Evidently Weber and Sartorius met with him then, as they sent a telegram requesting that the same artisan return to take a death mask. Sartorius wrote to inform Alexander von Humboldt, whereas Weber sent this message to Dirichlet:

> Dearest friend,
>
> This morning at 2 o'clock Gauss died. ... I learned of the death this morning and went over immediately with Sartorius; we found him sitting in his armchair as though he had fallen asleep, without any traces of pain. He suffered more during the previous days, partly from the asthmatic hardships with which his illness had begun already last Easter, and partly from the dropsy, which toward the end had spread over his entire body and in the last days hindered him from lying in bed, so that he could rest at night only by sitting in an easy chair. [Merzbach 2018, 177]

During the last phase of his life, Gauss's physician was his colleague, Wilhelm Baum, who joined the Göttingen faculty in 1849. He later treated Riemann,

[18] Bernhard Riemann to Julie Ebell, 16 May 1955, Riemann Nachlass, Staatsbibliothek Berlin.

[19] Ida Riemann to Bernhard Riemann, 6 November 1955, Riemann Nachlass, Staatsbibliothek Berlin.

advising him in 1862 to take a long vacation in a warmer climate (see Section 4.5).

Wilhelm Weber wasted no time in the urgent matter of appointing a successor. He began bombarding his good friend Gustav Lejeune Dirichlet with pleas to accept Gauss's former chair, hoping that such a high honor would entice him to leave Berlin, the city to which he and his wife Rebecka had been so deeply attached. Much had changed in the meantime, however. Rebecka's parents were no longer alive, and after their deaths the family home at Leipzigerstrasse 3 was sold. Her musically talented siblings, Fanny and Felix, had both suddenly died from strokes before reaching the age of 50. Homelife for the Dirichlets had also grown far quieter as their son Walter had taken up life as a country farmer, so only their daughter Flora still lived with them along with Dirichlet's mother, Anna Elisabeth, who would live to reach 100. Rebecka dreaded the prospect of moving to Göttingen as she made plain to her nephew Sebastian Hensel, likening the town to Jean Paul's fictive Kuhschnappel [Merzbach 2018, 179].

The Hanoverian Ministry moved nearly as quickly as Weber, so that by April Dirichlet already had received the expected offer. Berlin's bureaucrats, on the other hand, were slow to react; when they finally realized that Prussia was about to lose its second star mathematician (Jacobi had succumbed to smallpox in 1851), it was too late to make a counteroffer. After paying his respects to the aged Humboldt and his other good friends, he and his wife departed in September for a new adventure that proved far more rewarding than either could have anticipated.

Dedekind was especially delighted to find himself in the company of Dirichlet and his charming wife. As a talented pianist and avid cellist, he was a welcome guest in their home, where *Hausmusik* was nearly always in the air. References to that musical life abound in Dedekind's letters from this time, as when he wrote his sister Julie about a concert that Clara Schumann and Joseph Joachim would perform the following evening (Nov. 5 1856) [Scharlau 1981, 37]. Initially, Dedekind felt somewhat overwhelmed by Dirichlet's brilliance, but with time he felt more and more comfortable in his presence. In a letter to his sister Julie from the summer of 1856, he expressed his growing sense of self-confidence in these words:

> ...I really get on well with all the people here, and there are a lot of them, almost too many. The most useful thing for me though is the almost daily contact with Dirichlet, with whom I am really only just beginning to learn; he is always equably kind to me, tells me in no uncertain terms which gaps I still have to fill, and gives me the directions and means required to do so immediately. So I already owe him an infinite amount, and there will surely be much more to come. [Scharlau 1981, 35]

Dirichlet also helped to launch Ernst Schering's career by posing a prize problem that the latter won and then submitted as his doctoral dissertation. After taking his degree in July 1857 under Dirichlet, who was then Dean of the Philosophical Faculty, Schering had the extraordinary luck to solve a second prize problem, which he then submitted as his postdoctoral dissertation. Thus, in July 1858, only a year after his promotion to Dr. Phil., he was awarded the *venia legendi* for mathematics. Riemann, as noted earlier, had explicitly adhered to the faculty's formal requirement that Habilitation take place no earlier than two years after conference of the doctorate. Dedekind had done the same. Both he and Riemann presumably knew that Gauss took this two-year requirement seriously.[20] Dirichlet evidently took a more liberal attitude, which enabled Schering to begin offering courses in the winter semester of 1858, by which time Dedekind had already begun teaching in Zurich. These events, which affected Riemann's career prospects at a critical juncture in his life, require some additional elaboration.

Riemann interacted far less with Dirichlet, no doubt partly because of his natural shyness. Nevertheless, Dirichlet shared Dedekind's high opinion of Riemann's creative talents, even though he found him less approachable on a personal level. Dedekind, by contrast, possessed a sober-minded temperament and maturity that made him seem at times more like Riemann's older brother, though he was actually five years younger than his troubled friend. Under Dirichlet's guidance, Dedekind began to acquire a distinctive mathematical taste that reflected many of the same attributes found in his mentor's works. Over the next few years, Dedekind began to lay the groundwork for his formidable achievements in modern algebra.

## 3.5 Teaching Complex Analysis

With only a few exceptions, Riemann's lecture courses dealt almost exclusively with two subjects: complex analysis (no surprise) and mathematical physics (more specifically partial differential equations and potential theory). These were by no means mutually disjoint interests, as can be seen from his work on hypergeometric functions [Riemann 1857a].[21] Riemann's very first course taught in Göttingen – later published as [Riemann/Hattendorff 1869] – dealt with the role of partial differential equations in physics (see 3.3). He often offered this in conjunction with a second standard course on "Gravity, Electricity, and Magnetism," which was also edited by Karl Hattendorff for [Riemann 1876b].

During his lifetime, Riemann left his deepest imprint on the then emerging field of complex analysis. He taught this subject for the first time in a 3-hour course offered during the winter semester 1855/56 under the title "Functions, in particular elliptic and Abelian, of a variable complex magnitude." Riemann offered this over two semesters, drawing all of three auditors: the Norwegian expert on hydrodynamics Carl Anton Bjkernes, Riemann's colleague Richard Dedekind, and his soon-to-be colleague Ernst Schering (Fig. 3.1). Bjkernes afterward studied in Paris, but he found the experience somewhat less rewarding than the time he had spent in Göttingen. In a letter to Schering he remarked:

[20] In fact, on these formal grounds he had rejected another candidate's application in 1854, noting that he had received his doctorate only three semesters earlier.

[21] For the historical and mathematical context relating to this important paper, see [Gray 2008].

> I see that Göttingen has as much to offer as the great Paris, or perhaps even a little more, since here there is no one who comes close to the kinds of lectures Dirichlet delivers and no one treats mathematics in such an original way as does Riemann. ... In Paris there is no one who can hold a candle to Dirichlet as a lecturer. [Schering, K. 1909, 455–456]

On his way to habilitating, Schering took virtually all of Riemann's courses, which included Theory of Elasticity (SS 1856) as well as Complex Functions and Hypergeometric Series (WS 1856/57). According to the obituary written by Schering's brother Karl [Schering, K. 1909, 456], Riemann made use of Ernst Schering's elaborated lecture notes in preparing the written versions of the two great memoirs [Riemann 1857a] and [Riemann 1857b]. That may well have been the case, but Schering and Riemann were also rivals, though few of their contemporaries probably realized this. Riemann's relations with Dedekind, on the other hand, appear to have been very harmonious throughout their time together in Göttingen. It was hardly an accident that Riemann went to his grave hoping his friend Richard Dedekind would help to salvage some of his work for publication (see Chapter 12).

Fig. 3.1: Ernst Schering (1833–1897).

As for Schering, whose relationship with Riemann was to say the least ambiguous, we can be grateful that his transcriptions of Riemann's first three lecture courses on complex analysis have survived.[22] Considering how little is known about Riemann's evolving interests in complex analysis during the five-year period between completion of his dissertation and the publication of its dramatic "sequel" on the theory of Abelian functions, Schering's rendition of these lectures represents an invaluable resource for studying the course of Riemann's thinking on this subject. Here, we can only afford to touch on a few of the more striking aspects of these lectures, drawing on Erwin Neuenschwander's critical edition [Neuenschwander 1996].

It would appear that Riemann referred explicitly to Dirichlet's principle for the first time in his lectures from the winter semester of 1855/56. He had, of course, made use of essentially the same idea in his dissertation from 1851 and was well aware that others besides Dirichlet had utilized much the same idea. As an indefatigable reader, Riemann constantly scanned recent and sometimes even older mathematical literature in search of clues and inspiration for his own bold ideas. Neuenschwander published this undated passage from Riemann's *Nachlass*, which describes various sources related to this methodological tool:

> By the name Dirichlet's principle I have designated a method for proving that a function is completely determined by a partial differential equation and suitable linear boundary conditions, i.e., that the task of determining a function according to these conditions admits of one solution and one solution only. This method has been applied by William Thomson in his note Sur une équation aux différences ... and by Kirchhoff in his treatise on the oscillations of an elastic disk, after Gauss had previously in a similar manner treated a problem which may be regarded as a special case ([Gauss 1840a], Art. 29–34). I have named this method after Dirichlet, as I learned from Professor Dirichlet that he had already used this method since the beginning of the forties (if I am not mistaken) in his lectures. [Neuenschwander 1981b, 225]

Unlike his published works, Riemann's lecture course also included references to the French literature bearing on complex analysis, including works by Cauchy, Puiseux, and Hermite.[23]

Kronecker once remarked that Riemann let his general theta-functions rain down from the sky; indeed, he often left his readers in the dark regarding the origins of his theory when they came to that closing part of [Riemann 1857b]. Naturally, his contemporaries recognized that Riemann had extended the classical theory of theta-functions set forth by Jacobi in his *Fundamenta nova theoriae functionum ellipticarum* [Jacobi 1829], but he made no mention of the fact that Jacobi's inversion problem (see Section 7.2) for hyperelliptic integrals in the case $p = 2$ had already pointed in the same direction. This case was solved nearly simultaneously and independently by Adolph Göpfel and Georg Rosenhain, the latter a one-time student of Jacobi. Both authors employed theta-functions for this special case, but according to Adolf Krazer, they were discouraged from employing theta-functions to study hyperelliptic integrals of higher genus because these involve $\frac{1}{2}p(p+1)$ parameters, whereas hyperelliptic functions of genus $p$ only require $2p - 1$ [Krazer 1903, v]. Karl Weierstrass was the first to make

[22] Cod. Ms. B. Riemann 37, SUB Göttingen, available online.

[23] Cod. Ms. B. Riemann 37, 42–43.

this leap for the general hyperelliptic case, whereas Riemann famously introduced theta-functions in $n$ variables to construct Abelian functions that arise by inversion.

Jacobi's theta-functions are not themselves doubly periodic, but he used these to construct such functions. One can easily see that any functions defined by a series

$$\sum_{n=-\infty}^{+\infty} a_n e^{2\pi i n z}$$

has period 1. By forming suitable quotients, Jacobi could obtain a function with periods 1 and $\tau$, where

$$\theta(z) = \sum_{n=-\infty}^{+\infty} (-1)^n q^{n^2} e^{2\pi i n z},$$

and $q = e^{i\pi\tau}$.

What becomes clear from Schering's lecture notes is that by the mid-1850s Riemann was not only thoroughly familiar with Jacobi's theory of theta functions but also with Rosenhain's use of thetas in his study of the genus two case of hyperelliptic functions [Rosenhain 1851/1895], his prize-winning memoir. If the pagination in Schering's texts reflects the actual order of presentation, though, then Riemann clearly jumped back and forth between different topics. Evidently he had several things in mind and often got sidetracked before finding his way back to the main road. During the summer semester of 1856, his main theme was Jacobi's theory, which he presented in a fashion much like his former teacher's approach. This began with a fairly lengthy discussion of the various types of theta-functions in one variable along with the numerous identities they satisfied. Only somewhat later in this course did Riemann show how these could be used to construct the standard types of elliptic functions. As noted in Section 7.1, Riemann no longer adopted this standard presentation when he lectured on elliptic functions in the summer semester of 1861. By that time, he had streamlined his lectures so as to include only an abridged section devoted to the theta-functions [Neuenschwander 1996, 79–80]. Schering's notes also reveal that Riemann's earlier lectures from 1855/56 dealt briefly with theta-functions in $n$ variables; in fact, assuming that text reflects the original order of presentation, he dealt with these *before* taking up the case of a single variable.

In his lecture course for the winter semester of 1856/57 on complex analysis and hypergeometric functions, Riemann proved the key existence theorems using series expansions rather than appealing to Dirichlet's principle. Schering mentioned this in a letter to Bjerknes from 2 December 1856, thus roughly one month after Riemann completed his paper on the theory of hypergeometric functions [Riemann 1857a]. He added that Riemann promised to give a general solution of the problem of representing distributions of electricity on a sphere, but in the end he ran out of time for this topic. He also lost some of his audience over the course of the semester; only three were present by the time he finished [Schering, K. 1909, 456].

As was customary with major papers published by the Göttingen Scientific Society – to which Riemann now belonged as an Assessor – he published a brief announcement describing [Riemann 1857a] in the *Göttinger Nachrichten.* Regarding the background to this study, he wrote:

> This treatise is dedicated to a class of functions, which are useful for solving certain problems in mathematical physics. The series arising from these perform the same service for difficult [physical] problems as do the so often used series involving sine and cosine functions in the case of simpler problems. These applications, in particular to problems in astronomy, ... seem to have led Gauss to investigate them and to publish part of his findings in 1812 .... [Riemann 1876/1892/1902, 1892: 84–85]

Riemann then gave a brief account of results from previous investigations, including the unpublished findings in Gauss's posthumous papers. He indicated, however, that those findings had since been superseded by Kummer's paper from 1835, the most substantial contribution to this topic prior to Riemann's study. He had this to say about it:

> Kummer managed to modify Euler's method into a procedure for finding all transformations. However, this required such extensive calculations that he stopped short of finding third-degree transformations and restricted himself to those of the first and second degree and their compositions. (*ibid.*)

Riemann then contrasted Kummer's methods with his own, referring to § 20 of his dissertation and indicating how he could recover all known results almost without any calculation. He hoped to present the Royal Society with a sequel, but only two fragmentary works and some results from his lecture courses were later published posthumously in [Riemann 1876/1892/1902].[24]

Since Gauss had published his 1812 study as "pars prior," many mathematicians probably anticipated that a sequel would eventually appear. In fact, there is good reason to believe that Riemann chose to delay work on the theory of hypergeometric functions – the topic of [Riemann 1857a] – because he knew that Gauss had announced this sequel to his [Gauss 1812]. Only after his death did he gain access to that unpublished work, which he found in Gauss's scientific estate. Riemann had a longstanding interest in this topic, but he clearly did not want to tread on ground that Gauss may have already covered. Over and again, he found himself moving along tracks that Gauss had long ago envisioned and partially explored, as had Cauchy. The latter, however, published constantly, whereas Gauss, though prolific by any normal standard, often held back his findings. How Riemann got to read Gauss's sketch for a sequel is unclear, but Wilhelm Weber may have helped to intervene on his behalf. As we have seen in other cases, Riemann often worried that Gauss might have anticipated his whole approach, yet clearly nothing like his theory of $P$-functions can be found

[24] For an account of [Riemann 1857a], see [Klein 1894, 41–253]; for its place in the history of hypergeometric functions, see [Gray 2008, 1–41].

in Gauss's papers [Gauss 1863–1933, 3: 203–229]. What is more, this topic provided a wonderful playing field for the general approach to complex functions that Riemann had set out in his dissertation [Riemann 1851/1876].

This set of circumstances surely weighed heavily on Riemann when Gauss was still alive. Back then, it would have been far too audacious for him to proclaim that his work went beyond what Gauss had accomplished before him. Soon after the great man died, however, Riemann was proud to publish a study of hypergeometric functions in which he noted that his approach not only went far beyond the published literature but also surpassed what Gauss had written on the subject, though never put into print.

Psychologically, Riemann's intellectual relationship with Gauss brings to mind how Johannes Brahms struggled to compose his first symphony in the wake of Beethoven's ninth. "I'll never write a symphony," he allegedly said, "you have no idea what it feels like ... to hear the footsteps of a giant behind you."[25] Similar thoughts might once have been in Riemann's mind, except that first, unlike music, mathematics is not a performing art, and secondly, Brahms spent twenty years working on his first symphony and was already famous when he finished composing it. Riemann was still hoping to get a modest-paying permanent appointment.

When he first submitted this paper on 2 November 1856, Riemann still harbored doubts about whether it would be accepted for the *Abhandlungen*, a high honor for someone still at the beginning of their career. In a letter to his brother Wilhelm, written that same day, he explained these circumstances:

> I also hope that my work will bear fruit. As I already wrote to you, my treatise is now ready for printing, and perhaps the Society will publish it in its *Abhandlungen* – a great honor, since for the last 50 years it has contained only mathematical treatises by Gauss.[26] The mathematical section of the Society – consisting of Weber, Ulrich, and Dirichlet – will likely, at least according to Weber's assurances, request the publication of my treatise. [Dedekind 1876a, 552]

Weber chaired the mathematics/physics section that recommended Riemann's paper, but he probably made no assurances as to the outcome. He might well have anticipated resistance from senior colleagues, some of whom might question why this young man's paper, dealing with a topic earlier treated by the great Gauss, deserved to appear in the *Abhandlungen*. Riemann's status as Assessor gave him the right to publish in the society's organs, but he could have chosen to publish in the less prestigious *Göttingische Gelehrte Anzeigen* instead. If anyone raised such issues – about which we have no way of knowing – Dirichlet surely would have made a strong case for publication. In any event, the mathematicians' recommendation was accepted, to Riemann's great relief; clearly this sign of outward recognition meant a great deal to him.

[25] This saying appears without attribution in Max Kalbeck's biography of Brahms.

[26] This overlooks that Weber had also published in this series.

Once he began teaching complex analysis in Göttingen, Riemann invariably began his courses with an introduction to the subject.[27] Neuenschwander wrote that he probably taught this material no fewer than six times, taking up roughly a third of the course, before moving on to the special topic that constituted its main theme.[28] Riemann ended this introductory presentation with remarks explaining the main ideas behind Riemann surfaces. He realized that the brief discussion in his dissertation was not only inadequate but also quite incomprehensible without references to concrete examples. One can surmise the flavor of his presentation from the version recorded by Ernst Abbe,[29] who took the course on complex functions offered during the winter semester of 1861/62.

Riemann was, of course, not the first to tackle the problem of multi-valued functions. In his course, he discussed the familiar example of the logarithm function resulting from integration around a simple pole. Setting $z - a = re^{\phi i}$ so that $\log(z - a) = \log r + \phi i$ and writing

$$\log(z - a) = \int_C \frac{dz}{z - a},$$

where the closed curve $C$ wraps $k$ times around $z = a$, we get $\log(z - a) = \log r + 2k\pi i$, which for fixed $z$ has infinitely many values [Neuenschwander 1996, 52]. Riemann even discussed the simplest example illustrating his method, namely, the function $w^2 = z$, for which he made cuts from the origin passing along its two leaves to two corresponding points. He thus made no use here of the so-called Riemannian number sphere $\mathbb{C} \cup \infty$, though he often appealed to the function value $z = \infty$. Since $w^2 = z$ has branch points at $z = 0, \infty$, the simple modern construction joins those two points along two arcs that open up slits in the two spheres. Identifying their edges in the proper manner then produces another topological sphere, thereby showing that the Riemann surface has genus $p = 0$.[30]

For the purposes Riemann had in mind, though, he surely wanted to keep the picture as simple as possible. He summarized the general situation as follows:

> In general, for a multivalued function, we can cover the $z$-plane with a surface $T$ which, for each point on the $z$-plane, consists of as many leaves lying apart as the function has different values for this point. A point around which one leaf continues into another is called a branch value of the function. In the vicinity of such a point, the surface $T$ can be viewed as a helical surface of infinitely small height, the axis of which is perpendicular

27 The critical edition [Neuenschwander 1996] can be profitably compared with Weierstrass' approach in [Weierstrass 1988].

28 In four instances, the main topic was elliptic and Abelian functions (WS 1855/56, WS 1857/58, SS 1859, SS 1861), whereas in two others (WS 1856/57, WS 1858/59) it was hypergeometric series; see [Neuenschwander 1996].

29 Abbe later made his name as an inventor of optical instruments, later becoming co-owner of Carl Zeiss AG, which manufactured microscopes, telescopes, and other optical devices.

30 The same principle applies when dealing with elliptic functions, associated with polynomials of degree 3 or 4, $P_3(z)$ or $P_4(z)$. For polynomials of odd degree $z = \infty$ yields a fourth branch point, so the construction of the associated Riemann surface is the same in both cases, yielding a torus with $p = 1$.

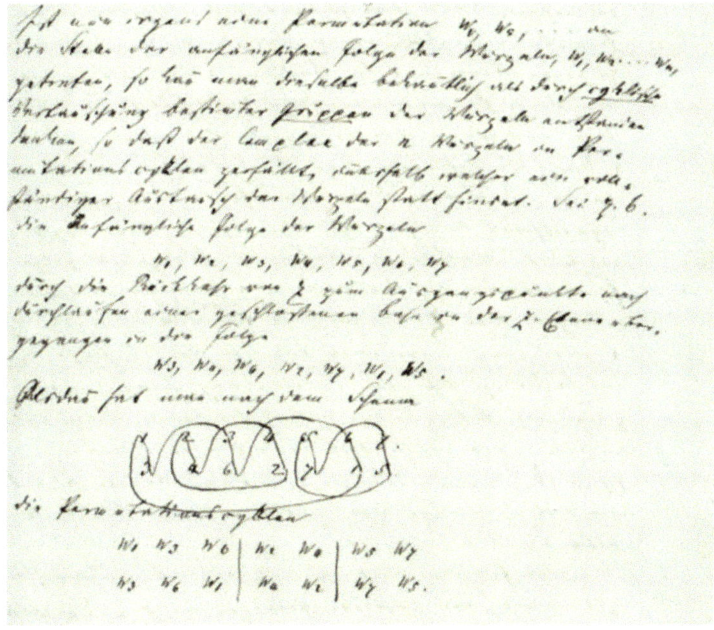

Fig. 3.2: A page from Ernst Abbe's text of Riemann's lecture course, 1861/62.

> to the $z$-plane at that point. If the function returns to its original value after several revolutions of z around the branch value, the top leaf passes through the others into the bottom one.
>
> The multi-valued function has only *one* definite value for each point of such a surface representing its branching behavior and it can therefore be viewed as a completely determined function of position on this surface. [Neuenschwander 1996, 74]

Riemann next considered the general case of an algebraic function $F(w, z)$ of degree $n$ in $w$ and $m$ in $z$. He discussed how the roots of $w$ will typically be permuted whenever $z$ moves continuously along a closed path, a phenomenon he illustrated by an easy example.

> If $z$ now traverses a closed curve, thus returning to the starting point, the expression $F(w, z)$ must again assume the same value, and thus must again have the same $w$ values as roots. Thus, either the individual roots have each regained their initial value, or a permutation has taken place among them: and the latter will be the case if the trajectory of $z$ corresponds to a branch point among the $n$ leaves, which must now be imagined to cover the $z$-plane. [Neuenschwander 1996, 75]

The example Riemann gave is shown in Figure 3.2, where the seven roots $w_1, \ldots, w_7$ are permuted in three groups of cycles. He writes these:

$$w_1\ w_3\ w_6\ w_2\ w_4\ w_3\ w_7$$

$$w_3\ w_6\ w_1\ w_4\ w_2\ w_7\ w_3$$

where the top row shows a cycle of length 3 – $(w_1\ w_3\ w_6)$ – followed by cycles of length 2: $(w_2\ w_4)$ and $(w_3\ w_7)$. Riemann then described these in connection with the branching behavior of the surface:

> Each cycle is separate from the others and indicates a particular branch point. The order of the indices of such a cycle indicates how the leaves of surface $T$ pass into one another. Geometrically speaking, in this case, at the branch point, surface 1 will merge into surface 3, surface 3 into surface 6, and this into surface 1. . . . [Neuenschwander 1996, 76]

These few passages should make evident that Riemann's lecture courses had a strong didactic character, an aspect he tended to ignore entirely in his published works. Over time, he attracted a devoted group of students, some of whom allowed others to make copies of their texts. In this way, knowledge of these new geometrical methods circulated on a small scale within Germany and also in Italy (see Section 11.1). Late in his life, Riemann gave Gustav Roch permission to publish material from one of his courses on complex functions; his article appeared in two parts in Oskar Schlömilch's *Zeitschrift für Mathematik und Physik* [Roch 1863a] and [Roch 1865a].

During the years they taught together, Dedekind regularly attended Riemann's lecture courses as well as those taught by Dirichlet. He took particular interest in Riemann's first course on Abelian functions from the mid-1850s. This can be seen from a letter he sent to Riemann in September 1856, congratulating him on his thirtieth birthday. Dedekind began with the salutation "Most honored Herr Assessor!" acknowledging the title Riemann (as well as Klinkerfues) had been accorded by the Göttingen Scientific Society:

> As I have not yet received any news from you, I now take the liberty of sending you my best wishes on your birthday and repeating my thanks for the rich instruction I owe to you during the past year; hopefully next winter there will be many more occasions to increase this debt of mine. At the end of the semester you expressed the intention of finally completing and writing down your theory of Abelian functions, and I asked you to send me installments of the finished work; if you really have used the vacation for this work, then I repeat my wish and I do so, as you can certainly be assured, out of *my own* interest. For I have a great desire during the remainder of this vacation to delve into this material more deeply. [Dugac 1976, 210]

One month later, Riemann and Dedekind were together again in Göttingen, where they met with an important visitor from Berlin. This was Carl Wilhelm Borchardt, the newly appointed editor of "Crelle's Journal." Dedekind later sent him two short papers for publication, but Borchardt was particularly keen to receive a major paper on Abelian functions, a topic that had been on Riemann's mind very much of late. In a letter to his sister Julie, Dedekind commented:

> I meet a great deal with Dirichlet, who constantly gives me new proof of his benevolence and interest . . . I also spend a great deal of time with my excellent colleague Riemann,

> who is without doubt the most profound living mathematician after or even with Dirichlet, and will be recognized as such if his modesty allows him to publish certain things. These, however, will only be understandable to a few for the time being. This contact with both of them is invaluable and will hopefully still bear fruit. [Scharlau 1981, 37]

Riemann was initially heartened by Borchardt's visit, about which he wrote his brother on 2 November 1856:

> From now on, I will devote all my free time to the work on Abelian functions that I told you about. Shortly before my return here to Göttingen, the editor-in-chief of the mathematical journal, Dr. Borchardt from Berlin, was also here and, through Dirichlet and Dedekind, sent me a request to send him, as soon as possible, a presentation of my investigations on Abelian functions, however rough it might be. Weierstrass is currently very active, but his recently published work, which Scherk[31] told me about, contains only the initial preparations for his theory. [Dedekind 1876a, 520]

## 3.6 Crushed Hopes

For Riemann, the year 1857 was an eventful one, but also a highly stressful time. Mathematically, this was a highpoint of his career, as it saw the publication of his two most important papers: the first on hypergeometric functions [Riemann 1857a] and the second on Abelian functions [Riemann 1857b]. And yet personally he could take little pleasure in these accomplishments, as much of the time he was overcome by dark thoughts that severed him from nearly all human contact. One source of his troubles stemmed from his fixation on the idea of marrying Laura Weber, even though he barely knew her. Laura was one of Wilhelm Weber's nieces, which makes it very likely Riemann encountered her years before during one of the Weber family's gatherings in Göttingen. Riemann was a frequent guest of his erstwhile patron, a bachelor whose household was usually run by one of his nieces.

In March 1857, Riemann's secret thoughts about Laura Weber led him to undertake a trip to Leipzig in order to visit with her family. Remarks in a letter to his brother Wilhelm suggest that Bernhard went there mainly hoping to speak with her father, Ernst Heinrich Weber, about his desire to marry Laura. Nothing more is known about their conversation, but a passage from that letter makes evident that Riemann had been dreaming of her for a very long time. When he saw her in Leipzig, "Laura's appearance . . . reawakened previous feelings. I had anticipated this moment with some trepidation, because I hadn't seen her in several years and feared that the impression I might have would not correspond to the image I still had of her" [Neuenschwander 1981a, 114]. Reading this, one can easily sense the desperation running through Riemann's mind.

[31] This was Heinrich Scherk, under whom Kummer took his doctorate in Halle. Scherk later left for Kiel, but lost his professorship in 1854 when Schleswig-Holstein was annexed by Denmark. He thereafter taught at a business school in Bremen, where both Riemann brothers apparently knew him.

Some weeks later, after Riemann had returned to Göttingen, he received a letter from his sister Ida. She was now living in Dockenhuden, a village in Holstein outside of Hamburg, while her younger sisters, Helene and Marie, continued living with Wilhelm in Bremen. Bernhard sent her these wishes: "May you always be as content with your situation there [in Dockenhuden] as you have been so far and not take too seriously whatever is unpleasant and distressing to you, something to which I believe we two among our siblings are most inclined." As so often, Bernhard began composing this reply on 22 April 1857, but then broke it off, adding a postscript almost two weeks later. By the time he sent this, he was already staying with his siblings in Bremen. This letter further reveals that while Bernhard was eager to read news from Ida, as conveyed to the family in Bremen, he did not expect her to write regularly to him. Very likely, he recognized that his habit of leaving letters unanswered discouraged his sisters from corresponding with him.

Nevertheless, he had sent an earlier letter to Ida concerning the events which had transpired during his trip to Leipzig. In the meantime, after reading one of Ida's letters to their sister Helene, he learned about how she had reacted to that news:

> In your letter to Bremen, if I'm not mistaken, you write to Helene that the news about me had so shaken you that you burst into tears. I don't know what I wrote to you about it; I was in a very gloomy mood at the time; certainly many things could have happened in a better, happier way than they did. I was still very restless on the journey to Leipzig, but after everything I now know about Laura, I believe I would be very happy with her. [Neuenschwander 1981a, 115]

Whether or not Ida felt reassured by these words, no one can say.

Whatever transpired in Leipzig, Bernhard evidently still held out hopes this dream would have a happy ending. Given that his financial situation was anything but secure, his urge to marry at age thirty must have struck some as a misplaced desire. In any event, after his trip he sent his brother a message, written in a desperate state of mind. Two days later, having calmed down somewhat, Riemann wrote him again about his difficult situation. He was now hoping to gain professional advancement in Göttingen, for which he was counting especially on Wilhelm Weber's support. In the letter to Ida, Bernhard wrote that during April he visited Wilhelm Weber's home in Göttingen almost every day. Laura's sister Sophie ran the household for her uncle, who also hosted her brother Julius, a law student. Riemann only once got to visit alone with these family members, however, since Kohlrausch had come over from Marburg to continue working with his Göttingen colleague.

The situation at the observatory had still not been resolved, raising Riemann's hopes of gaining employment there, but he shared that ambition with his former student, Ernst Schering. Riemann also sensed that his efforts to engage Weber's support may well have backfired, as he reported on 25 April to his brother:

> When I visited the Webers' ... at the beginning of this week, when Kohlrausch was also visiting at the time, a very unpleasant atmosphere prevailed, especially towards me.

> There was talk of a threatening thunderstorm and that a lot of rain could still come in May, etc. [Neuenschwander 1981a, 116]

Weber had spoken with a group of his colleagues about a preliminary decision affecting the vacancy at the observatory; this turned out to be unfavorable for Riemann. Bernhard now had the impression that predictions of a coming storm were quite misleading: it had already broken open at this meeting, during which Weber's colleagues put him on the defensive. It seems they had heard, perhaps through Schering, that Riemann had been lobbying Weber to gain this position.

This turn of events truly surprised and pained Riemann, as he sensed that Weber was angry with him for having brought about this unpleasant situation. By the following day, though, things had calmed down. Riemann realized that Weber's anger had already passed, even though he himself felt greatly disappointed. In the company of Schering and Klinkerfues, he also gained the impression that Schering was in line to gain this coveted position, despite the fact that he had only recently taken his doctorate. One can easily appreciate Riemann's dismay, but Ernst Schering had played his cards shrewdly. He had already been working with Klinkerfues in the observatory well before he gained this position as an *Assistent.* In February 1857, he took up residence in its east wing [Schering, K. 1909, 457].[32] As no one foresaw at the time, Schering's physical proximity to Gauss's scientific library and estate would later lead to considerable tension and conflict. These circumstances, which eventually affected Riemann as well, arose once work on the Gauss edition was underway.

Soon after Gauss died, key members of the Royal Society began making plans for the publication of his collected works [Gierl 2004, 152–153]. One of the key instigators was the Society's standing secretary, the mineralogist Friedrich Hausmann, who had, like Gauss, studied at the Collegium Carolinum in Braunschweig. The original plan called for a total of seven volumes, but eventually twelve would be published, some with extensive commentary. No editorial project of remotely comparable size had ever been undertaken before by the Göttingen Royal Society. Following Hausmann's death in December 1859, Wilhelm Weber was primarily responsible for overseeing the edition, supported by Riemann.

On 27 June 1857, Schering wrote to Bjerknes:

> Gauss's library has been purchased by the University, and Dirichlet has undertaken the publication of his posthumous writings. In the notebooks, one finds a large number of short, unfinished essays on a wide variety of subjects. [Schering, K. 1909, 457]

As noted above in Section 3.5, Riemann had obtained access to Gauss's papers already before November 1856. This suggests perhaps that Schering had been studying these unpublished documents for several months before writing this letter.

Schering went on to report about several new findings he had made since this acquisition took place:

[32] In October 1868, after Alfred Clebsch assumed Riemann's chair, Schering moved into Gauss's former apartment in the west wing of the *Sternwarte*.

> Gauss already possessed much that was only later discovered and published by others; among these: the fundamental theorems on $\theta$-functions, the determination of the class number of binary quadratic forms with negative determinant, entirely in accordance with the principles Dirichlet presented for number theory. Extremely interesting is his hand copy of Disquisitiones Arithmeticae, in which he noted the above and the dates of discovery of the fundamental theorems. From this, it is evident that he worked with tremendous speed at that time and that he must have found many things in a totally different way than appears in his written work ... [Schering, K. 1909, 457–458]

Schering's excitement over these new findings practically leaps off the page; nor was this merely a momentary reaction. For he spent much of his life working on and basking in the reflected glory of Gauss's works.

Initially, Dirichlet was placed in charge of this project, assisted by Stern, Schering, Riemann, and Dedekind, an arrangement fraught with problems from the start. Dirichlet never managed to take control of the edition, primarily because of his health. During their stay in Rome in 1844, Dirichlet contracted malaria, from which he never completely recovered. His final years in Göttingen turned out to be the least productive period in his otherwise remarkable scientific career [Merzbach 2018, 218–219]. In the meantime, the Royal Society charged Schering with the responsibility of putting Gauss's extensive scientific estate in order, as a first step toward preparing the individual volumes for publication.

Returning to the meeting held in Weber's home in April 1857, Riemann had another reason to feel embittered about what had transpired. On that occasion, he complained about provisional plans for the courses that were to be offered in the summer semester, in particular those he and Dedekind were scheduled to teach. He himself had yet to announce any course offerings, perhaps hoping to gain an unofficial leave of absence. He may well have been flirting with the idea of focusing all his energy on the paper he had promised for Borchardt's journal. The question arose, however, as to whether he or Dedekind would be teaching the standard course on the method of least squares, which both had taken under Gauss.[33] As Riemann reported to his brother, what took place next made him most unhappy. Immediately after this meeting, his complaints were passed on to Dedekind, who went to see him directly.

> He stayed with me for an entire afternoon and we agreed on how to arrange our lectures so as not to clash. However, I was very uncomfortable that this had come around again so quickly. Dirichlet, too, seemed to me, although still very polite, no longer as well-disposed toward me as usual, and this also tormented me. Dedekind was now also tasked with persuading me not to abandon my course and urging me to announce when it would begin; but I didn't want to do this until I knew it would take place. [Neuenschwander 1981a, 117]

Riemann's position was not just weak, it was actually indefensible. He knew, of course, that such a delay contravened the established procedure, which required instructors to announce their course offerings well before the semester began. This enabled students to plan their schedules and, if necessary, sit in for a few weeks before deciding on whether to sign up for a course officially. Faced with

[33] Dedekind agreed to do so in the coming semester; Riemann only taught this course once, some three years later.

Riemann's stubborn attitude, Dedekind decided he needed to mobilize some of the students, and he succeeded in doing so.

Not long after Riemann returned from visiting Weber, a student from his last course came to tell him that he and four others wished to sign up for his lecture course; they had been awaiting the announcement and would sign on as soon at it appeared. Although Riemann suspected an ulterior motive behind this interest, he replied that he would begin his lectures the following Monday. To his brother, however, he confided that he felt unwell and would have preferred spending the semester in Bremen, where he would have found it easier to concentrate on his work. In fact, he broke off this course early so that he could leave Göttingen and finish work on his paper there.[34]

Dedekind later wrote this about Riemann's ensuing efforts:

> ... he now devoted himself with all his energy to the completion of this work [which he finished on July 2]. However, his health had suffered greatly due to the excessive exertion, and by the end of the summer semester he was in a state of mental exhaustion that severely dampened his spirits. [Dedekind 1876a, 552]

Dedekind treated the severity of this episode rather gingerly, in fact, no doubt in deference to Riemann's widow and his sister Ida. Already in July, Riemann had written to his brother, who was desperate to know whether he should take an emergency break from his work in Bremen to visit him in Göttingen.[35] He urged him to write immediately and indicated he might need to bring their youngest sister Marie, who was sixteen. Both must have been seriously ill at this time, as Wilhelm died later that year and Marie followed him to an early grave only months later. Other contemporary documents also indicate that Dedekind was beside himself trying to deal with this situation. He realized that his friend was behaving in a strange manner, but he really had no idea what might be causing this. Dedekind also wrote to a physician in the family about Riemann, another indication that he was deeply worried about his friend's state of mind.

Since he was personally indisposed at this time, Dedekind wrote to his sister Mathilde, requesting that she look after Riemann and his companion August Ritter at the Dedekind family's second residence in Bad Harzburg:

> Instead of my person, may I recommend to you the two doctors Riemann and Ritter. The first, about whom I've spoken to you so often, is presently most unhappy. He spent the entire summer in Bremen preparing a highly important work for publication. But his lonely life and constant bodily pains have led to hypochondria and a deep mistrust of other people and himself, even if he appears to act in a friendly manner. He has in the meantime completed this work, and there is reason to hope that a quiet stay in Harzburg and a relaxed interaction with other people will have a beneficial effect on him. ... One must do all one can to pull such a valuable and scientifically distinguished person as Riemann out of his deep misery, but he must not take notice of this intention. It was always difficult to do him a favor, as he would only accept this if he could be convinced that you were doing this out of consideration for oneself as much as for him. He hates it when he causes trouble for others. Here he behaved in the strangest ways only for such reasons, because he believes no one can stand him, etc. [Scharlau 1981, 44–45]

[34] Dedekind to his sister Mathilde, 13 August 1857 [Scharlau 1981, 44].

[35] Cod. Ms. B. Riemann 19, 75v, 26 Juli 1857, SUB Göttingen.

This plan quickly backfired, as Mathilde Dedekind soon found their new house guest in a morbid state of mind. He must have begged her to write her brother, requesting that he come to Bad Harzburg as soon as possible. From other correspondence it appears clear that Riemann did not enjoy being in the company of Ritter, whom Dedekind described as "an interesting, but harsh and even repulsive person, who has seen much in the world." In any case, her letter evoked an unusually angry response from her brother:

> Impossible! I went through all that thoroughly in my last letter. ... I can hardly get away from here before Wednesday. Exactly that is what I told Riemann. What you're doing for him is certainly completely sufficient, so no extra trouble on his behalf. Greet him and Dr. Ritter heartily from me. I will travel directly from here to Harzburg and stay for one or two days. [Scharlau 1981, 45]

When he arrived, Dedekind found his friend in a despondent state of mind. Thus, the one or two days he thought to stay turned into one or two weeks, much longer than he had originally planned. He later wrote these lines describing this very dark period in Riemann's life:

> The excessive strain alone had taken a heavy toll on his health, and by the end of the summer semester he found himself in a state of mental exhaustion that darkened his mood to the highest degree. To refresh and strengthen his health, he spent a few weeks in Harzburg, where his friend Ritter accompanied him for a few days and where he was later joined by his colleague Dedekind, with whom he took many walks and also made longer excursions in the Harz Mountains. On such walks, his mood brightened, his confidence in others and in himself grew, and his harmless jokes and his unreserved conversation about scientific subjects made him the most amiable and stimulating companion. During this time his thoughts turned again to natural philosophy, and one evening after returning from a strenuous hike he picked up Brewster's Life of Newton and spoke at length with admiration about the letter to Bentley, in which Newton himself proclaimed the impossibility of direct action-at-a–distance.[36] [Dedekind 1876a, 520–521]

This lighter mood, to be sure, quickly passed, as nothing seemed to pull Riemann out of his deep depression; not even his astounding mathematical achievements.

## 3.7 Abelian Functions

In his biographical essay, Dedekind wrote that Riemann first sent three shorter papers to Borchardt on 8 May 1857, whereas the two-part manuscript containing his main results followed on 2 July [Dedekind 1876a, 520]. In the original publication of [Riemann 1857b] these appear as four separate papers, each concerned with a special aspect of Riemann's theory, whereas in the *Werke*, which most readers turned to after 1876, this structure can easily be overlooked, since there all four papers are combined.

[36] As discussed in Section 2.7, Riemann was long familiar with Newton's letters to Bentley. However, the key passage from the third letter, which Riemannn cited in [Riemann 1876/1892/1902, 1876: 498], does *not* appear in Brewster's biography.

Riemann knew full well that his approach to Abelian functions was bound to send off shock waves not only for analysts in Germany, but also in France. Some of Cauchy's followers may have even recognized that their master's death came at an ironic moment, namely, just before Riemann's paper came out. In the three preludes to "Theory of Abelian Functions," he summarized the main substance of his dissertation, after which came the introduction to the main text. Riemann took care to inform his readers about the prior evolution of this work, a precaution he took on no other occasion. He characterized its contents, with the exception of Sections 26 and 27, as forming a portion of the material presented in his two lecture courses held during 1855/56 and the following summer semester of 1856. He noted further, however, that he had obtained the results presented in Sections 1–5, 9, and 12 already in 1851/52, when he investigated the properties of conformal mappings of surfaces of higher genus (see Section 5.1). He dropped this work afterward – the next years were dominated by interests in physics and philosophy – only returning to complex analysis in the spring of 1855. Over the next year, he worked out the arguments in his paper up through Section 21, completing the remainder afterward.

Riemann also knew that Weierstrass was preparing a sequel to his famous paper solving the Jacobi inversion problem for hyperelliptic functions (Section 7.2). In fact, this paper had already been submitted to the Berlin Academy [Bottazzini 2003]. Weierstrass defined Abelian functions to be the single-valued, analytic functions of several complex variables arising from Jacobi's inversion problem. In Riemann's approach, Abelian functions were the integrals of algebraic functions introduced by way of Abel's theorem (Section 7.3). Riemann's paper [Riemann 1857b] was written in two separate parts. In the first, he laid out a general theory of algebraic functions and their integrals on a given surface of genus $p$, but without making any use of $\theta$-series. He introduced three classes of Abelian integrals according to their singularities. He then determined how these lead to bounds on the number of meromorphic functions on a surface. By reformulating Abel's theorem in terms of birational transformations, Riemann showed the significance of $p$ as an invariant for what became known as birational algebraic geometry.

The second part of Riemann's paper was devoted to the study of $\theta$-series for $p$ complex variables. Using these, he was able to give a complete solution of the Jacobi inversion problem. At the same time, he was careful to acknowledge Weierstrass's previous findings in a suitable way:

> Jacobi's inversion problem, which is solved here, has already been solved for hyperelliptic integrals in several ways by Weierstrass's ongoing work, which has been crowned with such beautiful success. An overview of this was published in the 47th volume of *Journal für die reine und angewandte Mathematik*. So far, however, only that part of this work outlined in Sections 1 and 2 and the first half of 3, concerning elliptic functions, has been published (vol. 52, p. 285). To what extent the later parts agree with the work presented here, not only with regard to results but also the methods leading to them, will to a large extent only be revealed by the promised detailed presentation of the same. [Riemann 1857b, 101–102]

When Weierstrass read the above passage, he realized that the challenge it posed required a completely new approach to the problem; he, therefore, promptly withdrew his submission.[37] Word of that decision must have spread quickly, thereby fanning the flames that led to Riemann's fame. In the years that followed, interest in Riemann's work grew. Some mathematicians tried to get their hands on his dissertation, realizing that he had already elaborated his novel geometric approach to complex functions in [Riemann 1851/1876]. Meanwhile, Weierstrass published nothing more on this topic, though he quietly reworked his theory after Riemann's death with the help of his now favorite pupil, the Russian mathematician Sofia Kovalevskaya.

In November 1857, Bernhard Riemann gained the title of associate professor and a salary increase from 200 to 300 Taler, still a very modest sum. The pleasure this must have given him, though, was overshadowed by the sad news that his brother Wilhelm had died in Bremen. This left Bernhard's three sisters without support, but even before they could join him in Göttingen, the youngest, Marie, also died. No medical records or other sources seem to have survived that might shed light on the causes of their deaths, but most likely this was due to tuberculosis.

It was around this time that the President of the Zurich Polytechnic (forerunner of the ETH), Johann Karl Kappeler wrote to Dirichlet asking for advice regarding suitable candidates for a vacant mathematics professorship at the Polytechnic. Among several others, Dirichlet especially recommended Riemann and Dedekind. He ranked Riemann slightly higher, but admitted that he had only taught courses for advanced students; at the same time, he praised Dedekind's abilities as a lecturer.[38] Realizing that he faced an important and difficult decision, Kappeler headed off to Göttingen in order to hear both of them lecture. He found Riemann's speaking style too introverted (*zu stark in sich gekehrt*) for the tastes of engineering students [Frei/Stammbach 1994, 37]. Clearly, this was a sound decision, though Dedekind would find it difficult to adapt his teaching style to the expectations of the engineering students at the Poly, especially the older ones.

[37] For the reasons Weierstrass gave some years later, see the quote at the end of Chapter 7.

[38] For a detailed account of these events as they unfolded in Zurich, see [Volkert 2017].

# Chapter 4
# Triumph and Tragedy, 1859–1866

Dedekind's departure for Zurich meant that the Göttingen faculty had only one mathematician to teach courses in higher algebra and number theory: Gustav Dirichlet. This circumstance may have been a factor in enabling Ernst Schering to habilitate and thereby acquire the *venia legendi* in speedy fashion; he was already teaching courses in the winter semester of 1858/59. Georg Westphal continued to offer his standard 3-hour course on mathematical astronomy, but this probably did not attract many students, perhaps not even a quorum. Schering's introductory course on astronomy drew five auditors, a reasonable number in those days. Like Dirichlet, Schering's competences extended to both pure and applied mathematics; his course offerings ranged from topics in astronomy and mathematical physics to number theory.[1]

## 4.1 Death of Dirichlet

During his last years in Göttingen, Dirichlet often complained about feeling weak and feverish. Doctors eventually determined that he had never fully recovered from the malaria he contracted back in 1843 during his sojourn in Italy. In the fall of 1858, around the time Sebastian Hensel and his new bride spent a week visiting with the Dirichlets, Gustav suffered a heart attack [Hensel 1911, 458–459]. At first, he seemed to be on the road to recovery, but then with little forewarning he lost his wife Rebecka, who died in December. After this shock, Dirichlet's health deteriorated rapidly until he, too, died in May of the following year. Their daughter Flora was not yet fifteen when her parents passed away.[2]

[1] For a detailed list of Ernst Schering's courses, see [Schering, K. 1909, 465–467].

[2] Her uncle, Paul Mendelssohn Bartholdy, became her guardian and she lived with him for several years in Berlin. At age twenty-five she married Wilhelm Georg Baum, son of the Göttingen surgeon. The Baum and Dirichlet families had earlier been neighbors, so this relationship had longstanding roots. The couple later settled in Danzig, where the younger Baum continued in his father's footsteps as a medical surgeon. They had six children together [Merzbach 2018, 224].



D. E. Rowe, *Bernhard Riemann: His Life and Wondrous Mathematical Legacy*,
https://doi.org/10.1007/978-3-032-25457-3_5

In retrospect, it might seem as though Riemann would have been regarded as the ideal candidate to assume Dirichlet's chair. Not only was he held in high esteem as a brilliant mathematician, but his work had deep connections with the Göttingen tradition exemplified by both Gauss and Dirichlet. From the perspective of the time, however, Riemann's age – he was still in his early thirties – counted against him, as he was far younger than those who typically held full professorships. Beyond this issue, the Göttingen faculty was reluctant to nominate Riemann as Dirichlet's successor for fear of the jealousies this might arouse. For although he had already been appointed Assessor of the Royal Society in 1856, he had nowhere near the seniority of Göttingen's two other mathematicians, Justus Ulrich and Moritz Abraham Stern.

Ulrich had already joined the faculty back in 1817; he was promoted in 1831 to a full professorship, a position he held up until 1879. Over the course of that time, Ulrich did the bulk of the teaching for courses in elementary applied mathematics. During his early career, he devoted much of his time and energy to writing textbooks, which he often used in these courses. Although he rarely published original research, in 1845 Ulrich nevertheless gained the honor of membership in the Royal Society's Mathematics Section, in which he was completely overshadowed by Gauss and Weber. His colleague Stern was, by contrast, an accomplished mathematician with numerous publications in Crelle and other periodicals. His career prospects, however, had long been impeded by religious discrimination: as an unbaptized Jew he had to wait thirty years before attaining an associate professorship during the liberal reforms of 1848.

In view of these circumstances, the faculty at first set its eyes on Karl Weierstrass, who only held an associate professorship in Berlin [Dugac 1973, 180–182]. The majority favored appointing an outsider, fearing that if they passed over Ulrich and Stern to promote Riemann they would be insulting these two older colleagues. This situation reflected a general principle that applied in all academic affairs: promotion always took seniority into serious account. Thus, only in rare instances did a man in his early thirties gain the coveted position of *Ordinarius*. In any event, this plan failed, because, to everyone's surprise, Weierstrass declined this prestigious chair, paving the way for Riemann's appointment.

At the time of Dirichlet's death, Dedekind remained in Zurich. Afterward, more than two months passed without Dedekind hearing any news at all from Riemann. Then, but to his great dismay, he learned from his sister Mathilde that Riemann had informed her of one of Dirichlet's last wishes. This concerned an unfinished manuscript on hydrodynamics, which Riemann likely knew about long before. On 31 July 1857, Dirichlet had presented an abstract for this paper on hydrodynamics at the Royal Society [Dirichlet 1857], but he later realized that he no longer had the strength to write it up for publication. What Mathilde Dedekind learned from Riemann was that in one of his last conversations with Dirichlet, the latter told him he would be grateful if Dedekind took on the task of completing that paper.

Dedekind's reaction to his sister's message, expressed in a letter to her from 17 July 1859, reflected his patience and understanding for Riemann's chaotic state of mind:

> Regarding the news as to Dirichlet's last request, which touched me very much, I ask that you remind Riemann so that he no longer hesitates to send me what he has. With any other person I would say they behaved very badly for having withheld this news for such a long time. But he is sick, and I know all too well his unfortunate dithering and hesitant manner, so I cannot raise even the slightest accusation against him. [Scharlau 1981, 48]

Once he received the drafts from Riemann, Dedekind was able to fill in the gaps and submit the text for publication as [Dirichlet 1860/1861]. Presumably he communicated this as well as his supplementary note [Dedekind 1861] directly to Riemann, who was thus informed about their contents from the beginning. Based on these, Riemann soon thereafter published an important sequel study [Riemann 1861a/1876], which he described in these words:

> In his last work published by Dedekind, Dirichlet broke surprisingly new ground for investigations into the motion of a homogeneous fluid ellipsoid, whose elements attract each other according to the law of gravity. The pursuit of this beautiful discovery has its own special appeal for mathematicians, quite apart from the question of the causes for the shape of celestial bodies, which originally prompted these investigations. Dirichlet himself only solved the problem he addressed completely in the simplest cases. For further investigations, it is convenient to give the differential equations for the motion of the fluid mass a form that is independent of the chosen initial time. This can be done, for example, by searching for the laws according to which the principal axes of the ellipsoid and the relative motion of the fluid mass relative to these can change. By treating the problem in this way, we will indeed follow Dirichlet's treatise, but, to avoid confusion, we have not been able to retain the notation he used. [Riemann 1861a/1876, 182]

During Easter vacation of 1860, Riemann and his two sisters, Ida and Helene, moved into their new quarters in the west wing of the astronomical observatory (see Fig. 0.2), where Gauss had lived during most of his long life. As Dirichlet's successor, Riemann assumed co-responsibility for preparing the initial volumes of the Gauss edition, an undertaking of great prestige that was very dear to Wilhelm Weber's heart, even though managing it caused him considerable difficulties. Ernst Schering, who had earlier outmaneuvered Riemann to obtain a lower-level position at the observatory, had in the meantime been given the task of organizing Gauss's papers in preparation for editing these volumes [Schering, K. 1909, 457–458].

Some may have felt that Schering, who had only habilitated in 1858, lacked the experience to play a leading role in this project. While the precise circumstances remain unclear, Schering later expressed his unhappiness over what eventually transpired [Schering, K. 1909, 462–463]. Initially, it had been expected that Dirichlet, as Göttingen's star mathematician, would play a leading role in this work, but his poor health and other matters prevented him from doing so. After his death, Schering complained that Dirichlet had never lifted a finger to support work on the project, a circumstance that would hardly improve with Riemann, whose health problems would eventually require that he spend long periods of time in Italy. Much later, in 1873, Schering registered another complaint: "My friend Riemann pointed out to me that I should insist my name be printed on the cover as editor of the first volume" (*ibid.*). Whether he ever tried to persuade members of the Royal Society to implement this request is unknown, but if so, nothing came of it.

In November 1859, Schering was offered a position as associate professor in Giessen. This outside offer set off a sense of panic in Göttingen, as Weber and others feared that work on the Gauss edition might come to a standstill were Schering to leave. An ad hoc committee filed a proposal that aimed to persuade the Ministry to help salvage this project. The ideal (though unrealistic) solution would have been to hire back Dedekind, while promoting Schering to a professorship. In any case, Weber's petition – supported by Wöhler, Sartorius, and Riemann – insisted that Schering's role was indispensable. Soon thereafter, in February 1860, Schering declined the offer from Giessen, and by the following month he was appointed associate professor with membership as an Assessor in the Royal Society [Dugac 1973, 183]. From this point onward, Schering began to take full control of the Gauss edition, despite the fact that the arrangement with the Royal Society called for the participation of three others: Riemann, Dedekind, and Stern. Over time, Wilhelm Weber came to realize that Schering had little interest in cooperating with other co-editors (see Section 12.1).

In 1859 Riemann became a corresponding member of the Berlin Academy, an honor he would likely have received even without having been named Dirichlet's successor in Göttingen. Gustav Dirichlet had been a regular member of the Academy from 1832 to 1855, the year in which Berlin mathematics underwent a major generational change. After 1855, Dirichlet was briefly made an honorary member of the Academy before becoming an external member in 1856. On 4 July 1859, Weierstrass wrote a glowing report nominating Riemann, in which he affirmed that his works belonged to the most impressive recent publications in mathematics. He recognized these as distinguished "by the importance of the results they contain but also by the originality and depth of the methods" [Biermann 2022, 27]. Furthermore, he predicted that they would exert an essential and durable impact in the future.[3]

Weierstrass's praise for Riemann is particularly striking in view of the fact that his paper [Riemann 1857b] employed radically new methods to solve a problem Weierstrass had been struggling with for years. Indeed, as some of his followers knew, Weierstrass decided to withdraw his own recent submission on the same problem once he saw what Riemann had accomplished. These circumstances would thus seem to suggest that Weierstrass's praise for Riemann's work was written in a spirit of graciousness and generosity that overlooked their differences of opinion. For, as became well known, he and Riemann stood poles apart when it came to methodological approaches, as the future would clearly reveal. Still, in the year 1859 the future course of complex analysis was anything but clear, making it likely that Weierstrass's views had not yet hardened at this time. Moreover, the recent death of Dirichlet was now on everyone's mind.

---

[3] Weierstrass's laudatory remarks are cited in the introduction to Chapter 5 and in Section 7.2.

## 4.2 Conversations in Berlin

In September, Riemann and Dedekind traveled together to Berlin, where they were warmly received by the new mathematical faculty. Some fifteen years later, Dedekind recalled the surrounding circumstances, namely the fulfillment of Dirichlet's last request (see Section 4.1).

> Immediately after receiving all the papers [from Riemann], I began drafting the hydrodynamic treatise, and after completing most of it, during the autumn break I traveled to Göttingen from Zurich with the manuscript and from there, accompanied by Professor Riemann, I went to Berlin to present it to Profs. Kronecker and Borchardt.[4] [Scharlau 1981, 51]

Dirichlet's passing meant that the earlier generation of mathematicians Riemann had known as a student in Berlin were now all deceased, so this trip introduced him to a new mathematical circle.[5] Ernst Eduard Kummer was Dirichlet's first choice as his successor, and he later became a major figure at both the University as well as the Academy. From 1863 to 1878 he served as secretary of the Mathematics-Physics Section of the Berlin Academy. His colleagues in the philosophical faculty twice elected him as Dean and during the academic year 1868/69 he held the office of Rector. An eloquent speaker, Kummer delivered a moving memorial lecture in honor of Dirichlet to fellow members of the Academy [Kummer 1860] (quoted in the Introduction).

As a Gymnasium teacher in Liegnitz, Kummer struck up a lifelong friendship with a young man named Leopold Kronecker. This circumstance likely influenced Kronecker's decision to move to Berlin in 1855, though not as Kummer's official colleague. In fact, he held no official position until 1860, when he was nominated by Kummer and Weierstrass for membership in the Berlin Academy. Thanks to this status, Kronecker could begin offering lecture courses at the University, the same arrangement that had enabled Jacobi to give lectures at the University earlier. As a wealthy man, Kronecker had no need for a university position, nor did Carl Wilhelm Borchardt, who married into the Oppenheim banking family in Köngisberg. Borchardt habilitated in Berlin in 1848, but it seems his and Riemann's paths only crossed much later. After assuming the editorship of Crelle's Journal in 1855, he joined the Berlin Academy as a full member. Later in Göttingen, Riemann met with Borchardt in his capacity as editor (see Section 3.5). This meeting in some sense paved the way for Riemann's only personal encounter with Weierstrass, who was a member of the Academy since 1856, but would only join the faculty in 1864.

Years later, in a letter to Georg Frobenius, Dedekind recalled the pleasant days he and Riemann spent together in Berlin. Weierstrass invited them both to dinner on the evening of 25 September at a Bavarian restaurant near the Gendarmenmarkt. Unfortunately, he did not mention the topics they discussed,

[4] Carl Wilhelm Borchardt, a former student of Jacobi in Königsberg, succeeded Crelle in 1855 as the principal editor of the *Journal für die reine und angewandte Mathematik*.

[5] Jakob Steiner, to be sure, was still alive, though by now a burned-out volcano; Riemann may have met him, but he had never taken an interest in Steiner-style synthetic geometry.

probably because he could only remember listening *con amore* to the conversation between these two great experts on function theory [Dugac 1976, 284]. These few details are about all that is known about their personal encounters in Berlin, but what resulted from them proved to be of great mathematical significance.

Two very different topics dominated these conversations: complex analysis and number theory. According to an anecdote Felix Klein recalled decades later, Riemann and Weierstrass discussed the status of Dirichlet's principle at this time.[6] Klein wrote as if both were aware that Riemann's appeal to Dirichlet's principle for proving his existence theorems was only a heuristic argument that lacked rigor. Riemann supposedly conceded this, but claimed his existence theorems were nevertheless correct. Since no corroborating evidence supporting this exchange of views exists, one can either dismiss it as mathematical gossip or consider its plausibility.

It was not for another ten years that Weierstrass, in 1869, published a note showing why Dirichlet's principle was unsound. Since Klein was in Berlin at this time and met occasionally with Weierstrass, it seems entirely plausible that they would have discussed this topic. It appears far less likely, though, that Weierstrass – meeting Riemann for the first time only two years after the publication of [Riemann 1857b] – would have suggested to Riemann that his reliance on this general principle was unsound. If this particular topic came up at all back at that time, then Klein probably misremembered or refashioned what Weierstrass then told him (he was hardly an unbiased witness). The larger issue, though, concerns the timing: at what point did misgivings about Riemann's arguments arise and who reacted to them? It seems Riemann's former student, Friedrich Prym, was the first to publish a counterexample that refuted the general line of proof Riemann set forth in [Riemann 1857b]. Prym's paper only appeared in 1871, around the time others had begun searching for ways to refine or get around Dirichlet's principle [Bottazzini & Gray 2013, 314–315].

What we know with more assurance is that Weierstrass was curious to learn how Riemann proved that a single-valued complex function in $n$ variables could never have more than $2n$ periods. Although this theorem accorded with all known cases, Riemann was the first to give a general proof. He presented this as one of the main results in his course on elliptic and Abelian functions from 1855/56. In a letter to Weierstrass, Riemann sketched a proof of this theorem, which found its way into print several years later, when Borchardt extracted part of the letter for publication in [Riemann 1869/1876]. The result itself was surely treated as folklore knowledge, but it turned out to have greater significance in light of another result Riemann unveiled the next year in Paris (see Section 4.3).

The other topic of conversation particularly piqued Kronecker's curiosity. The density of the prime numbers had long intrigued eminent mathematicians, even those who had never worked on the problem themselves. Kronecker learned from

[6] Klein based his information on a conversation he had with Weierstrass, which might have taken place during the winter semester of 1869/70 when he was a postdoc in Berlin; see [Klein 1926, 264] and [Klein 1895, 492].

Riemann that he had conceived of a new method for attacking this problem, one that promised to yield a better approximation of the density of the primes than those known theretofore. After returning to Göttingen he tried to write up this investigation for publication by the Berlin Academy. Roughly five weeks later, Riemann managed to submit this manuscript, but before doing so he sent a letter to Weierstrass explaining the circumstances. Though he admitted that his study contained significant gaps, he nevertheless expressed confidence that the results were sound.

Riemann's letter to Weierstrass had this to say about his recent visit as well as his subsequent efforts to tackle this elusive problem:

> First of all, my warmest thanks for our extremely friendly reception in Berlin; all the good things we enjoyed there and above all the pleasure we derived from your conversation will remain in my grateful memory for a long time to come. In the time that has passed since then, I have spent a great deal of effort investigating the frequency of prime numbers, but have not come to a proper conclusion. Without Dr. Kronecker's wish to learn more about this subject, I would probably not have decided to write anything about it. However, with the imminent resumption of lectures, which will perhaps prevent me from continuing this investigation, I have now dared to prepare a small essay about it for the Academy, which will be submitted to you and your colleagues for evaluation, and for which I hope for a favorable reception from your side. [Riemann 1859/1990, 823]

Since this manuscript was quite short (eight printed pages), no one apparently objected to publishing its tentative results. Thus, soon afterward "On the number of prime numbers below a given quantity" appeared in the monthly reports of the Berlin Academy [Riemann 1859/1876]. Few took note of its contents at the time, but by the 1890s it became famous as several eminent mathematicians began to study the properties of Riemann's zeta function [Edwards 1974] (see Section 8.2).

No evidence seems to have survived that would indicate Riemann had pursued this problem before arriving in Berlin, making it likely that his conversations with Kronecker ignited (or more likely reignited) this interest.[7] During the course of this visit, Riemann drew attention to a letter that Gauss had written in 1849 to the astronomer Franz Encke, which revealed Gauss's longstanding interest in the distribution of primes.[8] When or how Riemann became aware of this letter remains unclear; Kronecker, on the other hand, informed him of an interesting letter he had received from Dirichlet toward the end of the latter's life. Neither of these two letters had appeared in print, which adds some mystery to these stories, while partially illuminating these otherwise scarcely known discussions.

Less than a year before his death, Dirichlet spent a few days during the summer of 1858 vacationing with Kronecker in Ilsenburg. One of their topics

[7] As noted earlier in Section 1.2, the story that Riemann had devoured an edition of Legendre's book on number theory as a Gymnasium student should be set against the fact that he subsequently showed virtually no inclination to study that subject.

[8] This letter was first published in [Gauss 1863–1933, 2: 444–447]; the entire Gauss-Encke correspondence is now available in [Wittmann 2018]; see also [Tschinkel 2005].

of conversation had been Dirichlet's previous results on an asymptotic formula for the mean number of divisors for large $n$. Only a short time after his visit, Dirichlet wrote Kronecker on 23 July 1858, that he had since found a way to sharpen his previous result. He only regretted that he could not turn to this immediately, as he first needed to complete his unfinished hydrodynamical paper [Dirichlet 1889/1897, 2: 407]. As fate would have it, he would not live to accomplish that task either; for this was the paper Dedekind received from Riemann after Dirichlet's passing (see Section 4.1).[9] The fact that Kronecker informed Riemann about these circumstances, including the letter he received from Dirichlet, can be seen from Riemann's draft of a note to Kronecker asking that he send over a copy of that letter.[10]

On 22 January 1866, Kronecker wrote about the significance of Riemann's work in nominating him to be a corresponding member of the Berlin Academy (the closing summation appears in Section 12.4). In reference to the paper he sent to the Academy in 1859, Kronecker called this publication

> ...its greatest ornament, as Mr. Riemann provides a complete determination, in analytical form, of the number of prime numbers lying below a given magnitude. Legendre, Gauss, and Dirichlet had already turned their attention to this subject, and the latter had occupied himself with it anew just a few months before his death. However, judging by all that is known regarding the efforts and achievements of these three great mathematicians in this domain, their investigations generally aimed merely at obtaining the closest possible approximation of the number of primes, whereas Mr. Riemann sought – and indeed actually found – a precise analytical expression for it. [Biermann 2022, 29]

As noted in Section 1.4, Riemann had studied under Franz Encke in Berlin, though he surely had no idea that his professor was interested in the distribution of the prime numbers. By the time Encke wrote Gauss to convey his recent findings, Riemann had already returned to Göttingen. Gauss's letter was written in reply to Encke's long letter from December 1849, in which he mentioned a talk Dirichlet had recently given at the Berlin Academy.[11] On 9 August 1849, Dirichlet spoke there about his paper concerning an asymptotic expression for the mean number of divisors of a given number [Merzbach 2018, 192–194]. Evidently, Kronecker first heard about this topic almost a decade later in 1858, when he and Dirichlet in 1858 vacationed together. It was natural, of course, that Kronecker related his memories from that last visit to Riemann, now that their former teacher was no more. Riemann realized that what Kronecker had heard was directly related to the information Encke had conveyed to Gauss. Referring to Dirichlet's presentation, Encke reported that he had claimed Legendre was the first to find a relation between the natural logarithm function and the distribution of the prime numbers. Encke also seemed to remember an earlier claim made by Euler, which he mentioned as well.

Legendre's formula was no secret to mathematicians familiar with this question. Besides Dirichlet, Niels Henrik Abel and Pafnuty Lvovich Chebyshev had

---

[9] Following in Dirichlet's footsteps, the Ukrainian mathematician Georgi Voronoi found such a formula in 1903 [Merzbach 2018, 270–271].

[10] This draft can be found in Cod. Ms. Riemann 3, 21r, SUB Göttingen.

[11] Encke to Gauss, 4 December 1849, https://gauss.adw-goe.de/handle/gauss/2714.

referred to it in print by this time. Unlike Chebyshev's work from around 1850 – which Riemann knew (see [Edwards 1974, 3–5]) – Legendre's was only an empirical finding without any hint of a proof. Chebyshev was able to prove that for large enough $x$,

$$0.89 \int_2^x \frac{t}{\log t} < \pi(x) < 1.11 \int_2^x \frac{t}{\log t},$$

where $\pi(x)$ is the number of primes less than or equal to $x$. He also proved that if the ratio of $\pi(x)$ to this integral exists as $x \to \infty$, then this limit equals 1. This latter claim is the prime number theorem, first proved by Jacques Hadamard in 1895.

Encke's letter brought back some pleasant memories for Gauss, who sent him a lengthy and rather surprising reply, dated December 24, 1849. Remarkably, Gauss confessed that he did not know, or may have since forgotten, Legendre's conjecture, which he then confirmed by consulting the second edition of his *Essai sur la théorie des nombres* from 1808.[12] His letter began:

> Your remarks concerning the frequency of primes were of interest to me in more ways than one. You have reminded me of my own endeavors in this field which began in the very distant past, in 1792 or 1793, after I had acquired the Lambert supplements to the logarithmic tables. Even before I had begun my more detailed investigations into higher arithmetic, one of my first projects was to turn my attention to the decreasing frequency of primes, to which end I counted the primes in several chiliads and recorded the results on the attached white pages. I soon recognized that behind all of its fluctuations, this frequency is on the average inversely proportional to the logarithm, so that the number of primes below a given bound $n$ is approximately equal to
>
> $$\int \frac{dn}{\log(n)},$$
>
> where the logarithm is understood to be hyperbolic. Later on, ... I extended my computation further, confirming that estimate. ... I have frequently (since I lack the patience for a continuous count) spent an idle quarter of an hour to count another chiliad here and there; although I eventually gave it up without quite getting through a million. Only some time later did I make use of the diligence of Goldschmidt to fill some of the remaining gaps in the first million and to continue the computation according to Burkhardt's tables.[13] Thus (for many years now) the first three million have been counted and checked against the integral. [Gauss 1863–1933, 2: 444–447]

Gauss next compared the results one obtained from Legendre's small modification of the above function. Based on this he concluded that for large values of $n$ the deviations from the numbers given by Legendre's formula were even greater than those that arose when using the integral above.

---

[12] Gauss and Legendre vied for the honor of founding modern number theory as a discipline, for which see [Bullynck 2010, 195–197].

[13] These were factorization tables published decades earlier by Jean Carl Burckhardt, who grew up in Leipzig and taught himself astronomy as a boy. He then entered Leipzig University in 1792, where he studied mathematics under C.F. Hindenburg. By the late 1790s, he worked as an astronomer in Paris, taking French citizenship in 1799. In 1807, Burckhardt succeeded Jérôme Lalande as director of the observatory at the École militaire.

Although Encke was still alive at the time Riemann visited Berlin, Weierstrass and Kronecker likely knew nothing about Encke's amateur interest in prime numbers. They were, in any case, more recent arrivals, and both had never heard about the letter cited above until Riemann told them about it. This can be surmised from a partial draft of a letter Riemann sent to Weierstrass, stating that he was enclosing a copy of it and requesting that it be returned later.[14] One should keep in mind that, as Dirichlet's successor, Riemann had now assumed major responsibility for launching the Gauss edition, which published this letter in 1863. This suggests the possibility that Riemann's copy was based on a document in Gauss's estate, to which Riemann had ready access over the last three years.

That he knew the contents of this letter accords well with this key passage, which appears at the end of his paper:

> In fact the comparison of $\mathrm{Li}(x)$[15] with the number of primes less than $x$, which was undertaken by Gauss and Goldschmidt. This was undertaken up to $x =$ three million, showing that the number of primes is already less than $\mathrm{Li}(x)$ in the first hundred thousand and that the difference, with minor fluctuations, increases gradually as $x$ increases. The thickening and thinning of primes which is represented by the periodic terms in the formula has also been observed in the counts of primes, without, however, any possibility of establishing a law for it based on these fluctuations. [Riemann 1859/1876, 153]

Riemann may well have known that Benjamin Goldschmidt had extended Gauss's table of prime numbers even before the exchange took place between Encke and Gauss. Since he interacted a good deal with Gauss's protégé, Goldschmidt could have shown him the tables listing some three million primes during Riemann's first year in Göttingen. This was compiled to lend empirical support for Gauss's conjecture that the number of primes below $x$ is approximately $\mathrm{Li}(x)$. Since Riemann met Goldschmidt soon after he began his studies in Göttingen 1846, he may have learned about this Gaussian project during his first academic year. Were that the case, he might well have discussed this with Dirichlet in Berlin or possibly later in Göttingen. Riemann also knew some of the results published by Chebyshev, who spoke with Dirichlet about this when they met in Berlin [Edwards 1974, 3–5]. That meeting took place in October 1852, thus only a month after Dirichlet's visit in Göttingen. As for the letter, it seems quite likely that Encke would have shared its contents with Dirichlet, with whom he enjoyed good relations. If so, then Dirichlet would have surely imparted this information to Riemann at some point.

Riemann's paper will be discussed in Section 8.2, though not with an eye to its many lacunae. It remained for several decades a large black box for anyone who tried to understand its many subtleties.

---

[14] This appears in Cod. Ms. Riemann 3, 21r, SUB Göttingen.

[15] The logarithmic integral $\mathrm{Li}(x) = \int_2^x \frac{dt}{\log t}$, whereas $\mathrm{li}(x) = \int_0^x \frac{dt}{\log t}$. So $\mathrm{Li}(x) = \mathrm{li}(x) - \mathrm{li}(2)$, taking due account of the singularity at $x = 1$. This is the first term in Riemann's formula, which leads to a better approximation.

## 4.3 Conversations in Paris

Riemann undertook a longer trip during the semester break that began before Easter 1860, when he spent a month in Paris. He reported about some of his experiences there in a letter to his sisters, who were vacationing in Bavaria at the time. Writing from Heidelberg on 27 April 1860, Bernhard requested that either Ida or Helene return to Göttingen promptly, as he would begin teaching again soon. Now that he had assumed Dirichlet's chair, Riemann also took responsibility for teaching Gauss's standard course on the method of least squares, which he would only offer this once. Alongside this 2-hour course, he taught his standard 4-hour introduction to potential theory with applications to gravity, electricity, and magnetism.

The letter reflects Riemann's general dependence on his sisters at this time. Thus, he asked for their help in contacting the housekeeper who cared for their rooms in the observatory. He also thought his sisters might have taken the cabinet keys with them, in which case he requested that they send these promptly to their Aunt Julie, so that he could get them from her. Otherwise, he related that he had taken daily notes about what he saw and did while in Paris, so he would tell them all about this later; luckily, however, he conveyed a few things already in this letter.

Overall, he expressed satisfaction with his relatively brief trip. Unfortunately, the weather during his stay was often stormy and unseasonably cold. For days at a time, it rained and toward the end Paris experienced several days of snow showers and sleet. He thus regretted the lost opportunity to visit museums and other sites due to the inclement weather, even though staying inside proved very uncomfortable as well since the rooms lacked stoves. Riemann nevertheless felt compensated for these unlucky circumstances by the friendly reception accorded him by the many Parisian scholars he met for the first and only time.

On the Saturday after he arrived, he was invited to a social gathering hosted by Joseph Serret, who a few weeks earlier had become a member of the Institute. Here is how he described that setting:

> Such a tea or reunion differs quite a bit from our social gatherings. It begins at 9, often around 10, and lasts until 1. During this time, one can come and go as one pleases; some people are just coming from the theaters, which in Paris rarely close before 12:30. Nothing is served but tea, ice cream, and a variety of jams, i.e., sugared fruit and similar sweets. It cannot be denied that this informal way of socializing has a lot to offer. The group at Serret's consisted of 30 to 40 men and women, among them several Germans, or rather German speakers, with whom I mainly conversed. There was great anticipation about the outcome of the election of a foreign member of the Academy (associé étranger de l'institut) to replace Alexander von Humboldt;[16] there are only eight positions available, and it is therefore considered a great honor to be among those eight members. At that time, [the organic chemists] Wöhler and Liebig were in the

[16] Humboldt was nearly 90 years old when he died on 6 May 1859, just one day after the death of his former protégé Gustav Dirichlet.

> running,[17] and Wöhler seemed to have the best chance. ... After much hesitation, [the zoologist Christian Gottfried] Ehrenberg was finally elected at the last meeting (there is a meeting every Monday from 3 to 6 p.m.). Naturally, I was keenly interested [for the sake of] my old teacher Wöhler, to whom I had wholeheartedly wished this honor. [Neuenschwander 1981a, 119–120]

The following Tuesday, Riemann was the guest of Joseph Bertrand at a dinner that lasted from 6 to 11 PM. The atmosphere on this occasion was far more relaxed than at Serret's since this was a familial gathering:

> ...still, the group was quite large, as the couple has eight children and both of Madam Bertrand's parents are also still alive. When I paid a visit to Bertrand's some time later, I met a large group of children who had gathered to perform a comedy. However, I arrived a little late for the comedy and only had the pleasure of watching the tango that followed, which was quite hilarious. [Neuenschwander 1981a, 120]

On this occasion, Riemann had the opportunity to meet Charles Briot,[18] whose works he much appreciated. Briot invited him to visit his countryside home in Châtenay, a few train stations from Paris, where they spent a very pleasant day together.

Cauchy was no longer among the living, but his achievements in complex analysis had by now attracted the attention of leading French mathematicians. Riemann reported further that Briot was currently translating his work into French and planned to correspond with him about this. It appears, however, that nothing came of this plan. In any case, this trip left a strong impression on Riemann:

> We took a very pleasant walk in the area around Châtenay, visited [the writer] Chateaubriand's country estate and many other interesting places. Châtenay is the birthplace of Voltaire[19] and is not far from the main road that leads from Fontainebleau to Versailles. [Neuenschwander 1981a, 120–121]

Riemann further reported meeting a number of German professors while in Paris; practically every week a consular official brought him news of another arrival, though, like him, nearly all of them would leave by the end of April. Another invitation came from Charles Hermite, whose admiration for German mathematics was second to none in France:

> Hermite was also very friendly and courteous and invited me to lunch twice; he probably felt he could not forgo this, as he owes much to the German scholars, having been previously much supported by Dirichlet and Jacobi in particular; however, I would have gladly forgone it, as it must have been a burden to him given his wife's condition. [Neuenschwander 1981a, 121]

---

[17] Friedrich Wöhler and Justus Liebig had been collaborators in Giessen up until 1834, when Wöhler became full professor in Göttingen.

[18] Charles Briot and his friend Claude Bouquet had just published [Briot/Bouquet 1859], their first textbook on elliptic functions.

[19] Voltaire was the pen name of François-Marie Arouet. His birthplace was long believed to be in Paris, but in 1789 Nicolas de Condorcet published a biography, which claimed that the famous writer's illegitimate birth took place nine months earlier in Châtenay.

Although this letter offered no hints of the mathematical matters he discussed with Hermite, these conversations were surely the most important during his entire trip. Unfortunately, as so often, no details have survived, but on this occasion Riemann must have explained the gist of a fundamental new finding. For a short time later Hermite wrote about his "extremely remarkable and beautiful analytic discovery" [Hermite 1862/1905, 148]. This concerned properties that Riemann contended would characterize the periodicity matrices which define general Abelian functions, i.e., those that do not necessarily arise by inverting Abelian integrals. For whatever reason, Riemann never published this theorem, so what is known about it came from Hermite, who wrote about this result in a supplement to Lacroix's calculus textbook (how this later became known is yet another mystery). In his paper on Abelian functions (see Section 7.3), Riemann had already hinted that the methods he had used could be modified for the case of general $\theta$-functions [Riemann 1857b, 101].

As Tom Hawkins showed, Riemann's result was very close to another (unpublished) finding of Weierstrass, who was pursuing the very same problem [Hawkins 2013, 416–420]. In a somewhat modernized form, Hawkins formulated Riemann's finding as follows:

> (a) If $f$ is an Abelian function, then there is a period matrix $\Omega = (\Omega_1\ \Omega_2)$ for $f$ with the following properties: (i) $T = \Omega_1^{-1}\Omega_2$ is symmetric and (ii) Im $T$ is positive definite.
> (b) If $\Omega = (\Omega_1\ \Omega_2)$ is a $g \times 2g$ matrix whose columns are linearly independent over $\mathbb{R}$, then Abelian functions with period matrix $\Omega$ exist, if $\Omega$ satisfies (i) and (ii) of part (a).
> [Hawkins 2013, 418]

This result supplemented Riemann's earlier result, as conveyed to Weierstrass, namely that a single-valued complex function in $n$ variables could never have more than $2n$ periods (Section 4.2). In the meantime, he could prove that only $n$ of those periods were independent. Moreover, the matrix for the periodicity module had a special symmetric form, based on what today are called Riemann's bilinear relations for Abelian functions. The upshot of this new finding revealed that a single-valued $2n$-periodic function could be expressed as a quotient of general theta functions [Wirtinger 1895]. As so often, the process of clarifying all the various technical difficulties that arose from this took a very long time to resolve (see [Hawkins 2013, 416—431]).

Little else is otherwise known about Riemann's sojourn in Paris, but it seems likely that during his stay he learned that in 1858 the Academy had announced a prize competition for a problem connected with the conduction of heat. What led him to compete for this prize no one can say with any certainty. Quite possibly, the problem itself had an immediate appeal for Riemann, who was always on the lookout for physical questions that seemed amenable to mathematical treatment. For example, his interest in Helmholtz's experimental results in acoustics led him to model the propagation of air waves by a non-linear differential equation in [Riemann 1860], a paper written around this same time. Many commentators, however, have assumed that the problem itself held little interest for Riemann; they claim that what actually spurred his interest was the possibility of attacking it with mathematical methods he had described in the second section of his Habilitation lecture. For here, Riemann showed that the problem posed by the Academy could be reduced to another involving the equivalence of two differential forms. On the other hand, one should not overlook the prize itself – a gold medal valued at 3,000 Francs – considering Riemann's none-too-stable financial situation. Since his brother's death in 1857, he had been supporting his two sisters, Ida and Helene, neither of whom was likely to marry.

In any case, Riemann decided to submit an incomplete manuscript on 1 July 1861, in hopes of gaining recognition for his partial solution of this problem. Moreover, he wrote his "Commentatio mathematica" [Riemann 1876b] in Latin, apologizing to the Academy for his lack of facility in French. Since this prize was never awarded to Riemann or anyone else, this matter languished until after his death. Richard Dedekind was the first to study [Riemann 1876b], which will be discussed in Section 8.3.

## 4.4 Health Concerns

During the summer semester of 1860, Riemann taught his standard lecture course on mathematical physics, alongside a 2-hour public course on the method of least squares, a canonical offering during Gauss's day. This activity coupled with his research pursuits left him exhausted, as may be judged by comparing a photograph from around this time (Fig. 4.1) with another taken approximately five years earlier (Fig. 2.3). By August, Riemann was ready to leave Göttingen for an extended stay in Carlsbad (then in Bohemia, today Karlovy Vary in Czechia). This scenic town had been the site of the 1819 conference, called together by Prince Metternich, which led to the Carlsbad Decrees (discussed at the beginning of the Introduction). By 1860 it had become a popular touristic locale, known especially for the international celebrities who went there for spa treatments.

Along the way, Riemann accompanied Wilhelm Weber on a trip to Dresden, where they met the other two Weber brothers and Wilhelm's niece Sophie, all of whom came from Leipzig. Riemann described their activities in a letter to his sisters, written on 20 August. In Dresden, he had hoped to encounter the geometer Richard Baltzer, who taught at Dresden's famous Kreuzschule. Unfortunately, this surprise visit came to nothing, as the latter happened to be away. Riemann had met him once before at the home of the Dirichlets, during which Baltzer made a good impression on him.

His letter also mentioned an excursion that took place later that same afternoon:

> ...Weber had arranged a tour to Tharandt with Church Councilor Schwarz; all the participants, namely Schwarz with his daughter, the four Webers, Professor Zenker from Dresden, a cousin of Weber's, whom I had already met in Leipzig, and myself, met at Tharandt station at 4:00 p.m. The weather was initially fine, allowing us to enjoy the charms of Tharandt. From the center of the town, one has a view of three

valleys that branch off in different directions, and into which the small town extends picturesquely. Here, on a protruding hill, lie the church and the old castle ruins. We hadn't been there long when a thunderstorm arose, announced by subdued thunder. [Neuenschwander 1981a, 122]

Fig. 4.1: A rare photograph of Bernhard Riemann, ca. 1860.

Weber's cousin was Friedrich Albert von Zenker, professor of pathological anatomy at the Medical Academy in Dresden. Councilor Schwarz knew the pastor in Tharandt and Sophie Weber knew his wife, so they stayed behind while the others took a train back to Dresden in the middle of a terrible storm. The next day, Riemann, the astronomer Hansen, and the other three Webers took a train ride through Saxon Switzerland to Teplitz, from where they planned to hike through the Ore Mountains to Karlsbad. Sophie Weber stayed a bit longer with Schwarz before returning to Leipzig. Riemann related how "she seemed to avoid meeting me alone, but I found an opportunity once again to express my request that Laura come to Göttingen very soon" (*ibid.*). Clearly, Riemann still harbored thoughts of marrying Laura Weber, a perhaps realistic possibility now that he had become a full professor.

Toward the end of his stay in Karlsbad, Riemann wrote again on 11 September. He had little to say about the spa town: "I hope the treatment will do me good; it was terribly boring" [Neuenschwander 1981a, 123]. He mentioned three other guests with whom he socialized there: the aforementioned astronomer Peter Andreas Hansen from Gotha, Georg Julius Ribbentrop, professor for Roman Law in Göttingen, and Franz Seitz, professor of medicine and director of a hospital in Munich. He also related that he would return with Professor Hansen, first on foot to the Bavarian railway and then by train via Coburg, Eisenach, and Kassel to Göttingen. The most significant part of this letter described what did not happen:

> I haven't written to L[aura] yet, or rather, I've written several letters, but I've always torn them up. The memory of the painful and embarrassing things that happened between us in the past torments me incessantly. I hoped that we might be able to get over it more easily verbally, but I didn't have the opportunity to speak to her in Leipzig. Writing about it is impossible for me. (*ibid.*)

This seems to be the last time Laura Weber was mentioned by Riemann in the letters that have survived. Unless she visited Göttingen, they probably never saw one another again. Laura Weber apparently never married and lived at home with her family in Leipzig. Although eight years younger than Riemann, she only survived him by a year.

During the following winter semester of 1860/61, Riemann taught a 4-hour course on partial differential equations. He had only offered this once before, during his first semester of teaching, probably because this had been one of Dirichlet's standard topics. Around this time, Riemann worked on his hydrodynamics paper [Riemann 1861a/1876], which was inspired by [Dirichlet 1860/1861]. As noted in Section 4.1, Dedekind had completed that paper at Dirichlet's request, which led him to write his supplementary note [Dedekind 1861]. Thus, it was fitting that both he and Riemann took up this classical problem of describing the motion of a homogeneous fluid ellipsoid, whose elements attract each other according to the law of gravity. As Riemann emphasized, Dirichlet's investigations surprisingly broke new ground. He went on to say that "the pursuit of this beautiful discovery has its own special appeal for mathematicians, quite apart from the question of the causes for the shape of celestial bodies that prompted these investigations" [Riemann 1861a/1876, 182].

During the summer semester of 1861, Riemann taught two courses, the first at 9 each morning and the second meeting directly afterward at 10. The first course was "Mathematical Theory of Gravity, Electricity, and Magnetism," which he was teaching for the fourth time. One of his auditors was Karl Hattendorff, who would later edit these lectures for publication in [Riemann 1876b]. Riemann's second course was on complex analysis, this time with special emphasis on elliptic functions. Twenty-one-year-old Ernst Abbe – who later gained fame for his work on optical instruments and as founder of the Zeiss Foundation in Jena – attended both of these lectures. Thankfully, he also wrote about them in some detail to his good friend Harald Schütz, who was then attending Weierstrass's lectures in Berlin [Abbe 1986]. Others who heard Riemann's lectures included Friedrich Prym and Johannes Thomae, who commented that one could only understand Riemann's lectures on hearing him summarize what he had presented at the beginning of his next lecture [Liebmann 1921, 136].

Abbe had also attended Riemann's lecture course on elliptic functions from the summer semester of 1859. In the meantime, he took his doctorate under Wilhelm Weber with a dissertation on the equivalence of heat and mechanical work. Riemann was his examiner in mathematics, and he gave Abbe an "Eins," i.e., the highest grade. At the start of the semester, Abbe reported that twelve to fourteen auditors were taking Riemann's course. He also praised his introductory lecture on elliptic functions, which Riemann delivered without reading from a text. "The man is really making rapid progress," and Ernst Abbe was in a position to judge: he had already taken both of these courses just two years earlier.[20]

Abbe also observed how interest in Ernst Schering's parallel course on the method of least squares had dwindled to the point that Schering decided to cancel it. This left Abbe in no way despondent, since

> [Schering] promised to help me with my private study of this doctrine by handing over Gauss's treatises. That was fine with me. But my joy was almost spoiled by Riemann's strong desire to do the same with his course on elliptic functions. [Abbe 1986, 81–82]

He then related the following anecdote relating to Riemann's state of mind and body at this time in his life:

> Think of it! One fine morning, as he was walking out of the lecture after me, he spoke to me in a rich voice and stammered out excuses that he was lecturing so badly and repeating himself so often, etc. But he was always so exhausted by the second hour that he could not, with the best will in the world, do otherwise; and he was therefore thinking of giving up the lecture or reducing it to a two-hour audience. (It is indeed true that his fatigue and exhaustion were very noticeable; he was often completely unable to think and could not get the simplest things out). [Abbe 1986, 82]

Abbe was taken aback when he heard all this and tried to dissuade Riemann from giving up his course on elliptic functions. He reminded him that several of his listeners had come to Göttingen mainly to hear these lectures, and he even claimed that all of them "were completely enraptured by his lectures and would find it very painful if he gave these up." After Riemann explained why several of Abbe's suggestions for rescheduling the course were impracticable,

[20] Erwin Neuenschwander used his *Ausarbeitung* from the courses on complex analysis as the principal text for the critical edition he published in [Neuenschwander 1996].

> he finally decided to instruct me to arrange an exchange of hours between the two courses ..., so that he could at least deal with the more difficult subject, the elliptic functions, first. This will also be put into practice immediately after the break – to the great satisfaction of all that the imminent danger could be averted in this way. (*ibid.*)

As the summer semester of 1861 came to a close in August, Riemann suddenly withdrew a number of books on geodesy from the library. Besides works by Gauss and others ([Gauss 1828c], [Gauss 1844], [Gauss 1847]), he checked out a book by Bessel and Baeyer that presented their results from surveys in East Prussia [Neuenschwander 2022, 66]. This was not the only time Riemann suddenly took up a new field of interest, but in this case the impulse came from an external source: the Hanoverian Ministry of Education. Riemann probably never had any great interest in geodesy. Nevertheless, since he now held the chair previously occupied by Gauss, he took on formal responsibility for overseeing this research in Göttingen. For this reason, the Ministry turned to him, asking for a response to a proposal put forward by the Prussian geodesist Johann Jacob Baeyer in [Baeyer 1861]. This booklet of 130 pages was nothing like a conventional scientific research proposal; its often emotional language read more like a call to arms, a clear reflection of Prussian ambitions at this time. Baeyer sought to lead a large-scale scientific study of the topography of Central Europe, an undertaking not unlike the one his hero, Alexander von Humboldt, brought to life decades earlier to promote the study of terrestrial magnetism. General Baeyer, who had enjoyed a distinguished military career, was an influential figure in Berlin, someone who knew how to give marching orders. His book, on the other hand, was full of flamboyant propaganda that began by singing the praises of the recently deceased von Humboldt. By this time, Riemann's health naturally prevented him from taking part in this venture. He eventually delegated that task to Ernst Schering, who in 1864 attended a major conference in Berlin as a commissar of the Hanoverian government [Schering, K. 1909, 459]. From this time forth, Schering played an active role in geodetic research. This work no doubt gave him considerable satisfaction as he, once again, found himself walking in the footsteps of Gauss.

Meanwhile, Riemann's teaching career was approaching its end. The following winter semester of 1861/62, he taught only one course, namely a continuation of his lectures on elliptic and Abelian functions, which met for only three hours each week.[21] Word had by now spread, attracting more and more interest in his lecture courses, but time was not on Riemann's side. One of his last auditors was Paul Gordan, who came in 1862 after taking his doctorate in Berlin. He barely got to hear Riemann lecture, but went on to work with Alfred Clebsch in Giessen. Their book on Abelian functions [Clebsch/Gordan 1866] was sharply criticized by those who stood closer to Riemann's original vision (see Section 11.5).

---

[21] On the importance of this course for Riemann's *Werke*, see Section 12.3.

## 4.5 Marriage and First Visit to Italy

The last four years of Bernhard Riemann's life were the happiest and yet the saddest of all. Sickness and death had plagued his whole family, and in July 1862 he had to break off his lectures after becoming seriously ill. Just one month before then, though, he had the good fortune to marry Elise Koch (Fig. 4.2), who was nine years younger than he. Little is known about her except that, like the Riemanns, she grew up in the home of a Lutheran pastor. Elise was born in Körchow, a small village close to Wittenburg, lying roughly fifty kilometers due north of Quickborn.

It appears not unlikely that Elise Koch's father later took a position at a Lutheran church in Bremen, though direct evidence supporting this has yet to be found. If that were the case, this would account for how his daughter Elise came to know members of the Riemann family. Wilhelm had previously taken a job with the postal service in Bremen, and after their father's death in Quickborn he was joined by his two surviving sisters, Ida and Helene. If they thereafter began attending Pastor Koch's church, then this would explain how the two sisters became friends with Elise Koch. After Wilhelm's death in late 1857, both of them eventually moved to Göttingen to live with Bernhard. At the time, he earned a salary of only 300 Taler, but his income increased dramatically after his appointment as full professor in 1859. There seem to be no extant sources that might shed light on how his marriage to Elise Koch came about, but this plan was likely under discussion no later than January 1862. This conjecture hinges on a new reading interest that Riemann suddenly developed at that time: he began borrowing books on life insurance and the companies that issued such policies. Some of the literature he read was older, but he also borrowed a fairly recent book that addressed the solvency of life insurance companies. On three occasions, Riemann borrowed Charles Babbage's *A Comparative View of the Various Institutions for the Assurance of Lives* [Babbage 1826], the third time in September 1862, shortly before he and his wife departed for Italy. Thus, for a period of roughly one year, Riemann immersed himself in a topic that clearly had considerable importance for him personally.

At the time of their marriage, the Riemanns lived in the west wing of the astronomical observatory; Ernst Schering had been living in the east wing since 1857. The previous rivalry between Riemann and Schering, discussed in Section 3.6, was probably forgotten by now, at least in Riemann's mind. Presumably their personal relations had normalized somewhat during the intervening years. Nevertheless, and not notwithstanding Schering's later claims that he enjoyed a close friendship with Riemann, there is every reason to suspect this was a gross exaggeration. Riemann did not make friends easily, especially in Göttingen, where he often felt very much alone. Still, for a brief time, Schering had the opportunity to interact with Riemann on a fairly regular basis, which meant he could also observe how Elise dealt with her husband's moods and infirmities. About her ability to do so, Schering wrote this:

Fig. 4.2: Elise Riemann née Koch (1835–1904).

> She would be a life partner who, with understanding and ample patience, knew how to treat the peculiarities arising from his serious and protracted illness in a beneficial way. Furthermore, she contributed to alleviating his sorrowful mood by ensuring that the grief of leaving her did not [overwhelm] his thoughts and weigh so heavily on him that he would not be able to complete the work he had begun and which he had already mentally brought to completion. [Schering 1867, 167]

This passage and other remarks from Schering's obituary imply an intimacy with Riemann that can hardly withstand close scrutiny (see 12.1).

Soon after he married, Riemann contracted pleurisy, which caused him to experience sharp chest pains. His condition quickly worsened, forcing him to stay at home. Elise reported that he held his fall lectures on potential theory in their living quarters in the *Sternwarte*, but he soon gave these up as well. Although probably no one realized it at the time, Riemann's teaching career was already nearing its end: he would never again complete a full course of lectures. His two physicians – one of whom had treated Gauss during the last phase of his life – both urged Riemann to seek a long period of rest in a milder climate.

Fortunately, his older colleagues realized the severity of Riemann's illness and took initiatives to help him recover his health. The geologist Sartorius von Waltershausen (Fig. 4.3) advised the Riemanns to spend the forthcoming winter in Sicily, not only because of its warm weather but also to take advantage of his several connections there. Over many years, he had gotten to know prominent people on the eastern end of the island, particularly through his research expeditions to Mt. Etna. This suggestion immediately appealed to the newlyweds, who never dreamed they would have the opportunity to go on an extended honeymoon to Italy. Together with some other colleagues, including Weber and Wöhler, Sartorius was able to raise 500 Thaler to help pay for this trip. He also made several other arrangements for them as well; this included writing several letters of introduction that proved useful once they arrived.

Given the general fascination with Italy shared by so many Germans from this era, we should briefly take note of how this all began. In the life of Johann Wolfgang Goethe, his nearly two-year-stay in Italy during the 1780s represents a major turning point. Although he had achieved early fame as a writer, the 37-year-old Goethe had produced no noteworthy literary works for a decade before he set off on this momentous journey that changed his whole outlook on the world. German elites who later read his stylized autobiographical account of those experiences in *Italienische Reise* were among those who launched the Teutonic love affair with the enchantments awaiting them in the south. For many, Goethe's opening motto said it all: "Auch ich in Arkadien" (Even I was in Paradise). Rebecka and Fanny Mendelssohn both longed to visit Italy, and their letters from 1843 add sparkle to the trip the Dirichlets made in the company of Jacobi and Steiner. Riemann may well have read parts of Goethe's *Italienische Reise*. Yet whether or not he did, what he and his new bride experienced during their three trips to Italy clearly reflected a similar fascination with its rich culture, warm climate, varied geography, and lighthearted people.

Goethe managed to spin an impressive web of personal connections throughout his long life, so it should come as no surprise that one particular thread

Fig. 4.3: Wolfgang Sartorius von Waltershausen (1809–1876).

tied him to a professor in Göttingen, who happened to play an important role in the background to the story told in this chapter. The historian Georg Sartorius met Goethe in 1801 when the writer first visited the town. If Italian culture played a role in their shared interests, this took a back seat to their larger hopes for Germany. As a scholar, Sartorius made a name for himself with his three-volume history of the Hanseatic League, the trading organization that evolved from 1300 onward between numerous northern European cities. It extended as far south as Göttingen, which joined in 1350 to take advantage of the protections the Hanse offered. As a major factor in the transition period from feudalism to early forms of capitalism in Germany, the Hanseatic League has since been the subject of many studies since the time Sartorius first addressed its importance, which he saw as linked with the liberal movement of his time. After the downfall of Napoleon, Sartorius and Goethe hoped to promote this trend in the face of those who wished to restore the continent to the political conditions that prevailed before the onset of the French Revolution. The 1815 Congress of Vienna dashed these hopes, but liberalism continued to bubble up for decades afterward.

In 1809, Georg Sartorius's second son was born in Göttingen. His parents named him Wolfgang, in honor of Goethe, and their illustrious friend agreed to become his godfather. Shortly before his death, Wolfgang's father purchased an estate that came with a noble title. Thus, from 1826 onward his son's name became Wolfgang Sartorius Freiherr von Waltershausen. As a young man, Sartorius had helped support the Gauss-Humboldt project by carrying out a series of magnetic observations in various parts of Europe. By the time Riemann first met him, Sartorius was professor of geology and mineralogy in Göttingen and a close friend of Gauss and Weber. When Gauss died in 1855, he was one of the two speakers at his funeral, the other being Gauss's son-in-law Heinrich Ewald. Historians of mathematics know the name Sartorius von Waltershausen from the biography of Gauss he published just one year later [Sartorius 1856/2012].

Wolfgang Sartorius had first visited in 1834, which marks the beginning of many years of intensive study of the geological features associated with volcanic activity on Mount Etna. In 1838, he began a five-year stay in Italy and Sicily, during which he undertook several explorations of the region. Based on these, he eventually published an impressive study, *Atlas of Mount Etna*, with 57 copper plates. As a wealthy man of the world, Sartorius befriended Julius Ewald Jaeger, a German merchant in Messina with a large family. The Jaeger family provided a home away from home for the Riemanns during the winter of 1862/63, when they were happy to spend in a warmer clime. Sartorius's contacts in Sicily thus played an important role in making the Riemanns' first trip to Italy a delightful experience.

Toward the middle of November the couple left for Marseilles, where they boarded a ship that arrived in Messina on the 25th.[22] The seas were rough along the way, but Riemann felt enchanted by the whole experience and spent much of his time on deck. In Messina they at first had to stay in an expensive guesthouse, but later they were invited to reside with the family of the recently appointed Prussian Consul. Julius Ewald Jaeger had been living for several years as a businessman in Messina, where he developed many contacts with the city's leading citizens. He and his family lived in a villa in Gazzi about three kilometers south of the city center. There, his daughter-in-law took especially good care of the Riemanns, who got to know the whole family as well as a very large circle of their German friends.

One of these was Otto Hartwig, who spent five years in Messina as a Lutheran pastor for the local German community. Riemann may have known him already earlier, as he lived in Göttingen for a time when he was working on his dissertation in philosophy. Hartwig took his doctorate in Marburg in 1857 and he later

[22] Many of the details that follow concerning their experiences in Italy during the final years of Riemann's life are taken from Elise Riemann's notes for Ernst Schering, which were eventually published in [Schering 1909, 441–447].

returned there to work in the university library. During his years in Messina, he became passionately interested in Italian culture and made a name for himself later as an expert on the early history of Florence. He eventually left his position in Marburg in 1875 to assume the directorship of the University Library in Halle, where he became a pioneer in modern bibliometric studies and library science.

On Christmas Eve, the Riemanns went into the city to admire its cathedral. Originally a Norman construction from 1200, the Messina Cathedral had suffered major damage from earthquakes over the centuries. For the Christmas season it was illuminated during the evening hours. Christmas day was spent with the Jaeger family, who also invited them for New Year's Eve. Elise wrote that she and Bernhard felt a sense of enchantment in these unfamiliar surroundings. He was also feeling much better physically and enjoyed taking excursions into the countryside, often leaving at eight in the morning and not returning until dusk. Elise reported that he often spoke with people he encountered and listened to what they had to say about life.

Elise also recalled a particularly memorable trip with Dr. Julius Schubring.[23] He had recently arrived in Messina, where he tutored the children in the Jaeger family. As an adventuresome spirit, Schubring was keen to gain experience of the surrounding region, and he invited the Riemanns to join him and the Jaeger sons on a tour. Riding on donkeys, they climbed to the top of some of the highest peaks to take in breathtaking views and enjoy the clear fresh air. They also enjoyed the splendid views from the Real Cittadella, a fort built by the Spanish during the late seventeenth century. Over time, it had fallen into the hands of various European powers up until Italian unification. In 1860 it came under heavy bombardment by the Piedmontese army, but still withstood the attack and remained the last stronghold of the Kingdom of the Two Sicilies until the garrison surrendered the following year.

On a somewhat longer excursion, the Riemanns visited Torre Faro, the northernmost point on the island immediately across from Calabria over the Strait of Messina. Elise recalled how her husband particularly enjoyed spending time directly on the beach, especially when there was a mild Scirocco wind; he and Elise often sat there for hours, letting the foam of the waves splash over them. There were only a few cold days, but otherwise the weather was splendid. In fact, Consul Jaeger informed them that they had chosen a perfect time to visit: during his fifty years in Sicily, this was the mildest winter he had ever experienced. Nevertheless, Riemann still experienced problems with excess sputum, though otherwise his health seemed to be on the mend.

In early March he and Elise took a trip south to Syracuse, the fabled city where Archimedes held off Marcellus's Roman army before meeting with his death. They traveled by land, passing through Taormina and Acireale, a region

[23] To my delight, Gert Schubring recognized this name and easily identified its bearer as a family relative, whose connection with Sicily was well known later in life. As noted in the Introduction, Julius Schubring's father and namesake befriended Felix Mendelssohn Bartholdy in Berlin; later in his life, he published their correspondence [Schubring 1892/1973].

of dazzling beauty. Along the way they stayed in Naxos, then left the next day for Catania, where they visited relatives of the Jaegers that first evening. They also received an invitation from Baroness Bruca, whose husband had recently been appointed Senator in the newly formed Kingdom of Italy. She brought them news from their mutual friend Sartorius von Waltershausen, from whom they were pleased to hear. Looking inland the next day, they saw the geological wonder that drew Sartorius back to Sicily over many years: Mt. Etna, the largest active volcano in Europe (Fig. 4.4). Residents of Catania could never forget that they were living in a truly precarious locality.

Fig. 4.4: Mt. Etna, with Catania in the foreground.

At the university, the Riemanns met the mathematician Lorenzo Maddem, who took them through the library and other buildings. Maddem was not only director of the Meteorological Observatory, he also held the chair for rational mechanics and theoretical geodesy. As a devotee of the local culture, he had been working since 1845 on developing Catania's impressive botanical garden, which took advantage of the region's volcanic ash. Maddem took the Riemanns on a tour there and introduced them to his principal collaborator, the botanist Francesco Tornabene, who was prior of the former Benedictine Monastery located at the city's center.

A highlight during their four-day stay in Catania was a tour of San Nicolò l'Arena, which was still under reconstruction after more than 150 years. Their tour guide spoke good German, showed them all around the monastery, including the church, where they were able to hear an organist demonstrate its impressive instrument. What they otherwise heard about from their guide concerned the devastation the monastery suffered when Etna erupted in 1669, leaving the city of Catania surrounded with lava. The monastery building survived, but its walls were totally covered. Less than twenty-five years later, Sicily was shaken by a gigantic earthquake, setting off a tsunami that left the entire east coast ravaged. Afterward, little remained of Catania and its monastery, and so the rebuilding began.

During this stay, they took a guided tour to some of the ancient ruins, including the Roman theater still under excavation. Then it was on to Syracuse by ship. There a physician took them on a tour of the sights. This included the Cave known as the Ear of Dionysius, referring to the tyrant Dionysius I of Syracuse, though the name stems from the painter Caravaggio, who was struck by its shape. Legend has it that Dionysius used this cave to imprison political dissidents, since its near perfect acoustics made it easy to listen in on their plans. Elise wrote little else about their stay in Syracuse, perhaps because they needed to return soon to Messina, as they would soon depart on a longer journey that took them to Palermo and Naples. As before, on this shorter trip they left with letters of recommendation to persons in both cities.

In Palermo they spent one evening with the local consul, then set off the next day on a whirlwind tour. They visited the Palazzo Butera, a Baroque palace facing the Mediterranean, then took in the botanical gardens, as well as the Zisa, a magnificent twelfth-century Norman hunting lodge and summer palace in the western part of the city. After this, they toured the massive Norman Palace above the city, which served as the seat of power for the Kings of Sicily as well as subsequent Sicilian rulers. Its Palatine Chapel is still famous for its unusual mixture of Byzantine, Norman, and Fatimid architectural styles. The Riemanns then made a short trip to Monreale, where they took in the Cathedral of Santa Maria la Nuova and the Benedictine cloister, two magnificent buildings exemplifying the Norman-Arabic tradition.

In Naples they met the mathematician Giuseppe Albeggiani and the famous chemist Stanislao Cannizzaro, about whom Elise recalled that he spoke reasonably good German. Both were natives of Palermo, and so probably knew one another from there. In 1869, Albeggiani would succeed Cannizzaro as professor of physics and natural sciences in Palermo. The latter played a major role at the first international chemical conference held in Karlsruhe in 1860. There he presented strong evidence in support of Avogadro's hypothesis, which posited the existence of gases composed of diatomic molecules. As a result of Cannizzaro's work the conference adopted a standardized system for atomic weights in chemistry. Elise recorded all these distinguished names without further comment, but one must imagine that they had received many tips from Wolfgang Sartorius about whom they might contact throughout Italy.

From Rome, they traveled northward, passing through Civitavecchia and Livorno before reaching Pisa. There, Riemann was greeted with enthusiasm by Enrico Betti, who recalled their meeting from five years earlier.[24] Betti was three years older than Riemann and shared similar interests in mathematical physics, having studied in Pisa under Ottaviano-Fabrizio Mossotti and Carlo Matteucci. Mossotti had been exiled from Italy for his liberal political views. After a period in Switzerland and England, he was appointed professor of experimental physics at the University of Buenos Aires in 1827. He returned to Italy in 1835 and five years later accepted a professorship at the University of Pisa, where he taught courses on mathematical physics, celestial mechanics, and geodesy. Riemann never met him, however, as Mossotti died on 20 March 1863.

Betti thought it likely Riemann would need to live in southern Europe, so he asked Riemann whether he might agree to accept the professorship vacated by Mossotti's death. Riemann replied that he could not decide about this without first consulting with his friends in Göttingen when he returned. With this understanding, Betti spoke with the Minister of Education, who made arrangements for Riemann to be offered this position.

Betti also accompanied the Riemanns when they stayed in Florence, where they met a number of interesting people. One of these was the mathematician Placido Tardy, who was born in Messina, but left Sicily in 1848 for political reasons. Since 1851, he held a professorship in Genoa, where he later served as rector. Betti later informed Tardy of his conversations with Riemann on certain aspects of higher-dimensional topology, the inspiration for what later became known as Betti numbers [Weil 1979]. Another was the physician Corrado Tommasi, who had studied pathology under Rudolf Virchow in Berlin. During the Second Italian War of Independence, he volunteered as an army physician and was posted among the Hunters of the Apennines under the command of Giuseppe Garibaldi. In 1864 he was Professor of Anatomy at the Institute of Florence. Two years later, Tommasi became a public figure owing to his role in stemming the outbreak of cholera in Palermo. A third intellectual who joined this group was Pascual Villari, professor of history at the Institute of Higher Studies in Florence and later a prominent political figure in Italy. As an author, Villari was especially known for his books about the lives of Savonarola and Machiavelli. In the evenings, the Riemanns and Betti often got together with some of these people. Occasionally, they were also joined by the director of the Florence Astronomical Observatory, Giovanni Battista Donati, a pioneering figure in spectroscopic studies of stars and comets. Usually they gathered at the Gran Caffè Doney, a fashionable meeting place located on Via Tornabuoni near the British Consulate. It was originally founded by a French nobleman who had previously served in Napoleon's army.

---

[24] In 1858 Betti, Francesco Brioschi, and Felice Casorati traveled to Germany and France, hoping to establish lasting scientific relationships with leading mathematicians there. They met with a friendly reception in Berlin, where Weierstrass had recently begun teaching, and also in Göttingen. Brioschi was primarily an algebraist with limited interest in Riemann's work on complex analysis, but both Betti and Casorati were deeply influenced by his novel ideas. Betti was one of the distinguished members of the Risorgimento generation who fought during the wars of independence; others included Brioschi and Luigi Cremona [Tazzioli 2012, 467–470].

Fig. 4.5: The Splügen Pass between Italy and Switzerland.

Elise Riemann reported nothing about her husband's condition during this entire time, from which one might cautiously conclude that his health remained reasonably good throughout this trip. Unfortunately, however, things took a turn for the worse on the trip back. Dedekind's account relates that "he caught a severe cold while crossing the Splügen Pass, where he carelessly walked for a stretch through the snow before arriving in Göttingen on June 17" [Dedekind 1876a, 556]. They had set off from Lake Como, traveling north to reach the Splügen Pass, which lies nearly 2,000 meters higher in elevation. Probably they knew that Elise was roughly ten weeks pregnant at this time. Even though it was late spring, the Splügen Pass (Fig. 4.5) between Italy and Switzerland could still be treacherous, and on future trips they avoided this route. Riemann returned in terrible shape and could only speak with a hoarse voice.

Over the summer, the offer from Pisa hung in the air for several weeks. This eventually prompted Betti to write on July 22, urging Riemann to make a decision:

Dear friend!

> I have arranged the matter in such a way that, once your acceptance has been granted, there can be no further difficulties with your appointment as professor at the University of Pisa. However, speed is absolutely necessary, since if, contrary to my wishes and those of my colleagues, you do not wish to accept, we must know immediately, as we would then have to think about another way to fill the position. Therefore, I feel compelled to ask for your decision, which I hope will be favorable, though the delay makes me somewhat fearful.

Riemann may well have seriously pondered this possibility, as on July 11 he borrowed Mossotti's study of molecular forces [Mossotti 1838] from the university library [Neuenschwander 2022, 69]. After consulting with his Göttingen friends – presumably Weber and Sartorius – he learned that both had serious misgivings about the wisdom of such an adventure. Realizing that his delicate health had to be his prime concern, Riemann reluctantly turned down the offer:

> Mon tres cher ami!
>
> Yesterday evening I received your letter from July 22, and I hasten to write you in response. Although the offer you have made opens a beautiful perspective for me, I can only respond negatively. The state of my health is not good enough to undertake giving lectures next winter. I hope you will forgive me for not having let you know my decision more promptly ....

Mossotti's chair, a professorship in geodesy, was later offered to Eugenio Beltrami, who accepted.

## 4.6 Second Trip to Italy

Riemann's doctors once again recommended that he leave Göttingen before the weather turned cold and humid. Thanks to the support of Sartorius von Waltershausen and Weber, the family now made plans for a second stay in Italy. They departed already on the 21st of August, 1863, taking Riemann's sister Helene along with them. This time their destination was Pisa, where Enrico Betti offered to help them get settled. Along the way southward, though, they visited Munich and Innsbruck, before staying in Meran and Bozen in South Tyrol. After this, they passed through Verona on their way to Venice, where they spent the week of September 20–27, before traveling on to Florence. Betti found a dwelling for them in Pisa and his housekeeper made arrangements for them to hire a girl to help them. All seemed well, but unfortunately the winter months soon turned far colder than usual, in fact, so cold that the Arno River froze over. Riemann and his sister Helene both became ill, but the family took great joy in the birth of baby Ida on 22 December, 1863.

The timing for their visit was, in another respect, particularly well chosen. Indeed, they found the people of Pisa in a particularly festive mood on the 15th of February, 1864, when the city celebrated the two hundredth anniversary of the birth of its most famous citizen: Galileo Galilei. Italian nationalists had swept away the French and Austrian occupying powers, uniting the provinces of the country under King Victor Emmanuel II and Prime Minister Camillo di Cavour. A prominent Piedmontese patriot from this time was Salvatore de Benedetti, who grew up in a Jewish family in Novara. In 1862, he came to Pisa as professor of Hebrew studies, a position he held until his death in 1891. For the Galilei celebrations, Benedetti did not miss the chance to tie this famous name to the recent momentous political developments:

> In our day we have seen the blessed hour arrive, prepared precisely by the wisdom and martyrdom of our greatest men, the desire and hope of the centuries, in which all of Italy, shaken from its age-old lethargy, felt itself. And where did the first stirrings of the new life appear? In the brotherhood of national knowledge, and in acts of reverence for the memory of the great ancients, neglected by the servitude of the times. Among them, Galileo Galilei was among the most eminent [...] And the name of Galileo was the first to be honored by a resurgent Italy. [De Benedetti 1864, 5]

Elise Riemann only noted that she and her husband were present at the scene, so to speak, though what exactly they saw or experienced remains unknown. In any event, they enjoyed an active social life in Pisa, where they befriended a German family from Augsburg by the name of Jakobi. Through Betti they also got to know a number of mathematicians at the university. Elise recalled the kindness of the family of physicist Riccardo Felici (remembered today for Felici's Law in electrodynamics). The Riemanns also met his student, Emilio Villari, who later taught experimental physics in Bologna. During the later part of their stay, Eugenio Beltrami came to Pisa as professor of geodesy and thus got to meet Riemann for the first time. A few years later, Beltrami's famous work on non-Euclidean geometry would be decisively influenced by reading Riemann's Habilitation lecture (see Section 6.3). They also socialized with the families of the geometer Giovanni Novi and the chemist Paolo Tassinari. Elise also recalled meeting Leopold Kronecker's younger brother Hugo, who had contacts in Pisa with whom he arranged to form an informal German club.[25]

For the summer months of 1864, they moved into a villa just outside the city walls of Pisa. Riemann was at first feeling fairly well, which enabled him to work on his difficult last paper on the zeroes of theta-functions; he only completed that article, however, in October of the following year [Riemann 1866/1876]. In August, fate dealt the Riemann family yet another blow. Bernhard's sister Helene had contracted tuberculosis, like her mother and his other now deceased siblings. Over the summer, her condition steadily worsened, until she died in Pisa on the 28th of August. At some point, her older sister Ida came to care for her, but to no avail. She and Bernhard were now the only surviving siblings from their family.

Arranging a Lutheran funeral in Italy was virtually impossible, but the Riemanns had gotten to know the head of the Waldensian community in Pisa, a lawyer by the name of Chiesi. This older Protestant sect, whose roots predated

[25] Hugo Kronecker had briefly studied medicine in Pisa before taking his doctorate in Berlin under Emil du Bois-Reymond. In a letter to Leopold Kronecker from October 1875, Elise Riemann sent his brother greetings and recalled meeting him during their stay in Pisa (see Section 12.4).

Fig. 4.6: Bernhard Riemann in Florence, 1864.

Fig. 4.7: Tribuna of the Uffizi Gallery in Florence.

the Reformation, somehow managed to survive in Piedmont under the harsh persecution of the Catholic Church. The French Revolution gave the Waldenses of Piedmont their religious freedom, and in 1848 King Charles Albert of Sardinia, the ruler of Savoy, granted them civil rights. Afterward, the Waldensians began migrating from their homelands, so that by the time of Italian unification they had congregations throughout the peninsula. Elise Riemann was full of praise for the help Senor Chiesi offered them in securing a burial site and making arrangements for Helene's funeral.

Ida remained in Italy after her sister's death, and in early September she and Bernhard spent some days together in Florence, at which time the well-known photograph (see Fig. 4.6) was presumably taken. During that time, Riemann sent a long letter on the 9th of September to his wife back in Pisa, one of the few such documents that has survived [Neuenschwander 1981a, 124–127]. He was delighted to have received two letters from her in the meantime. Much of what he had to say revealed that he had been running around quite a lot, though also paying heed to his physical endurance. On the first day, he and Ida went to see the Tribuna of the Uffizi, the stunning octagonal exhibition hall in the Uffizi Gallery (see Fig. 4.7). Already in the 1770s its splendor had attracted wealthy tourists from throughout Europe.

Ida returned the following day to see more of the Uffizi Gallery, but this time she went alone as her brother went to meet with Enrico Betti. The two of them went to see a friend, then ran some errands, and finally spent some time looking at books in the library. Afterward, Riemann felt too exhausted to climb the stairs at the Uffizi, so he returned to their apartment and began writing his letter, while waiting for Ida to return. Later that day, he met again with Betti, who took him to a meeting of the Congresso pedagogico. Riemann was astounded by the plans being made for a new type of school with a modern curriculum that no longer emphasized foreign languages. Since he stayed until the end of their session, this already made for a long day.

Riemann's letter touched on the issue of his health, including the possibility of returning to Göttingen. On the basis of a doctor's report, his leave of absence had been extended through the next winter semester, but Riemann claimed he was "feeling quite well again since May" and thus wanted to try living in Göttingen again. These thoughts were now much on his mind, in part because of an unanticipated visitor: his medical colleague from Göttingen, Karl Ewald Hasse, who planned to examine him later in Pisa. Encouraged by this, Riemann went to see him already in Florence. They probably hardly knew one another, but Riemann was hopeful that Hasse's diagnosis would annul the earlier one he had received.

The next day, Bernhard and Ida accompanied Hasse on a sightseeing tour of Florence. Since the museums were closed, they visited the Church of San Marco and its monastery. Afterward, they admired the frescoes in the Basilica della Santissima Annunziata, and then went to see the tombs of the Medici in the Basilica of San Lorenzo. It was an enjoyable outing, but also exhausting for Riemann. He was sure that Hasse noticed how tired he became when walking around the monastery of San Marco. Riemann wrote that Hasse was extremely courteous; he left them around two that afternoon, but offered to take Ida to the Pitti Palace the next day. The remainder of Riemann's letter reveals that he was quite torn about whether or when they should leave Pisa:

> As far as my going back to Göttingen is concerned, I believe September would be the most favorable month for moving, since September in northern Germany is roughly the same as the second half of November and the first half of December here. So perhaps going back in September would be preferable to going back next spring, in the event that I had to be back in Göttingen in April or May.
>
> Yesterday evening I spoke to Betti about this: he was very passionately against it and said that once I was really cured, it wouldn't matter whether I returned to Göttingen in the spring or the autumn. He strongly advised me to stay here for the winter. However, I now also fear that Hasse will not take on the responsibility of letting me go to Göttingen for the next winter.
>
> Actually, it seemed to me as if he believed that I could live in Göttingen at least as well as Keferstein, who at first also suffered a lot there, but is now getting along quite well. [Neuenschwander 1981a, 125]

The zoologist Wilhelm Moritz Keferstein had, for reasons of his health, spent the winter of 1858–59 in Naples. Riemann could not have anticipated that Keferstein would later contract a lung disease that ended his life at age 37.

Riemann took kindly to Hofrat Hasse, but he worried about the influence of gossipmongers who spread rumors about the life he was enjoying in Italy.

"He may also have heard many such comments in Göttingen from people who consider me to be overly favored." Nevertheless, he continued:

> [Hasse] gives me the impression of a very conscientious and good-natured man who deserves complete trust, and I would really like it if he spoke out completely openly. Perhaps he expects *me* to make the decision to return, but after the experiences of last year, I don't want to take responsibility for it. I believe it would be neglecting my duty toward you and our child if I went back without needing to return and without a specific explanation from a doctor.
>
> If Hasse weren't coming to Pisa, I might have been tempted to take Ida to Milan. Betti would probably have traveled with us, while informing Brioschi, Casorati, ... about our trip in advance, so that I would be able to partly renew and partly make new acquaintances along the way ... [Neuenschwander 1981a, 126]

This single letter is virtually all that survives from Riemann's hand during their several months in Italy. Nevertheless, it fully confirms what Dedekind wrote about his often confused state of mind and his inability to take decisive action in important matters. Nothing more is known about the diagnosis he received from Prof. Hasse, nor do we know whether Betti's arguments finally won the day. Perhaps Riemann quickly reconciled himself with the decision to remain in Italy throughout the coming winter. It would also be interesting to know whether he met with Hasse again after he returned to Göttingen in the fall of 1865. Karl Ewald Hasse was director of the medical clinic in Göttingen and a colleague of the anatomist Jakob Henle. One of their students from this time was Robert Koch, who would become famous along with Louis Pasteur for their pioneering research on the germ theory of disease. In 1905 Koch was awarded the Nobel prize for his investigations and discoveries in relation to the disease that plagued so many families during that era: tuberculosis.

One month after the sojourn in Florence, Sartorius von Waltershausen visited the Riemanns for two days in Pisa. He was then returning from Sicily on the second of three trips to study the volcanic activity of Mt. Etna. During his two-day stay, he found Riemann feeling reasonably well, but years later he recalled:

> On the evening of October 20, as well as the following day, I had the joy of spending some wonderful hours with this great and noble man, gaining a look into the mind of one of the deepest thinkers of our century. His health was satisfactory in those days and I still held on to the silent hope that the mild Italian climate would have a favorable impact in prolonging his life. When I saw him again in Göttingen the following year, the last spark of his immortal genius was in expiry. [Sartorius 1856/2012, 159–160]

Riemann introduced their guest to Betti, who then traveled together with Sartorius to Florence and accompanied him on to Turin, the next leg of his journey.

Despite all his ailments, Riemann continued to enjoy his friendships with Betti and Felici over the winter. In March 1865, Enrico Betti brought with him another distinguished visitor, Francesco Brioschi, whom Riemann had met back in 1858 when these two Italians as well as Felice Casorati visited the mathematicians in Göttingen. At that time, Brioschi held a professorship in Pavia and worked primarily on topics in algebra. During the events leading up to the unification of Italy, he was a member of the Official Commission of the Ministry of Public Education, which aimed to reform the entire system of public education. Its proposals were implemented in 1861 when the Kingdom of Italy was founded. More recently, Brioschi had moved to Milan, where in 1863 he founded its Politecnico, the Technical University of Milan. He spent the remainder of his career as director and professor of mathematics and hydraulics at this institution. The Politecnico became a leading center for research and development in electrotechnology, paving the way to harnessing hydroelectric power as Northern Italy's primary energy source during the era of modern industrialization.

During this time, the Riemanns also got to know two other professors at Pisa University: the medical researcher Onorato Bacchetti and his close friend Carlo Matteucci, a leading expert on biophysics. As a student, Bacchetti had fought against the Austrians and afterward served as secretary of the Pisan University Brigade. Thus, he was one of the many Italian academics Riemann met who had played an active part in the Resorgimento. Matteucci had earlier studied under François Arago in Paris; Betti was one of many who took his courses on experimental physics in Pisa. Riemann almost surely knew about Matteucci's fundamental experimental work on bioelectricity, which was based on the pioneering studies of Luigi Galvani, Professor of Anatomy at the Bologna Academy of Sciences during the last three decades of the eighteenth century. Galvani's experiments on a detached frog's leg aimed to prove the existence of intrinsic animal electricity, a contention challenged by Alessandro Volta. During the early 1850s, Riemann had familiarized himself with current research in this field, which led to his own contribution on the phenomenon of Nobili's rings [Riemann 1855].

In the 1830s and 40s, Matteucci conducted electro-physiological experiments demonstrating that the nervous systems of higher organisms do, in fact, utilize measurable electrical impulses to send messages throughout the body. Mateucci's experiments were confirmed and further refined by the Berlin physiologist Emil du Bois-Reymond. In recognition of this work, in 1844 the London Royal Society awarded Matteucci the Copley medal. Afterward, he was also active in Italian politics, becoming a senator in 1860, at which time he was appointed inspector general of the Italian telegraph lines. Two years later, he became the country's Minister of Education.

In March of 1865, Friedrich Prym came to Pisa and stayed for several weeks with the Riemann family, including their daughter Ida, now fourteen months old. At first, Prym found his teacher's condition to be quite a bit better than he had imagined. He reported about this in a letter to Felice Casorati from 17 March 1865 [Neuenschwander 1978, 58]. Despite the attractions of Tuscany, Prym only rarely went out, as he was thrilled to be in Riemann's company and was meeting with him for about an hour each day. This situation changed quite quickly, however, as Riemann suffered badly whenever he spoke for a longer period of time. Only ten days later, Prym tried a different mode of communication, which involved sending his queries in writing (see Section 11.3). One month later, on 21 April, 1865, Prym wrote again to Casorati, letting him know that Riemann's health had taken a turn for the worse. During the preceding four weeks he had not felt well, though Prym thought this might have been partly due to the turbulent spring weather [Neuenschwander 1978, 58–59]. The following January of 1866, Betti imparted similar news to Casorati:

> Many things he had already done in his mind would be lost. It was so tiring and damaging for him to talk that little could be gained from it in the conversations one had with him. When Prym was here and made him talk a little more than usual, he suffered an immediately very serious deterioration. Lately he has been occupied with applications of Abelian functions to geometry and number theory. As for geometry, he told me that Clebsch still had a long way to go. [Bottazzini 1977, 31]

The family spent May and June of 1865 in Livorno on the Mediterranean just south of Pisa. From there, they departed northward to Lago Maggiore by way of Genoa. During the months of July and August, they stayed in two small villages near the town of Intra: Selasca and San Mauritius. The peaceful atmosphere in that serene locale left a lasting impression on Riemann, who later longed to see this part of the world one last time. By September, however, they returned to the Mediterranean coast, staying in Pegli just outside Genoa. Elise recalled the friendly reception they received from the Consul in Genoa, but her husband was by this time suffering from gastric fever. Originally, they had planned to spend the winter in Sardinia, mainly in the town of Cagliari, but Riemann now urgently wanted to return to Göttingen. The trip back went by way of Marseille, Lyon, and Geneva, during which Riemann suffered greatly before arriving in Göttingen on October 3, 1865. Elise was now pregnant with their second child, and both she and her husband were relieved that the winter in northern Germany was relatively mild. They then suffered yet another tragedy when her baby boy arrived stillborn. Elise recalled how her husband suffered deeply from this sad turn of events.

Although Riemann never had the opportunity to teach in Italy, he clearly took a serious interest in the country's language and culture. During his final stay in Göttingen, he studied the famous Renaissance epic poem by Torquato Tasso, *La Gerusalemme liberata.*[26] The life of the tormented artist Torquato Tasso, who may have suffered from a bipolar personality, inspired Goethe to write his play *Tasso*. For Goethe, this was a natural by-product of his enchantment with Italian culture, whereas in Riemann's case we have no clue as to what led him to this famous work.[27]

Only about two months later, Riemann's interest in Tasso's poem gave way to a new fascination that would occupy his attention for the remainder of his life. Ernst Schering later wrote about this, suggesting that Riemann now knew that his long-tormented life was about to end:

---

[26] Riemann twice borrowed this book in the Edizione critica riveduta e corretta da G.G. Orelli, published in Zurich in 1838 ([Neuenschwander 2022, 69–70]).

[27] In all likelihood he knew the German version, *Das befreite Jerusalem*, in the translation of Diederich Gries, a work Goethe and others found worthy of praise. Quite remarkably, Riemann's interests extended to the translator himself. Gries had died some twenty years earlier in Hamburg and was by now a largely forgotten figure. Riemann read about his life in the biography published in 1855 by Elisabeth Campe, née Hoffmann, daughter of a Hamburg bookseller. In fact, her marriage to the publisher August Gottlob Campe led to the founding of the well-known publishing house of Hoffmann & Campe.

> Fully anticipating his imminent death, he repeatedly and urgently requested from the doctor an indication of how much time he had left to live, in order to choose work he could complete within that time frame. [Schering 1867, 166]

If this was true, then Riemann's choice was all the more remarkable: he began studying human anatomy to begin an investigation of the mechanical properties of the human ear. Acoustics had been one of Riemann's many interests in natural philosophy, a tradition he often associated with Newton. In this case, however, he was surely inspired to tackle this problem by reading Helmholtz's *Lehre von den Tonempfindungen* (Sensations of Tone) [Helmholtz 1863]. He borrowed that famous study on 24 March 1866, less than two months before he and his wife set off on their third and final trip to Italy. During the intervening period, Riemann threw himself headlong into the literature on the anatomy of the ear. He even obtained the heads of freshly slaughtered cattle in order to study their oral apparatus directly.

As always, Wilhelm Weber was willing to lend a helping hand. On 18 April, Riemann replied to Weber, who had offered to contact their colleague, the anatomist Jakob Henle, to ask about obtaining specimens of animal heads suitable for study. This, Riemann answered, would not be necessary, but he had another request he wished to make of Hofrat Henle, namely,

> ... to proofread my treatise before printing and point out to me places where I may have suggested an hypothesis that, according to his observations, was incorrect or improbable. From my observations, I believe I have shown more completely than had before been possible, the conditions that the apparatus and its parts must meet. However, when searching for the individual devices by which these conditions are fulfilled, it is extremely easy to make mistakes if the anatomical information for this purpose is not precise enough, and especially if one knows as little about this as I do.

Riemann also referred to the difficulties he had encountered when trying to examine the hearing apparatus in cows:

> Regarding the attachment of the hammer, older preparations can provide no insight at all, only those from freshly slaughtered animals. It is easy for me to have heads of freshly slaughtered cattle brought to me; and in this way I have gained some insight into the stapes. However, the attachment of the hammer was always destroyed by the crude instruments available to me for breaking the bones containing the ear. Should your nephew visit me again, I ask you to let him know where I can buy a small saw or another instrument suitable for my purpose here in Göttingen, and under what name I can most easily obtain the appropriate item. [Gabke 2016, 9–10]

Riemann never completed this work, but his colleagues Henle and Schering published the incomplete manuscript in Poggendorff's *Annalen*. Although Riemann had the bad habit of rarely citing other authors, he made an exception in this case by noting that his provisional hypotheses differed from those in Helmholtz's masterful study. This prompted Helmholtz to publish a lengthy rebuttal [Helmholtz 1869], in which he noted at the outset that Riemann's study had an entirely different objective than his own.

During his last two months in Göttingen, Riemann's former student Karl Hattendorff met with him regularly in an effort to complete an older work on minimal surfaces (see Section 11.2). According to Hattendorff's recollection of that time together, Riemann was no longer able to speak, even in a whisper, without breaking out into violent coughing. Their work went by far more slowly than Riemann had originally imagined, nor did Hattendorff have any inkling that it would soon end. In June, however, he learned that Riemann had been seized by an irresistible longing to return to Italy. Perhaps this was his last impulsive decision in a life filled with hardships and grief. He now longed to spend his last days in the small village he and his wife had visited the previous summer: Selasca on the shores of Lago Maggiore.

## 4.7 Final Journey

Riemann's decision to leave for Italy came at a dramatic moment in time. In June, Prussian armies had begun to encroach on Hanover as war between the two great German powers, Austria and Prussia, had now become imminent. Hanover's King Georg V had a deep antipathy for Prussia and refused to join the other German states in their alliance against the Austrian coalition. His decision to join the Austrian side was finally reached on June 15, the day Riemann left Göttingen with his wife and baby daughter on a long and difficult journey. The next day Georg V arrived in Göttingen, where an army of 20,000 had assembled in preparation for the battle to come. Two weeks later they were totally outnumbered by the Prussian army near Langensalza, forcing the King to capitulate and go into exile. Only days later, the Prussians soundly defeated the Austrians at the Battle of Königgrätz. Bismarck made no territorial demands of Austria, whereas Prussia annexed Hanover outright, much to the anger and dismay of its royal family and their supporters.

During those fateful weeks that reconfigured the political map in Central Europe, the Riemanns faced any number of obstacles on their journey. The train that was to have taken them from Göttingen to Kassel broke down or was damaged, forcing them to take another route with a horse-drawn wagon that brought them to Giessen. This grueling trip only ended two weeks later, when they arrived with their daughter Ida (named after Riemann's sister) in the tiny village of Selasca, an idyllic spot overlooking Lago Maggiore. The previous year, they had become friendly with members of a small German community there. The fact that he was willing to undergo an arduous journey so that he could return to die in Italy says much about his attachment to the country and its people, feelings Elise Riemann later fully confirmed.

She herself shared these same sentiments, except of course when the conversation turned to mathematical matters, as was frequently the case when they stayed in Pisa. Riemann became very close with Enrico Betti there, but he got to know many others as well. This third and final trip, though, was nothing like the previous two. As Elise reported, Bernhard was only seeking peace and quiet, while hoping to complete his manuscript on the mechanics of the ear. This was not to be: Riemann died on July 20, less than one month after he and Elise arrived in Selasca (Fig. 4.8). His last days have been recounted many times, drawing on the account in Dedekind's biographical essay, but actually this was Elise's story, not his. Both of them knew Riemann intimately, but in different ways, making it all the more interesting to reconstruct how this biography came to be written (about which, see Section 12.4).

In his obituary for Riemann, Ernst Schering alluded to the series of family tragedies that preceded the fatal illness that ended Riemann's life:

> ...he perceived within himself the traces of another illness, which had already robbed him of his mother in his youth, then a sister, and after his father's death, his younger brother, who at that time bore the burden of caring for the family, and almost simultaneously another sister; finally, shortly before his own death, a third sister. One cannot without a sense of sorrowful pity imagine the moods that must have swept over him during the probably rare moments when he was not occupied with his mathematical and philosophical problems. A significant improvement in his mood occurred when, from 1858 onward, his two siblings who were still alive at that time provided him with constant companionship. [Schering 1867, 166]

Riemann probably never really shared his grief with those he knew in Göttingen; certainly Ernst Schering would have been the last person to whom he would have confided such feelings. The information cited above was very likely imparted to Schering by Riemann's sister Ida and his widow Elise (see Section 12.1). They also reported to Dedekind that Riemann's previous depression gradually lifted after March 1858, when his sisters Ida and Helene came to stay with him [Dedekind 1876a, 553].

What transpired after Riemann's death can only be reconstructed on the basis of fragmentary evidence.[28] After returning to Göttingen, Elise Riemann began to put her husband's scientific estate (*Nachlass*, literally "left afterward") in order. The German term is particularly apt in this case, as they had left Göttingen in great haste and Riemann's work habits were anything but orderly. He often filled whatever paper happened to be at hand with all kinds of calculations; often enough, these had no apparent relation with one another. A later letter from Dedekind to Heinrich Weber indicates that he visited Elise Riemann at this time [Scheel 2015, 81]. She wanted to speak with him about Riemann's last wishes, which surely reminded Dedekind of the news he only learned about indirectly after Dirichlet's death (Section 4.1). In that case, though, Dedekind had only to complete an unfinished paper, whereas Riemann's request concerned his entire scientific estate. His widow now conveyed that her deceased husband had hoped to speak with Dedekind during his final stay in Göttingen, but his declining health discouraged him from trying to contact him. It was now left to her to ask his old friend – whose career was still very young, to be sure – if he would be willing to go through Riemann's papers, in order to assess which of these, if any, might be publishable. How Dedekind responded- and in what ways this eventually led to the publication of Riemann's *Werke* in 1876, is a story taken up in Chapter 12.

---

[28] Various stories surely circulated as well, some kept current in accounts such as [Du Sautoy 2003].

Fig. 4.8: This plaque is all that remains from Riemann's grave. It is located at the entrance to the cemetery of Biganzolo near Lago Maggiore.

# Part II
# Riemann's Works

# Chapter 5
# Riemann's Doctoral Dissertation

Riemann's *Grundlagen* [Riemann 1851/1876] was a programmatic work that set the stage for much of what he accomplished afterward. Many mathematicians had worked with complex numbers, but before 1850 these "imaginaries" never acquired the full legitimacy accorded the real numbers. The latter were taken to represent idealized points on a linear continuum, which itself was seen as rooted in the idealized world of plane geometry. The number $i = \sqrt{-1}$, on the other hand, was taken to be merely a formal device for solving algebraic equations. For Riemann, however, both parts of the complex-valued function $u(x) + v(x)i$, $x \in \mathbb{R}$ had equal legitimacy.

By anchoring his theory of functions in the complex number plane, Riemann suggested he was only adopting Gauss's position, which by this time was well known. This new field of research owed even more to the work of Augustin-Louis Cauchy and his followers (for which, see [Bottazzini & Gray 2013, 131–213]), though Riemann's approach broke free from their earlier conceptions. Cauchy largely avoided appealing to geometrical arguments, whereas Weierstrass later turned this avoidance into a steadfast rejection [Ullrich 1990]. His influential views long prevailed among German analysts, though several were nevertheless attracted to parts of Riemann's program as well.

In view of Weierstrass's lofty place in German mathematics, his opinion of Riemann's dissertation is well worth citing. Probably, like most of his contemporaries, he only read it after the publication of its sequel [Riemann 1857b], the paper that launched Riemann's fame. Two years later, in nominating Riemann to be a corresponding member of the Berlin Academy, Weierstrass offered these remarks about [Riemann 1851/1876]:

> ...one only regrets that it remained almost entirely unknown until recently, for here Mr. Riemann gives an exposition of the principles on which his later investigations are based. The essence of his work can be briefly stated as follows. If one considers two quantities that can only be changed in one dimension or, what is the same thing, can be represented by positive and negative numbers, and one knows nothing more about them than that one of them is a continuous function of the other, no conclusions can be drawn from this alone with regard to the law that governs their connection. But things are completely different if the argument of a function is a complex quantity.



D. E. Rowe, *Bernhard Riemann: His Life and Wondrous Mathematical Legacy*,
https://doi.org/10.1007/978-3-032-25457-3_6

> Mr. Riemann shows that the concept of such a function already contains a general determination of the manner in which it changes with that quantity, and thus one can only arbitrarily control its behavior in the vicinity of individual values or a continuous sequence of values forming but one dimension of its argument. Functions that are subject to the same boundary conditions will then necessarily be identical. As a result, hitherto completely unknown, highly fruitful, and yet very simple methods are now established for comparing functions and especially for finding algebraic relations between transcendental quantities. While previously in such investigations, in order to alter an expression connecting the quantities under consideration, one had to accomplish this by carrying out an explicit transformation or by showing the agreement between two expressions for all values of those quantities. Mr. Riemann's principles reduce what is needed for this process to a bare minimum, completely eliminating purely mechanical calculation, while replacing the expedient of cleverly devised tricks by a truly scientific method. [Biermann 2022, 28]

Reading this, one cannot escape the impression that Weierstrass, too, regarded Riemann's dissertation as a profoundly important work. Mathematicians who accepted its main principles for dealing with complex numbers would henceforth grant full recognition to imaginary quantities, rather than treating these as mere auxiliary constructs. Beyond this, his dissertation introduced Riemann surfaces into complex analysis.[1] This key innovation involved a type of covering surface over the complex plane, which he used to represent multivalued functions. Before Riemann, this had posed an awkward problem that led some mathematicians to employ curves that dissected the complex plane into pieces. Thus, Cauchy's "lignes d'arrêt" produced various determinations for a function in the distinct regions outside these curves. Riemann's far more flexible approach utilized a connected surface lying over the plane.[2] This conception had the advantage that, for any point in the plane other than the branch points of a function, the number of leaves above it agreed with the number of function values. These leaves then join each other at the branch points forming a connected surface that corresponds to the determinations of the function. In the closing sections of his dissertation, devoted to his mapping theorem, Riemann only alluded to this general conception, while promising to develop it on another occasion. This was entirely in keeping with the spirit of his work, which pointed the way to a new field for future research.

## 5.1 Context

That physical ideas played a central role in Riemann's approach to function theory has never been disputed. The disagreements begin when commentators have tried to pin down precisely which works influenced him and how he assimilated the ideas of others. In their authoritative report on the development of algebraic functions, Alexander Brill and Max Noether underscored

[1] The more detailed description of Riemann surfaces in Section 6 of [Riemann 1857b] reveals that these objects were by no means easy to define; see the remarks about this in Section 7.3.

[2] For Weierstrass' alternative construction, see [Ullrich 2003]; see also [Elstrodt/Ullrich 1999].

the importance of works by George Green, Gustav Kirchhoff, and Hermann von Helmholtz [Brill/Noether 1894, 251–256]. These three authors were also cited in Felix Klein's influential booklet, *Über Riemanns Theorie der algebraischen Functionen und ihrer Integrale* [Klein 1882/1923], which argued that Riemann's classification of complex functions was rooted in different types of stationary flows on closed surfaces (see Section 11.6). Aside from the works by Gauss and Dirichlet on potential theory (Section 9.1), one can point to [Kirchhoff 1848] and [Helmholtz 1853] as two important papers that may well have influenced Riemann's approach to complex functions.

Section 2.5 described Riemann's situation at the time he submitted his dissertation. He had hoped to receive an exemption from having to hold forth in a disputation, still a formal requirement for promotion to Dr. Phil. After the faculty denied his request, however, he began preparing a list of theses he was prepared to defend. Initially, he came up with a list of six, five of which dealt with physical topics. Furthermore, all five reflect Riemann's highly critical views of various current ideas and technical advances [Riemann 1876/1892/1902, 1902: 112–113]:

1. Magnetic fluids do not exist;

2. Faraday's "induction in curved lines" is untenable.

3. A reversible pendulum [or Kater pendulum] is not the best means for determining the length of a pendulum.[3]

4. The theory of conservation of force has not been sufficiently proved experimentally.

5. The concept of voltage in the theory of electricity has not yet been formulated sharply enough.

For a few of these assertions, one can cite documents that shed light on Riemann's views, but in several cases we are left groping in the dark when it comes to understanding how he arrived at these various positions (see the introduction to Chapter 9).

Riemann's fourth thesis brings to mind Hermann von Helmholtz's famous lecture "Ueber die Erhaltung der Kraft," which he may have heard about when he took Jacobi's course on analytical mechanics in Berlin (Section 1.4). Jacobi had attended Helmholtz's lecture, which probably influenced his ideas about mechanics (see Section 1.5). As matters turned out, though, Riemann never did defend these theses in his disputation. He may have discussed some of them with Wilhelm Weber, with whom he worked closely, but for whatever reason these topics in physics were dropped. Weber had expressed skepticism regarding Riemann's chances of finding one or more opponents – normally fellow students with or without a doctorate – so he may well have thought it best to devise a different plan. In the end, it was decided that Riemann should defend the general approach to complex analysis he had advocated in his dissertation.

[3] This presumably refers to the use of a seconds pendulum, which measures the strength of gravity by the length of the pendulum's arm.

Riemann apparently made no reference to the role of physical ideas during his defense (see below), but for many years after his death leading mathematicians thought about this very connection, which appeared so striking. In the preface to [Klein 1882/1923], Klein referenced a conversation he once had with Friedrich Prym, whom he knew to be someone intimately acquainted with Riemann's ideas. According to Klein's recollection, Prym informed him that

> ...originally Riemann's surfaces were not necessarily many-sheeted surfaces over the plane but rather, on the contrary, complex functions of position which could be studied on arbitrarily given curved surfaces in exactly the same way as on surfaces over the plane. [Klein 1882/1923, iv]

Klein thus interpreted this to mean that Riemann's inspiration came not merely by way of potential theory in the plane, but rather by imagining stationary flows on arbitrary closed surfaces. Although Klein claimed he had found the original source that had led Riemann to his theory, he could not point to many hints in support of this view in the latter's dissertation. Indeed, Riemann's key idea in [Riemann 1851/1876] was a method for proving the existence of analytic functions with given boundary conditions. He thus had almost nothing to say about functions on Riemann surfaces that were closed and orientable, the topic he developed later in [Riemann 1857b] (Section 7.3). As discussed in [Bottazzini & Tazzioli 1995], Klein's influential interpretation was based on a misunderstanding, though it provoked interesting responses from leading mathematicians, including Enrico Betti, Luigi Bianchi, and Max Noether. Particularly noteworthy was the rebuttal Klein received from Prym, who wrote Klein in April 1882 to say that, although he could no longer remember the context of his remarks, he never held the viewpoint Klein had ascribed to him [Klein 1921–23, 3; 479].

Prym only knew Riemann during the final years of the latter's life (see Section 11.3), but like many others he had a longstanding interest in how his ideas had evolved. He offered his reflections about this in a lengthy letter to Klein from 6 February 1882, which contained this passage:

> According to a communication that Riemann imparted to me in the spring of 1865 during my stay in Pisa, he arrived at his theory of functions of a variable complex quantity through the observation that relationships between functions obtained by expanding the relevant functions in series persisted even when one went beyond the regions of convergence of the series representing them, and that in many cases one obtained correct results when operating with divergent series, as Euler, for example, repeatedly did. He then asked himself what actually enables this extension of functions from one domain into another, and he came to the conclusion that this resulted from the partial differential equation. Dirichlet, with whom he discussed the subject, fully agreed with this view; thus, this idea probably dates back to Riemann's student years, before he wrote his inaugural dissertation.

Prym thus took this to mean that Laplace's equation (or the Cauchy-Riemann equations) were at the very heart of Riemann's initial work. Based on this, he concluded that Riemann drew on

> ...all the resources offered by potential theory, in which he was Dirichlet's pupil, for the development of his theory as well as Cauchy's theory of functions of a variable

complex quantity. In this respect, it is correct to say that Riemann arrived at his investigations through mathematical physics. The inaugural dissertation is, in some parts, nothing more than the potential theory of Gauss and Dirichlet reduced to two dimensions. Finally, the perspectives Riemann opened up in the last sections of his inaugural dissertation can be traced back to studies of mappings of surfaces, to which Riemann was inspired by the relevant Gaussian work. [Bottazzini & Tazzioli 1995, 6]

Writing about a decade later, Brill and Noether argued that the core ideas behind Riemann's dissertation could be found in an undated note, which they assumed predated his *Grundlagen*. This note, "Equilibrium of Electricity on Cylinders with Circular Cross-Section and Parallel Axes," was first published by Heinrich Weber in [Riemann 1876/1892/1902, 1876: 413–416]. It deals with a type of problem that was well known, having already been studied by Pierre Simon Laplace, the great French mathematician. Laplace recognized that if an electrical force remains constant along lines parallel to the generators of a cylinder, and thus perpendicular to its circular cross sections, this leads to a reduction of the Laplacian equation to two variables instead of three. This, of course, conforms very nicely with Riemann's theory for analytic functions $w = u+iv$. These require two harmonic functions $u$, $v$, which satisfy this Laplacian equation, as does the "logarithmic potential" $v = \sum m \log r$ (as opposed to $v = \sum \frac{m}{r}$ for the potential in 3-space).

Otherwise, however, the situation remains murky, since we have no solid clues indicating that Riemann had already studied this case prior to writing his dissertation, as Brill and Noether maintained. What we have instead is merely Heinrich Weber's reconstruction of Riemann's ideas based on undated formulas he found in the estate.[4] Weber himself assumed these were lecture notes [H.M. Weber 1877, 150], in which case they would obviously postdate Riemann's dissertation. Furthermore, Weber noted that the general method employed here, which applies to regions of arbitrary genus bounded by lines and circles, sheds light on Riemann's work on minimal surfaces [Riemann 1868/1876], a topic he first took up in the early 1860s. On the other hand, Riemann himself noted that some of the results in Sections 1–5, 9, and 12 of [Riemann 1857b, 102] dated back to 1851/52, when he studied conformal mappings between surfaces of higher genus. Finally, as observed in [Bottazzini & Gray 2013, 304], the very sophistication of Riemann's argument makes it implausible that he found it before writing [Riemann 1851/1876].

Following Weber's reading, Riemann immediately translated this setting into another, in which the cylinder is replaced by a surface $S$ bounded by $n$ curves, e.g., circles. He then postulates the existence of a function $\zeta = \xi+i\eta$, with $|\eta| > 0$ on $S$ and which takes the circles bijectively to the real axis ($\mathbb{R} \cup \infty$), while $\zeta$ assumes every value in the upper half-plane $n$ times. This yields a conformal representation of $S$ as an $n$-fold covering surface $T$ of the upper half-plane. Since $S$ and $T$ have the same connectivity, $T$ will have $2n-2$ branch points. Riemann then showed that a certain second-order differential equation in $\zeta$ leads to an

[4] Emmylou Haffner reproduced three of the 27 pages written in Riemann's hand that served as the basis for Weber's reconstruction [Haffner 2021, 33].

algebraic function $s$ with the same branch structure as $T$. The ratio of two of its solutions than yields a conformal representation of $T$ with circles as boundaries. This surface has $4n-4$ simple branch points and thus genus $p=n-1$. Weber saw this mapping as the truly significant feature in the note, which he subtitled "Conformal Mapping of Figures bounded by Circles."

In his dissertation, Riemann only alluded to such cases, as he restricted his attention to simply connected regions. Nevertheless, his closing section indicated that one could easily extend the result he had derived – the Riemann mapping theorem – to conformal images of regions on surfaces of higher genus. In this connection, he cited two related works by Gauss, his study of conformal mappings of surfaces [Gauss 1825a] and section 13 of [Gauss 1828b], which presents the Gauss map for determining the local curvature on a surface.

## 5.2 Text

Part of what makes Riemann's dissertation difficult to read is that he only occasionally offered hints to motivate what his study was all about. A few appear at the very beginning, but otherwise one must jump ahead to Section 20, where he finally pauses to make some illuminating general remarks. He began there by saying:

> The introduction of complex numbers into mathematics has its origin and primary purpose in the theory of simple laws of dependence between variable quantities, expressed through operations on these quantities. If these laws of dependence are applied in an extended scope, by assigning complex values to the variable quantities to which they refer, a harmony and regularity that would otherwise remain hidden becomes apparent. The cases in which this has occurred so far encompass only a small domain – they can almost all be traced back to laws of dependence between two variable quantities where one is either an algebraic function of the other or a function whose derivative is an algebraic function – but almost every step taken here has not only given the results ... a simpler, more coherent form, but has also paved the way for new discoveries, as evidenced by the history of research on algebraic functions, circular or exponential functions, elliptic and Abelian functions. [Riemann 1851/1876, 37–38]

The terrain leading to the theory of elliptic and Abelian functions, treated as *functions of a complex variable*, remained indeed a topic for the future. Riemann was thus fully aware that his dissertation not only broke new ground, but also pointed the way to vast new developments in modern analysis. As Detlef Laugwitz emphasized, Riemann was the first to conceive of a complex function as a mapping, in this case between two complex planes $A$ and $B$.[5] A key passage from Section 2 reads:

> Both the quantity $z$ and the quantity $w$ are considered to be variable quantities that can assume any complex value. The assumption of such variability, which extends to a

[5] "What distinguishes Riemann's approach to the theory of functions from the approaches of Cauchy and of the Weierstrass school is the idea of a mapping. It is this idea that makes for the closeness between Riemann's view and our own" [Laugwitz 1996/1999, 72–73].

continuous area of two dimensions, is greatly facilitated by linking it to spatial concepts. Consider each value $x + yi$ of the quantity $z$ represented by a point $O$ on the plane $A$, whose rectangular coordinates are $x, y$, and each value $u + vi$ of the quantity $w$ represented by a point $Q$ on the plane $B$, whose rectangular coordinates are $u, v$. Any dependence of the quantity $w$ on $z$ will then be represented as a dependence of the position of the point $Q$ on that of the point $O$. ...One can therefore imagine this dependence of $w$ on $z$ as a mapping of plane $A$ onto plane $B$. [Riemann 1851/1876, 5]

Riemann immediately deduced in Section 3 that, so long as this mapping remains finite, it will be conformal:

Similarity therefore occurs between two corresponding infinitely small triangles and consequently generally between the smallest parts of plane $A$ and its image on plane $B$. [Riemann 1851/1876, 6]

At this point he cited the key work that had guided him in this direction, namely Gauss's prize essay on conformal mappings [Gauss 1825a]. There Gauss showed that by introducing curvilinear coordinates $(p, q)$ on a surface alongside Cartesian coordinates $(P, Q)$ with line element $dP^2 + dQ^2$, one can define a mapping $(p, q) \mapsto (P, Q)$ where the metric $ds$ can be expressed as $ds^2 = k(dP^2 + dQ^2)$. The connection with complex functions arises from factoring the metric, thus $dP^2 + dQ^2 = (dP + idQ)(dP - idQ)$. For a complex-valued mapping, defined by $f(p + iq) = P + iQ$ or alternatively $g(p - iq) = P - iQ$, Gauss showed that the first mapping determines the second, and vice-versa [Gauss 1863–1933, 4: 197].

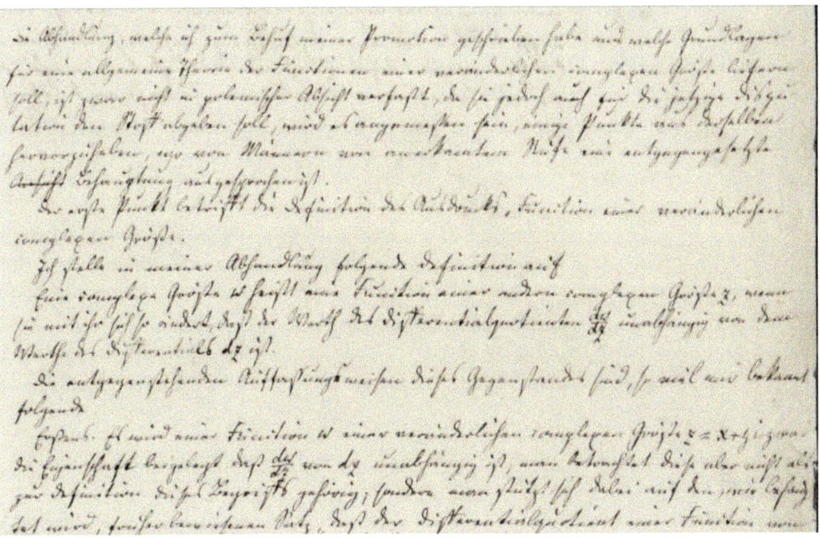

Fig. 5.1: From Dedekind's transcription of Riemann's disputation text.

Riemann's estate contains a draft of three Latin propositions for a disputation (Fig. 5.1), all of which concern the properties of a complex analytic function as defined in Section 12 of the dissertation. A rough English translation of the introduction reads:

> Although we did not write the preceding dissertation as a polemic (let us use that word), but rather with reasoned argument, nevertheless, since it will also provide material for a public disputation, we set down, in place of theses, certain propositions about which contradictory opinions are found among authors of distinguished reputation" (Cod. Ms. B. Riemann 13, 96r, 96v).

Another document in Riemann's papers makes it quite certain that these propositions were, indeed, those he defended in his disputation. In it he briefly noted the issues at stake by contrasting his definition of an analytic function with two alternatives, one of which was advocated by Cauchy. Interestingly, the name Augustin-Louis Cauchy only arises here, but nowhere in the dissertation itself. The text translated below is taken from a transcription prepared by Richard Dedekind (Fig. 4):[6]

> The treatise I wrote for my doctorate, which is intended to provide the foundations for a general theory of functions of a complex variable, was not written with polemical intent. However, since it is also intended to provide material for the current debate, it will be appropriate to highlight some points from it where men of recognized reputation have set forth contrary claims.
>
> The first point concerns the definition of the term "function of a complex variable."
>
> In my treatise, I provide the following definition: A complex quantity $w$ is called a function of another complex quantity $z$ if it changes with it in such a way that the value of the differential quotient $dw/dz$ is independent of the value of the differential $dz$.

In other words, if $w = f(z)$, where $dz = dx + idy$, then $dw/dz$ is independent of $dy/dx$, so the derivative of $w$ is independent of the direction in the $xy$-plane. In order to make clear that this was an essentially new conception, Riemann named two opposing definitions of a complex function:

> First, a function $w$ of a variable quantity $z = x + yi$ is attributed the property that $dw/dz$ is independent of $dz$. However, this is not considered to be part of the definition of this concept. Instead, one relies on the previously proven theorem, which claims that the differential quotient of a function of $a+b$ with respect to $a$ is equal to the differential quotient with respect to $b$.
>
> Second. One considers a complex quantity $w$ (generally of a specific type) as a function that varies with $z = x + yi$ but in any arbitrary way. This view was expressed by Cauchy, who among the French was the first to deal extensively with the theory of complex quantities. He propounded this at [a recent] session of the Paris Academy, when he presented a report on a work by Puiseux, and he has further developed this in several subsequent lectures.
>
> Of these two ways of understanding the concept of a function of a complex variable, the first is opposed to mine in that it shares the property that I use to define this concept, but does not consider this to be part of the definition; the other does not attribute this property to complex functions at all.

---

[6] Available online from the Göttingen Digitalisierungszentrum (GDZ): (Cod. Ms. B. Riemann 13, 100r, 100v).

In this connection, it is noteworthy that Riemann's dissertation began with a loose definition for continuous real-valued functions. The editors of his *Werke* found a more precise formulation in his papers, which they printed in the endnotes to the dissertation [Riemann 1876/1892/1902, 1876: 46]:

> By the expression: the magnitude $w$ varies continuously with $z$ between $z = a$ and $z = b$ we mean: within this interval an infinitesimally small change in $z$ corresponds to an infinitesimally small change in $w$ or, more concretely, any arbitrary magnitude $\epsilon$ always allows for choosing a magnitude $\alpha$ such that whenever the interval about $z$ is smaller than $\alpha$ the difference between any two values of $w$ is never greater than $\epsilon$.

This, of course, is just a somewhat wordy rendering of the familiar definition of continuity usually attributed to Weierstrass. Riemann added the remark that a continuous function is everywhere finite.

In Section 4, Riemann derived the familiar Cauchy-Riemann differential equations, which he claimed gave necessary and sufficient conditions for a function to be holomorphic in a domain,[7] where for $w = f(z) = u + iv$ :

$$\frac{\partial u}{\partial x} = \frac{\partial v}{\partial y}, \quad \frac{\partial v}{\partial x} = -\frac{\partial u}{\partial y}.$$

Assuming these derivates are themselves differentiable, it then follows that

$$\frac{\partial^2 u}{\partial^2 x} + \frac{\partial^2 u}{\partial^2 y} = 0 \quad \frac{\partial^2 v}{\partial^2 x} + \frac{\partial^2 v}{\partial^2 y} = 0,$$

so $u$ and $v$ are harmonic functions.

Riemann studied their behavior not just in the complex plane but also on surfaces over it. He introduced what came to be known as Riemann surfaces in Section 5:

> For the following considerations, we limit the variability of the quantities $x$ and $y$ to a finite domain by considering the location of point $O$ to be not the plane $A$ itself, but rather a surface $T$ spread out over it. We choose this formulation, in which it will be unobjectionable to speak of surfaces lying on top of each other, in order to leave open the possibility that the location of point $O$ extends multiple times over the same part of the plane. However, in such a case, we assume that the overlapping parts of the surface are not connected along a line, so that a folding of the surface or a division into overlapping parts does not occur.
>
> The number of surface parts over each part of the plane is then completely determined if the boundary is given in terms of its position and orientation (i.e., its inner and outer sides); its position, however, can still vary. [Riemann 1851/1876, 7]

Riemann then defined a branch point $P$ of order $m-1$ as one requiring a point $Q$ to encircle $P$ $m$ times before it returns to its original position on the surface $T$. He then asserted that the surface $T$ is fully determined, up to a finite number of different shapes, once the position and orientation of its boundary is given as well as the position and order of its branch points. Furthermore, a "variable quantity,

[7] His assumption that they constitute a sufficient condition, however, requires additional conditions (see [Bottazzini & Gray 2013, 266]).

which ... takes on a specific value that changes continuously with its position, can obviously be regarded as a function of $x, y$ ..." [Riemann 1851/1876, 9].

Before indicating how one can introduce analytic functions on such surfaces, Riemann first made a number of remarks about their connectivity properties. In his dissertation, the surfaces of interest usually have a closed boundary, a precondition for applying Dirichlet's principle. In order to discuss their connectivity properties, Riemann decomposed such a surface $S$ with boundary $\partial S$ by means of cross-sections (*Querschnitte*). These are lines drawn through the interior of $S$ from $P_1 \in \partial S$ to $P_2 \in \partial S$ without crossing itself. The latter can also lie in the part that is added to the boundary, i.e. in an earlier point of the cross section. A connected surface is said to be simply connected if it breaks into components with every cross section; otherwise it is said to be multiply connected. Riemann then introduced as a first topological invariant: if $n$ cross-cuts lead to a surface with $m$ components, then its *order of connectivity* is $n - m$. These first considerations, however, had little relevance for the topological concepts he developed later in [Riemann 1857b], which dealt mainly with closed surfaces.

For the remainder of this discussion, I will focus on the two topics which did have great import for Riemann's subsequent research. These later became synonymous with the method he called Dirichlet's principle and the central result known today as the Riemann mapping theorem. Both appeared for the first time in his dissertation, though not by these names. In fact, the term conformal mapping had not yet been coined, so Riemann (and before him Gauss) spoke of "mappings similar in the smallest parts," thus local similitudes, rather than mappings that preserve angle measure. In Section 21, Riemann offered a proof that in the complex plane a region topologically equivalent to an open disk is, in fact, conformally equivalent to the same. Naturally, this first version of the Riemann mapping theorem took no account of the difficult topological problem of deciding which curves bound a topological disk (Jordan's curve theorem). William Fogg Osgood is generally credited with having given the first rigorous proof in 1900 by establishing the existence of Green's function on arbitrary simply connected domains.

Before turning to Section 21, though, we should first take up what Riemann wrote about "Dirichlet's principle," beginning with the following result from Section 16:

> Theorem: If $\alpha$ and $\beta$ are any two arbitrary functions of $x, y$ for which the integral
>
> $$\int \left[ \left( \frac{\partial \alpha}{\partial x} - \frac{\partial \beta}{\partial y} \right)^2 + \left( \frac{\partial \alpha}{\partial y} + \frac{\partial \beta}{\partial x} \right)^2 \right] dT,$$
>
> extended over all parts of the surface $T$ spread arbitrarily over $A$ assumes a finite value, then this integral, when altered by a continuous function, or one that is only discontinuous at individual points and vanishes on the boundary, always attains a minimum value for one of these functions and, if discontinuities that can be eliminated by modification at individual points are excluded, for only one. [Riemann 1851/1876, 30]

This minimizing solution for $\alpha$ and $\beta$ turns out to be harmonic on the interior.

Riemann then let $\lambda$ be a continuous function that vanishes on $\partial T$, remains continuous except at finitely many points, and for which the integral

$$L = \int \left( \left( \frac{\partial \lambda}{\partial x} \right)^2 + \left( \frac{\partial \lambda}{\partial y} \right)^2 \right) dT,$$

is finite. Letting $\omega = \alpha + \lambda$, he introduced

$$\Omega = \int \left[ \left( \frac{\partial \omega}{\partial x} - \frac{\partial \beta}{\partial y} \right)^2 + \left( \frac{\partial \omega}{\partial y} + \frac{\partial \beta}{\partial x} \right)^2 \right] dT.$$

Riemann now asserted what he intended to prove:

> The totality of the functions $\lambda$ forms a closed connected domain in which each of these functions can continuously transition into every other but cannot approach one that is discontinuous without $L$ becoming infinite. For each $\lambda$ we get a finite value for $\Omega$ which becomes infinite with $L$, ... but can never sink below zero. Consequently, for at least one function $\omega$ the integral $\Omega$ assumes a minimum. [Riemann 1851/1876, 30]

Sections 17 and 18 are devoted to a series of technical results needed to prove this theorem. Riemann then turned in Section 19 to the main application he had in view. Here he asserted that an analytic function $w = u + vi$ of $z$ defined on a bounded and simply connected surface $A$ will be completely determined if the following conditions are met:

1) The function $u$ is continuous on all boundary points;
2) the value of $v$ is given arbitrarily at some point;
3) the function $w$ is finite and continuous at all points.

Riemann elaborated on the freedom that arises in connection with possible boundary conditions in an interesting passage, where the notion of a manifold comes into play. Others certainly used this term as well, but Riemann explicitly connected it with analysis situs (topology) in his celebrated 1854 lecture (discussed in Section 6.2). The freedom to choose functions $u$ (or conversely $v$) on the boundary meant that each point $P \in \partial A$ had to satisfy an equation that determined a one-dimensional manifold of values identified with $P$. The discussion that followed, however, became quite opaque:

> These conditions, the totality of which forms a continuous manifold and which are expressed by equations between arbitrary functions, are, however, in order to be permissible and sufficient for the determination of a function continuous throughout the interior of the region, to speak generally, and are subject to a limitation or supplementation by individual conditional equations – equations for arbitrary constants – because the accuracy of our estimate obviously does not extend to these. For the case where the area of variability of the size z is represented by a multiple connected surface, these considerations do not suffer any significant change ... [Riemann 1851/1876, 36]

In Section 20, cited earlier, Riemann summarized the advantages of his approach to function theory as opposed to the conventional methods based on analytic formulas. One might characterize his remarks as propaganda, though

Weierstrass only echoed these views in the passage cited at the top of the chapter. To illustrate his principles, Riemann gave a striking example in Section 21, where he presented his mapping theorem:

> Two given simply connected plane surfaces can always be mapped onto one another in such a way that each point of the one corresponds to a unique point of the other in a continuous way such that the correspondence is locally similar [conformal]; moreover, the correspondence between [pairs of] interior and boundary points may be arbitrarily given, and when this is done the correspondence is completely determined [Riemann 1851/1876, 40].

In modern language, this theorem is often formulated as follows: Any two regions homeomorphic to an open disc are conformally equivalent. This latter formulation has the flavor of an existence theorem, whereas Riemann's proof was partially constructive in nature, except when he invoked Dirichlet's principle to obtain a unique conformal mapping.[8] Riemann took a simply connected open domain $T$ and showed how this could be mapped conformally onto the interior of a small disc $D \subset T$ with center $z_0$. Using polar coordinates $(r, \phi)$, one can define a on $D$ by $f(z) = \log(z - z_0) = \log r + \phi i$. Taking a radius as a branch cut $\ell$, $f$ will jump by $-2\pi i$ when a loop passes over $\ell$. Riemann extends this cut by an arc that ends when it meets $\partial T$, and he extends $f$ to a function $\hat{f}$ on $T \cup \partial T$ for which $\hat{f}(z_0) = -\infty$ and points on $\partial T$ the real part of $\hat{f}$ vanishes. Riemann then claims the real part of $\hat{f}$ along the arc $\ell$ running from $z_0$ to $\partial T$ will increase monotonically $-\infty$ to 0. He further argued that each value $a \in [\infty, 0]$ corresponds to a simply closed curve $C_a \subset T$, and so by ensuring that $\partial T$ can be mapped to $\partial D$ each curve $C_a$ will map to a circle in $D$. In oversimplified form, this was the argument Riemann gave, naturally for a well-behaved domain $T$ bounded by, say, a smooth simple closed curve.

Standards of rigor constantly change, of course. As noted in Section 2.5, Gauss lauded Riemann's dissertation, and even Weierstrass, writing in 1859, only sang its praises (see the introduction to this chapter). Lars Ahlfors, one of the leading experts on complex analysis of his time, wrote in 1953 that "Riemann's mapping theorem is ultimately formulated in terms which would defy any attempt of proof, even with modern methods."[9] What struck Riemann's younger contemporaries was not so much the vague character of Riemann's argument but rather the fact that he failed to give any examples whatsoever. Writing nearly two decades later, Hermann Amandus Schwarz recalled how frustrated he felt when, as a student of Weierstrass he knew about Riemann's mapping theorem, yet he could not produce a single conformal mapping taking a simple rectilinear region in the complex plane onto a circle [Schwarz 1869, 105]. Around this same time, E.B. Christoffel was making progress in this direction, but he too underscored how few concrete results illustrating Riemann's mapping problem had been brought to light in nearly twenty years [Christoffel 1870a, 286–287].

---

[8] The idea behind his argument together with suggestive pictures is presented in [Gray 2021, 155–157]; another summary can be found in [Gray 2015, 172].

[9] Quoted from [Gray 2015, 173].

## 5.3 On Reading Riemann

According to university regulations, dissertations had to be published, though probably few were ever read. Riemann's thesis was in this respect no exception, at least initially. Only after the appearance of his monumental study on Abelian functions [Riemann 1857b], in which he referred directly to his dissertation, did the latter text begin to attract readers. Of course, the faculty and especially a candidate's adviser had to vouch for the quality of work submitted for a doctoral degree. Here, one might even wonder whether Gauss actually studied Riemann's text carefully, especially since he was already generally familiar with its methods and ideas. As noted in Section 2.5, Riemann took far longer than Dedekind to prepare his dissertation, which he finally submitted to the faculty in November 1851. Gauss praised both works, though he clearly recognized that [Riemann 1851/1876] was far more original. Characteristically, though, Gauss offered not even a hint of the topic Riemann handled so brilliantly. What he did say, though, was that most readers would find the text difficult due to its somewhat obscure arrangement (see Section 2.5).

Today Riemann's "Grundlagen" ("Foundations for a General Theory of Functions of a Complex Variable") is considered one of his most important works. Gauss and Cauchy both made fundamental contributions to complex analysis before Riemann, to be sure. Still, [Riemann 1851/1876] pointed to an entirely new approach to analytic functions by describing their fundamental properties in terms of singularities on Riemann surfaces. One no longer needed to focus on their local behavior by means of calculations with power series, as now the existence of uniquely determined analytic functions could be derived in global terms using Dirichlet's principle. Riemann then hit a high note toward the end of his dissertation with his famous mapping theorem, a result he linked to Gauss's results on conformal mappings in [Gauss 1825a]. These are the standard themes one encounters over and again in the secondary literature as an encapsulated summary of Riemann's text.

What about the opinions of those who have actually read his dissertation carefully? In his biography of Riemann, Detlef Laugwitz wrote at length about its content, quoting key passages, while summarizing the most important findings [Laugwitz 1996/1999, 96–107]. Since Laugwitz's goal was to distill the main content in a readable form, he focused entirely on fleshing out the standard picture familiar to contemporary mathematicians. His account thereby evades all the obscurities, but also many of the mysteries that other readers found so striking.

The distinguished analyst Lars Ahlfors, writing in 1951, gave a far more critical appraisal. Ahlfors underscored the difficulties that not only he, but also Riemann's contemporaries faced when trying to unravel this text:

> The most astonishing feature in Riemann's paper is the breathtaking generality with which he attacks the problem of conformal mapping. He has no thought of illustrating his methods by simple examples which to lesser mathematicians would have seemed such an excellent preparation and undoubtedly would have helped his paper to much earlier recognition. On the contrary, Riemann's writings are full of almost cryptic messages to

> the future. For instance, Riemann's mapping theorem is ultimately formulated in terms which would defy any attempt of proof, even with modern methods. [Patterson 2017, 265]

Was Riemann purposely obscure, perhaps in this respect taking his cue from the great Gauss?[10] This seems highly unlikely when we take into account his earnest efforts to explain the fundamental ideas behind his work, often illustrated by the kinds of examples Ahlfors missed in the published works. Mathematicians are bound to disagree about the degree to which Riemann's works are understandable or truly opaque. Moreover, since he wrote about many different topics, including physics as well as philosophy, the quality of the exposition tended to vary a good deal. Still, he was nearly always a terse expositor, one of the reasons [Riemann 1868/1876], his essay on the foundations of geometry, remains a challenging text to this day.

S.J. Patterson, referring primarily to Riemann's short paper on the zeta-function [Riemann 1859/1876], considered his writing clear enough, or at least no more obscure than other mathematicians from his era. Beyond that, Patterson emphasized that to read Riemann's work with understanding requires familiarity with the mathematical literature he himself knew so well:

> Riemann was a mathematician, or, perhaps more precisely, a mathematical physicist, who was mathematically ambitious and employed the most modern tools at his disposal. The present-day reader is confronted with the task of reconstructing the knowledge and thought-patterns of a sophisticated mathematician of the 1850s and 1860s. It is not possible to understand Riemann in isolation . . . both the group of older mathematicians in Göttingen and Berlin who influenced him and the published literature of the time are the context of Riemann's work. [Patterson 2017, 266].

Indeed, Riemann had many teachers – Gauss, Dirichlet, Jacobi, Weber, Stern, Goldschmidt, et al. – who influenced his thinking in a variety of ways. Yet he had several other older teachers – Newton, Euler, Lagrange, Cauchy, et al. – whose works he studied over long periods of time. Riemann could well have proclaimed that he stood on the shoulders of giants.

[10] Gauss was never obscure in the mathematical sense, of course, but he was a famously austere and reticent author. Even his admiring biographer freely admitted that Gauss's fellow mathematicians often complained about how his writing style made it extremely difficult to read his works with understanding [Sartorius 1856/2012, 82–83]. A critique of Gauss that Abel heard in Berlin likened him to "a fox who erases his tracks in the sand with his tail" [Rowe 2018, 24–25].

# Chapter 6
# Riemann's Habilitation Texts

Thanks to Dedekind's efforts, soon after Riemann's death the mathematical world could read for the first time the two texts he had prepared for his Habilitation. Both appeared together with Riemann's reconstructed paper on minimal surfaces in Volume 13 of the *Abhandlungen* of the Göttingen Royal Society, not a particularly accessible publication, especially outside of Germany. For that reason alone, one might plausibly conjecture that very few mathematicians in other countries read Riemann's post-doctoral thesis (*Habilitionsschrift*) [Riemann 1868/1876] during its initial years of reception. Within Germany this work was part of a flurry of activity on the foundations of real analysis that I will not describe here. Of the more than 550 papers Walter Purkert identified as works that cited Riemann's publications, most named the *Habilschrift*, his only contribution to the field [Riemann 1990, 870 ff.]. Not until the last quarter of the twentieth century could one read about this work in a now familiar context: the emergence of Cantorian set theory.[1] Today nearly everyone with a background in mathematics has heard of the Riemann integral, even if few are likely to know when or where it first appeared, much less why he introduced it.

As recounted in 3.1, Riemann delivered his most famous lecture – "On the Hypotheses underlying the Foundations of Geometry" [Riemann 1868/1876] – on 10 June 1854. His audience consisted of six full professors, only two of whom – Carl Friedrich Gauss and Wilhelm Weber – had any idea of the subject matter. Since the text Riemann read to them that day had a formal purpose, he apparently never contemplated submitting it for publication. Nor did he ever refer to it in any of his other published works.[2] In short, over the course of Riemann's brief lifetime very few people even knew of this essay and probably no one had the chance to read it. Its inception and reception thus form a prominent chapter in Riemann's mysterious life.

[1] Among the many studies from that period were [Dauben 1979], [Purkert/Ilgauds 1987], [Purkert 1989], and [Ferreirós 2007].

[2] One should note, however, that he may have intended to allude to one of its key results in his "Commentatio" of 1861; see Section 8.3.



D. E. Rowe, *Bernhard Riemann: His Life and Wondrous Mathematical Legacy*,
https://doi.org/10.1007/978-3-032-25457-3_7

Riemann's original manuscripts for his Habilitation – both for his thesis [Riemann 1868/1876] as well as his qualifying lecture [Riemann 1868/1876] – apparently no longer exist. They may have already disappeared soon after Dedekind prepared the transcriptions he submitted in July 1867 to the Göttingen Scientific Society.

## 6.1 Riemann's Postdoctoral Thesis

Given the ubiquity of the Riemann integral today, it should come as some surprise to learn that the paper in which he introduced it, [Riemann 1868/1876], was not a study of integration theory at all. Riemann's postdoctoral thesis dealt with functions that can be represented by a trigonometric series, a topic he may have learned about in Berlin through Dirichlet. As noted in Section 3.1, Riemann later received help from Dirichlet regarding the historical context that led to this area of research. Why he never returned to it remains unclear, but his teaching activity indicates that he was totally absorbed with problems stemming from complex analysis, including differential equations. Real analysis and number theory were outliers for him, despite the fact that two of his most famous innovations – the Riemann integral and his extension of Euler's zeta function to the complex domain – belong to these research fields.[3]

The representation of functions by infinite trigonometric series was a famous problem that attracted great interest in the wake of Joseph Fourier's fundamental work on heat conduction. Stated very simply, the problem at stake was to determine the precise conditions a real-valued function had to satisfy in order to have a Fourier series representation. Cauchy and Dirichlet had both attempted to tackle this question, and Dirichlet succeeded in finding sufficient conditions for functions that admitted at most finitely many discontinuities [Dirichlet 1829]. Riemann then generalized these results not only by extending the problem itself but also by introducing the so-called Riemann integral in attacking this more general problem. This called for finding necessary and sufficient conditions for a function to be representable by some trigonometric series, not necessarily a Fourier series [Dauben 1979, 6–19].[4]

In his paper from 1829, Dirichlet indicated that he hoped to return to this topic again, though he never did. This makes it likely that Riemann learned about this topic directly from Dirichlet while studying under him in Berlin. Were that the case, then Dirichlet would have likely informed him of another on the same topic by Friedrich Wilhelm Bessel. As noted in Section 2.6, Riemann was apparently familiar with [Bessel 1839] even before he wrote his *Habilschrift*. In it,

[3] Laugwitz wrote about [Riemann 1868/1876] at considerable length in [Laugwitz 1996/1999, 181–213], since that paper was an important contribution to integration theory; it also served as a key stepping stone that eventually led to Georg Cantor's theory of infinite point sets, for which see [Dauben 1979, 11–19].

[4] Dauben and other historians have conjectured that Riemann chose not to publish this work because he could only give a partial answer to this question [Dauben 1979, 14, 317].

he argued that Bessel's supposed improvement on the results in [Dirichlet 1829] failed to do so [Riemann 1868/1876, 238].

Riemann's study was unusual in that it began with a lengthy historical account bearing on the concept of a real-valued function in which he highlighted the eighteenth-century debates connected with the vibrating string problem. After this, he turned to Fourier's work, which set the stage for all the followed:

> Almost fifty years passed without any significant progress being made on the question of the analytical representability of arbitrary functions. Then a remark of Fourier's threw new light on this subject; a new epoch in the development of this part of mathematics began, which soon became apparent externally in great extensions of mathematical physics. Fourier noticed that in the trigonometric series
>
> $$f(x) = a_1 \sin x + a_2 \sin 2x + \cdots + b_0 + b_1 \cos x + b_2 \cos 2x + \ldots$$
>
> the coefficients can be determined using the formulas
>
> $$a_n = \frac{1}{\pi}\int_{-\pi}^{\pi} f(x) \sin nx \, dx, \quad b_n = \frac{1}{\pi}\int_{-\pi}^{\pi} f(x) \cos nx \, dx.$$
>
> He saw that this method of determination would also remain applicable if the function $f(x)$ was given completely arbitrarily; he set for $f(x)$ a so-called discontinuous function (the ordinate of a broken line for the abscissa $x$) and thus obtained a series which in fact always gave the value of the function. [Riemann 1868/1876, 232]

Another unique feature of this work, when compared with Riemann's other publications, has to do with its far more elaborate scholarly apparatus. In his dissertation, for example, he only mentioned the relevance of two works by Gauss for his study, which was literally devoid of any concrete references. The first part of his postdoctoral thesis, on the other hand, was peppered with footnote citing specific pages. As with his other early works, Riemann also attached a detailed summary of the contents, though here again a striking difference appears. Rather than simply plunging in, Riemann first devoted three sections to the prehistory of the question he undertook to study, namely, under what conditions a function can be represented by a trigonometric series. Section 4 to 6 then discuss his new concept for definite integrals, which enable Riemann to extend the scope of real-valued functions relevant for his study. Section 7 then describes, as usual, the "plan of the investigation," which he laid out afterward in two parts.

> For the prehistory, Riemann gave this brief synopsis:
>
> 1. From Euler to Fourier: The origin of the question lies in the dispute from 1753 over the scope of d'Alembert's and Bernoulli's solution to the problem of a vibrating string. Views of Euler, d'Alembert, Lagrange.
>
> 2. From Fourier to Dirichlet: Correct view of Fourier, opposed by Lagrange. 1807. Cauchy. 1826.
>
> 3. Since Dirichlet: Dirichlet's resolution of the question for functions occurring in nature. 1829. Dirksen. Bessel. 1839. [Riemann 1868/1876, 265]

What is sometimes called Riemann's rearrangement theorem for divergent series stems from a passage in Section 3, in which he attributed this idea to Dirichlet. There he described how in [Dirichlet 1829] the question of representation was resolved with all rigor for a large class of functions, namely integrable functions without an infinite number of maxima and minima. About this, Riemann went on to say:

> Knowledge of the path that needed to be taken to resolve this problem arose from the insight that infinite series fall into two essentially different classes, depending on whether or not they remain convergent when all the terms are made positive. In the first class the terms can be arbitrarily rearranged; in the second, on the other hand, the value is dependent on the ordering of the terms. Indeed, if we denote the positive terms of a series in the second class by $a_1, a_2, a_3, \ldots$ and the negative terms by $-b_1, -b_2, -b_3, \ldots$ then it is clear that $\sum a$ as well as $\sum b$ must be infinite. For if they were both finite, the series would still be convergent after making all the signs the same. If only one were infinite, then the series would diverge. Clearly now an arbitrarily given value $C$ can be obtained by a suitable reordering of the terms. We take alternately the positive terms of the series until the sum is greater than $C$, and then the negative terms until the sum is less than $C$. The deviation from $C$ never amounts to more than the size of the term at the last place the signs were switched. Now, since the number $a$ as well as the numbers $b$ become infinitely small with increasing index, so also are the deviations from $C$. If we proceed sufficiently far in the series, the deviation becomes arbitrarily small, that is, the series converges to $C$. [Riemann 1868/1876, 235]

The laws of finite sums are only applicable for the first type of series, not to the second. Riemann took for granted that Cauchy was unaware of the fact that some Fourier series fall in the second class, which undermined his attempt to derive these from a law that described the decrease of their individual terms.

Dirichlet was able to prove that every $2\pi$-periodic function on an interval can be represented by a trigonometric series if

1) it is integrable throughout,

2) does not have infinitely many maxima and minima, and

3) at jump discontinuities, it assumes the mean value between the mutual limit values.

> Through Dirichlet's work a large number of important analytical investigations were given a solid basis. By bringing into full light the point where Euler erred, he succeeded in settling a question that had preoccupied so many distinguished mathematicians for more than seventy years (since 1753). In fact, for all the cases of nature, which with it was alone concerned, [the question] was completely settled. For no matter how great our ignorance is about how the forces and states of matter change by infinitely small amounts according to place and time, we can safely assume that the functions to which Dirichlet's investigation does not extend do not occur in nature.

Riemann gave, nevertheless, two reasons why he wished to undertake an investigation for functions that, in his view, could not play a role in physical studies. First, such functions have general relevance for the conceptual foundations of the infinitesimal calculus. In particular, his re-examination of the concept of integration aimed to bring greater clarity and certainty to this aspect of the theory. Second, Riemann emphasized the applicability of Fourier series to parts of mathematics outside the realm of physical investigations. Here he highlighted its importance for analytic number theory, a field in which his own skills in handling Fourier series led five years later to one of his greatest achievements, [Riemann 1859/1876], though it too was full of "cryptic messages for the future."

In Section 4, Riemann prepared the ground for his investigation by introducing his new concept for the integral. Here he might have mentioned the name of Cauchy or others, but instead he indicated that the time was ripe to clarify the concept of a definite integral and the extent of its validity. So he asked simply: What do you mean by an integral? For Cauchy, a definite integral required a continuous function $f(x)$ defined on an interval $[x_0, x_n]$. One then formed sums

$$S = \sum_{k=1}^{n} \delta_1 f(x_1) + \ldots \delta_n f(x_n),$$

where $\lim \delta_i \to 0$. When the values of $S$ converge, the integral

$$\int_{x_0}^{x_n} f(x) \quad \text{is well defined in the sense of Cauchy.}$$

The Cauchy integral employs only one variable parameter $\delta_i$ to create ever smaller subintervals, whereas the Riemann integral couples $\delta_i$ with another infinitesimal parameter $\epsilon_j$:

$$S = \sum_{k=1}^{n} \delta_1 f(x_0 + \delta_1 \epsilon_1) + \ldots \delta_n f(x_{n-1} + \delta_n \epsilon_n).$$

Using this definition, Riemann could handle functions $f(x)$ with isolated singularities on the interval of integration, so long as the sums converged to a limit $S$ as the $x$-values approached one of the bad points. In Section 5, he gave necessary and sufficient conditions for the limit $S$ to exist, after which he gave an example of a Riemann integrable function $f(x)$ that is discontinuous on a dense set of points. He let $(x)$ denote the difference $|x - n|$ between $x$ and its nearest integer $n$ or $(x) = 0$, if $x$ lies midway between two integers. For $2n$ and $p$ relatively prime, define the series

$$f(x) = \frac{(x)}{1} + \frac{(2x)}{4} + \frac{(3x)}{9} + \ldots,$$

which converges for all $x$, but is discontinuous when $x = p/2n$.

Finally, in Section 7, Riemann turned to the task at hand. He noted that previous investigations dealing with Fourier series for cases occurring in nature began by studying completely arbitrary functions. One then sought additional conditions, such as those found by Dirichlet, that enabled one to prove that the function can be represented by its Fourier series. Riemann took up the corresponding question in the reverse order by asking: if a function can be represented by a trigonometric series, what can we conclude about how its value changes under a continuous change of its argument? He thus took an arbitrary series

$$f = \frac{1}{2} b_0 + a_1 \sin x + b_1 \cos x + a_2 \sin 2x + b_2 \cos 2x + \ldots, \quad \text{where } a_i, b_i$$

become arbitrarily small for large $i$. Writing this as

$$f(x) = A_0 + A_1 + A_2 + \ldots,$$

and integrating twice, one obtains

$$F(x) = C + C'x + A_0\frac{x^2}{2} + A_1\frac{x^3}{3} + A_2\frac{x^4}{4} + \ldots.$$

This function $F(x)$ now took center stage for a number of technical results that culminate with the theorems in Section 9:

> For a function $f(x)$ that is $2\pi$-periodic on an interval to be represented by a trigonometric series whose terms, for every value of $x$, ultimately become infinitely small, there must exist a continuous function $F(x)$ that depends on $f(x)$ such that
>
> $$\frac{F(x+\alpha+\beta) - F(x+\alpha-\beta) - F(x-\alpha+\beta) + F(x-\alpha-\beta)}{4\alpha\beta}$$
>
> converges to $f(x)$ when $\alpha$ and $\beta$ become infinitely small and their ratio remains finite. Furthermore,
>
> $$\mu^2 \int_b^c F(x) \cos \mu(x-a)\lambda x \, dx$$
>
> must become infinitely small as $\mu$ becomes infinitely large, if $\lambda(x)$ and $\lambda'(x)$ are zero on the boundaries of the integral and continuous between them and $\lambda''(x)$ does not have infinitely many maxima and minima. [Riemann 1868/1876, 251]

Conversely, if these two conditions are fulfilled, then there is a trigonometric series in which the coefficients ultimately become infinitely small and which represents the function wherever it converges.

In Section 12, Riemann dealt with a class of functions that become infinite at finitely many points; these he showed are still amenable to Dirichlet's methods, so long as they meet the second of his three sufficient conditions. He therefore assumed that $f(x)$ does not have an infinite number of maxima and minima, and went on to show that for such a function to be represented by a trigonometric series with coefficients decreasing to infinity, it is necessary and sufficient that at each point $x = a$, where $f(a)$ becomes infinite, the values of $f(a+t)t$ and $f(a-t)t$ become infinitely small with $t$ and that the function $f(a+t)+f(a-t)$ can be integrated up to $t = 0$.

Finally, in Section 13, he gave examples of various oscillating functions with an infinite number of local maxima and minima, which are integrable but cannot be represented by a convergent Fourier series. His first case was $f(x)$ defined on the interval $(0, 2\pi)$ by

$$\frac{d\left(x^\nu \cos \frac{1}{x}\right)}{dx}, \quad \text{with} \quad 0 < \nu < \frac{1}{2}.$$

This was surely one of the first appearances of the infinitely oscillating function, $\cos \frac{1}{x}$, a forerunner of its cousin, the topologist's sine curve. Riemann immediately generalized this example by considering

$$\int f(x)\,dx = \phi(x)\cos\psi(x),$$

where for $x \to 0$, $\phi(x) \to 0$ and $\psi(x) \to \infty$, and where these functions and their derivatives are continuous and have only finitely many local maxima and minima.

Alongside these examples of integrable functions which cannot be represented by a trigonometric series, Riemann showed that the opposite phenomenon can also occur. To show this, he merely needed to modify slightly the function he defined in Section 5 (see above): instead of writing $n^2$ in the denominator, he now considered:

$$g(x) = \sum_{n=1}^{\infty} \frac{(nx)}{n} = \frac{(x)}{1} + \frac{(2x)}{2} + \frac{(3x)}{3} + \ldots.$$

Citing a passage from Gauss's *Disquisitiones Arithmeticae*, he noted that for rational numbers $x$, $g(x)$ has a well-defined representation as the trigonometric series

$$\sum_{n=1}^{\infty} \frac{\sum_{n=1}^{\theta} -(-1)^{\theta}}{n\pi} \sin 2n\pi x,$$

where $\theta$ are the divisors of $n$. Since this series is unbounded in every interval, however small, $g(x)$ cannot be integrated.

As we know, Riemann's manuscript never saw the light of day during his lifetime; thus, its impact on other mathematicians only made itself felt in the early 1870s. It was during those years that a number of German mathematicians began to make an intensive exploration of the foundations of analysis, beginning with the properties of the real number line. In all likelihood, this fervor of activity motivated Dedekind to write up and publish his booklet *Stetigkeit und irrationale Zahlen* [Dedekind 1872], a work based on ideas he first developed more than a decade earlier while teaching in Zurich. Four years later, Dedekind entered into a lengthy correspondence with Rudolf Lipschitz, who found Dedekind's radical construction of the real numbers from the rationals, via so-called Dedekind cuts, a pointless exercise. The letters they exchanged in private ([Scharlau 1986]) might well be seen as one of the major turning points leading to what would later be called modern mathematics, a major theme in Laugwitz's biography of Riemann. Others who were inspired by Riemann's work in their studies of real analysis included Hermann Hankel, Eduard Heine, and most important of all, Georg Cantor. Thus, in a very real sense, the problems that Dirichlet and Riemann struggled with long before became the soil out of which Cantor created a new theory of the infinite, his transfinite set theory [Dauben 1979, 6–29].

## 6.2 Riemann's Habilitation Lecture

As noted in Section 3.1, Riemann's thesis and qualifying lecture were prepared to fulfill the formal requirements for Habilitation. Neither was published during his lifetime, nor did Riemann refer to them in his other works. These were occasional works that stood apart from his main research interests (see Section 2.7). Still, especially in the case of his Habilitation lecture, one has to wonder why he never took the opportunity to publish this or otherwise promote the remarkable ideas he set forth in 1854.

One can hardly view Riemann's attitude toward publication as akin to Gauss's famous motto "pauca sed matura" (few but ripe). He had no reservations about publishing the sketchy arguments in his paper on the zeta function (Section 8.2) or the incomplete results in his Paris prize paper (Section 8.3). Once Gauss had passed on, Riemann took the opportunity to publish his paper on generalized hypergeometric functions (Section 10.3), in which he rightly pronounced that he had gone beyond his predecessor. When Gauss was still alive, Riemann often expressed worries in letters to his family that the great man was working in the same direction and might suddenly publish a paper that made all of his own work worthless. Such fears were exaggerated, but one should not discount their psychological importance. Competition and secrecy were both strong factors motivating Riemann's research, even though he was also genuinely interested in imparting his ideas to others. His encounters with Eisenstein made him pull back from him, and Dedekind described his mental state in 1857 as bordering on paranoia. Riemann no doubt also experienced happy moments in Göttingen, but he just as surely felt a great deal of pressure there, in contrast to the relaxing atmosphere in Pisa. Why he never made an effort to promote his work on Riemannian manifolds we will never know; it is one of the major mysteries he took with him to his grave.

At the time he prepared for his Habilitation, Riemann was very busy pursuing his physical researches. That larger program, as he himself made clear, was his principal research goal [Ferreirós 2006]. Some of its foremost features will be taken up in Chapter 9. Riemann's essay has often been viewed in the context of non-Euclidean geometry, which raises the puzzling question as to why one finds no mention of the property of parallelism anywhere in his text. When Riemann did speak about Euclidean geometry, his focus fell on the free mobility of rigid bodies, which explains why Helmholtz was later able to connect his own reflections on the foundations of geometry with those of Riemann [Helmholtz 1868]. Yet Riemann's entire framework was far broader, not only because he treated higher-dimensional differentiable manifolds but also because he took invariant line elements rather than 3-dimensional bodies as the starting point for his analysis. Furthermore, he clearly distinguished between topological and metric properties of manifolds (see [Scholz 1980] and [Scholz 1999]).

Riemann's initial remarks alluded to the need to clarify these fundamental matters:

> It is known that geometry assumes as given both the notion of space and the first principles of constructions in space. The definitions for these are merely nominal, whereas the essential determinations appear in the form of axioms. The relations underlying these assumptions remain, however, in darkness; we neither perceive whether and how far their connection is necessary, nor whether they are a priori possible. [Riemann 1868/1876, 272]

According to Riemann, philosophers as well as geometers from Euclid to Legendre have left us in the dark because they have neglected to examine the general conception of multiply-extended magnitudes. He thus took this as his starting point, noting that such multiply-extended magnitudes are capable of different measure relations, so that space can be seen as a particular case of a three-fold extended magnitude. Its properties can then only be deduced from experience. This then poses the problem of finding the simplest facts which enable one to determine the measure relations of space. There may be several such systems of facts, but the most important system was the one laid down by Euclid when he codified Greek geometry *circa* 300 B.C. in the thirteen books of his *Elements*.

This claim would seem to fly in the face of what practically every commentator on Riemann's text has chosen to emphasize. Its reception history will be discussed below in Section 6.3, but here we should note what he wrote immediately after this, since this places the goal of his investigation in a clear context:

> These facts, like all facts, are never necessary, but rather they depend on the limitations of empirical observation; they are thus hypotheses. We may nevertheless investigate their probable validity beyond the limits of observation, whether in the direction of the infinitely great or the infinitely small. [Riemann 1868/1876, 273]

So what did Riemann mean, when he wrote that Euclid and his followers left him and his contemporaries in the dark? If Euclidean geometry, as passed down over the centuries, is truly the most important system one can study, why should anyone fiddle with its hypotheses? Traditionally, mathematicians identified Euclid's geometry with his theory of parallel lines, which many saw as a stumbling block. After many vain attempts to prove the fifth postulate, or some equivalent statement for the existence of uniquely parallel lines, by Riemann's time some had come to believe that these efforts would always be futile. For Riemann, whose philosophy of space was rooted in empirical findings, the existence of parallel lines could never be confirmed in nature, as parallelism was a property of curves extended to infinity.

To finesse that problem, Greek geometers worked with line segments, which could be extended to an arbitrary length. Their work adopted a constructive approach; H.G. Zeuthen, a leading expert on Greek mathematics around 1900, went so far as to maintain that the constructions carried out in Euclid's *Elements* were tantamount to existence proofs. In Book I, Euclid only invoked the parallel postulate when it was needed to prove that the angle sum in a triangle always equals two right angles (180°). He then ended Book I by proving the famous Pythagorean theorem and its converse: in any right triangle with side lengths $a, b$ and hypotenuse $c$, one has $c^2 = a^2 + b^2$, a condition that characterizes right triangles. Moreover, like the theorem for the angle sum in any triangle, the Pythagorean theorem holds only in Euclidean plane geometry. That hypothesis could be formulated for any surface, or in Riemann's language for any two-dimensional manifold. Indeed, he alluded directly to Gauss's theory of surfaces in setting forth his general conception of a differential form for distances on manifolds with any given number of dimensions $n$. A Pythagorean geometry is based on a specific type of metric, in Cartesian coordinates $ds = \sqrt{\sum dx_i^2}$. Since the geometry of space depends on a system of measurements, Riemann saw no advantage in reverting to global synthetic principles. Curvilinear objects required, in any case, locally defined metrical relations.

As he had done with his dissertation and postdoctoral thesis, Riemann appended a synopsis of the topics in his Habilitation lecture. In setting these out, he employed a tripartite structure, preceded by a section summarizing the "plan of the investigation." These three sections are differentiated in the body of the text, but afterward he presented them in more detail in order to outline the entire argument. The sections are:

I. Concept of an $n$-fold extended magnitude;

II. Metric relations admitted by a manifold of $n$-dimensions under the assumption that lengths of lines are independent of position, so that every line may be measured by any other;

III. Application to Space.

Section II presents Riemann's brief account of differentiable manifolds and curvature. Dedekind was the first to struggle with what came to be called Riemannian curvature or the curvature tensor, an object Riemann only vaguely described in the middle section of his lecture as a generalization of Gaussian surface curvature. Dedekind's efforts mark the beginning of a complicated story that has received much attention in the secondary literature. Yet, Riemann's interest in natural philosophy – apparent in the first and third sections of the text – has often been overlooked [Bottazzini & Tazzioli 1995]. Dedekind and Weber were keenly interested in utilizing relevant parts of Riemann's unpublished Paris prize essay [Riemann 1861a/1876] to account for what was missing, a complicated story that is taken up in Section 8.3.[5]

Riemann proceeded to analyze the concept of an $n$-fold extended magnitude, citing two texts by Gauss and others by the philosopher Herbart as relevant earlier studies. He first distinguished between discrete and continuous manifolds, calling the objects points in the first case and elements in the second. Discrete manifolds were familiar from everyday experience, whereas one rarely encountered continuous manifolds except in higher mathematics. In Subsection I.3, Riemann showed how by taking a constant real-valued function defined on an $n$-dimensional manifold one can construct a submanifold of dimension $n-1$. From here, he passed over to Section II, beginning with the special case of

[5] For detailed historical analyses, see [Cogliati 2014] and [Darrigol 2015]; [Spivak 1979, 135–153] presents an English translation of [Riemann 1868/1876] followed by detailed mathematical explanations.

an $n$-manifold equipped with a Euclidean metric. He conceded that the theory he was about to describe could only be developed using formulas based on abstract measure concepts. Under certain conditions, however, these abstract relationships can be given a geometrical interpretation. In both cases, Riemann referred to "the famous study by Privy Court Counsel Gauss on curved surfaces" [Riemann 1868/1876, 276].

This was Gauss's classic study on the differential geometry of surfaces, which introduced the concept of Gaussian curvature [Gauss 1828b]. A key passage deals with the role of bending invariants in surface theory:

> ... When a surface is regarded, not as the boundary of a solid, but as a flexible, though not extensible solid, one dimension of which is supposed to vanish, then the properties of the surface depend in part upon the form to which we can suppose it reduced, and in part are absolute and remain invariable, whatever may be the form into which the surface is bent. To these latter properties, the study of which opens to geometry a new and fertile field, belong the measure of curvature and the integral curvature, in the sense which we have given to these expressions. To these belong also the theory of shortest lines, and a great part of what we reserve to be treated later. From this point of view, a plane surface and a surface developable on a plane, e. g., cylindrical surfaces, conical surfaces, etc., are to be regarded as essentially identical; and the generic method of defining in a general manner the nature of the surfaces thus considered is always based upon the formula
>
> $$\sqrt{F\,dp^2 + F\,dpdq + G\,dq^2},$$
>
> which connects the line element with the two variables $p, q$. [Gauss 1863–1933, 4: 344–345]

Those who had read and grasped the significance of these ideas would have easily understood that Riemann had found a way to generalize this approach to higher dimensions.[6] The physical import of his approach, however, remained altogether unclear, as there seemed to be no conceivable way to test small curvature effects, whether in the large or the small.

Riemann's lecture was clearly conceived and delivered for the benefit of Gauss, who, if we are to credit Wilhelm Weber, found it extremely fascinating [Dedekind 1876a, 517]. Yet Gauss never published any of his various musings about the principles underlying our conception of geometry, thoughts he often imparted to those in his inner circle [Sartorius 1856/2012, 81]. Nor did Gauss ever contemplate the possibility of studying curvature properties of higher-dimensional manifolds. To do so, Riemann introduced geodesic coordinates, often called Riemannian normal coordinates. He alluded to these in the following passage:

> In order to give a tangible meaning to the curvature of an $n$-fold extended manifold at a given point and in a given surface direction through it, one must assume that a shortest line emanating from a point is completely determined if its initial direction is given.

[6] Riemann did not refer to Gauss's prepublication announcement, in which this statement appears, so we cannot be sure he ever read it. Darrigol, citing the passage above, wrote that it "invited generalization to manifolds of any dimension" and that "Riemann got the hint" [Darrigol 2015, 56], claims that hardly accord with the circumstances described in Section 3.1.

Accordingly, a certain surface will be obtained if all the initial directions emanating from the given point and lying in the given surface element are extended into shortest lines, and this surface has a certain measure of curvature in the given point, which at the same time is the measure of curvature of the $n$-fold extended manifold in the given point and of the given surface direction. [Riemann 1868/1876, 261]

This passage gave the first description of what came to be called sectional curvature.

Riemann highlighted the case of manifolds of constant curvature, whether positive or negative:

The manifolds whose measure of curvature is everywhere equal to zero may be regarded as a special case of those manifolds whose measure of curvature is constant throughout. The common characteristic of these manifolds ... can also be expressed by stating that figures within them can be moved without distortion. For, evidently, figures within them could not be arbitrarily translated and rotated unless the measure of curvature were the same at every point and in every direction. On the other hand, however, the metric relations of a manifold are completely determined by its measure of curvature; consequently, the metric relations surrounding any given point in all directions are precisely the same as those surrounding any other point; thus, the same constructions are executable from any point, and it follows that, in manifolds with a constant measure of curvature, figures may be assigned any arbitrary position. The metric relations of these manifolds depend solely upon the value of the measure of curvature $\alpha$. With regard to the analytic representation, ... the expression for the line element takes the form

$$\frac{1}{1+\frac{\alpha}{4}\sum x^2}\sqrt{\sum dx^2}.$$

[Riemann 1868/1876, 264]

When applying these notions to the physical world, Riemann emphasized the fundamental distinction between topological properties (extensional relations) and metrical properties (measurable relations). He also noted an important difference between discrete and continuous manifolds.

[In] a discrete manifold, the statements of experience are never entirely certain, though not imprecise, whereas in ... a continuous manifold, every determination from experience always remains imprecise – no matter how great the probability that it is nearly correct may be. This circumstance becomes important when extending these empirical determinations beyond the limits of observation into the immeasurably large and immeasurably small; for the latter can obviously become ever less precise beyond the limits of observation, whereas the former cannot. [Riemann 1868/1876, 284]

With regard to the curvature of physical space, Riemann drew some startling conclusions. If cosmologists today seem to be convinced that we live in a flat universe, they must surely be aware that this consensus opinion remains subject to what Riemann had to say about this:

When spatial constructions are extended to the immeasurably large, a distinction must be made between boundlessness and infinity; the former belongs to relations of extension, the latter to relations of measure. That space is an unlimited, three-fold extended manifold is a presupposition applied to every conception of the external world, according to which the realm of actual perceptions is supplemented at every moment and the possible locations of a sought-after object are constructed, and which is continually confirmed in these applications. The boundlessness of space therefore possesses a

> greater empirical certainty than any external experience; nevertheless, infinite extent by no means follows from this. Rather, if one assumes the independence of bodies from their location and thus ascribes to space a constant curvature, it would necessarily be finite as soon as this curvature had even the smallest positive value. If one were to extend the initial directions lying in a surface element to the shortest possible lines, one would obtain an unlimited surface with a constant positive curvature, that is, a surface which, in a planar, threefold extended manifold, would assume the shape of a spherical surface and which is consequently finite. [Riemann 1868/1876, 284]

The physical program Riemann was then pursuing was based on a revision of mechanics that incorporated ether effects that he related to gravity and magnetism. These physical speculations only came to light with the publication of Riemann's *Werke* in 1876, but they loom in the background of his qualifying lecture, particularly in those passages where he wrote about the microphysical world. The speculative character of the preceding quotation should be set against what Riemann afterward wrote:

> Questions about the infinitely large are moot for the interpretation of nature, unlike questions about the infinitely small. Our knowledge of the causal relations between phenomena depends essentially on the exactness with which we follow these into the infinitely small. The progress in our knowledge of mechanics in recent centuries has depended almost entirely on the exactness of the constructions. These have been made possible through the invention of the infinitesimal calculus as well as through the simple principles discovered by Archimedes, Galileo, and Newton, and which modern physics employs. Yet in the natural sciences, which still lack simple principles for such constructions, one seeks to recognize the causal relations by studying the phenomena in minute spatial regions, so far as the microscope allows. Questions about the measure relations of space in the infinitesimally small are therefore by no means superfluous [Riemann 1868/1876, 285]

Riemann's central concern in Section III was to show how his new conceptual approach to geometry could shed light on the properties of physical space, in particular on the smallest levels. For space at large, ordinary Euclidean geometry was wholly adequate, even though Riemann pointed to the possibility that space might have a very small positive curvature that gave it the structure of a 3-sphere. Since both cases assume space to be unbounded, Riemann considered this topological hypothesis to be far more certain than the narrower hypothesis that assumed its metric relations extended to infinity. Regarding the empirical basis supporting these considerations, he wrote:

> If we suppose that bodies exist independently of position, the curvature is everywhere constant, and it then results from astronomical measurements that it cannot be different from zero; or at any rate its reciprocal must be an area in comparison with which the range of our telescopes may be neglected. [Riemann 1868/1876, 285]

This, again, indicates that Riemann was advancing a general approach to geometry that subsumed the special case of a flat metric (Riemann used the term plane manifold).

> In a flat $n$-fold extended manifold, the measure of curvature is zero at every point in every direction; However, according to the earlier investigation, in order to determine the proportions, it is enough to know that it is zero at every point in n surface

> directions whose curvature measures are independent of one another. The manifolds whose measure of curvature is = 0 everywhere can be viewed as a special case of those manifolds whose measure of curvature is everywhere constant. The common character of these manifolds, whose curvature is constant, can also be expressed by saying they allow figures to move within them without stretching. Because obviously the figures in them would not be able to be moved and rotated at will if the measure of curvature were not the same at every point in all directions. [Riemann 1868/1876, 281–282]

This and other passages underscore the importance of the manifold concept in Riemann's (unpublished) text. Indeed, this single work became famous for launching the theory of Riemannian manifolds, a concept, however, that its author never publicly promoted.

In the closing lines of his lecture, Riemann brought out more clearly the relevance of his reflections for the theory of space in the small. In his view, the validity of the assumptions of geometry for the infinitesimally small was connected with the internal basis for the metric relations. If the ultimate constituents of space formed a discrete manifold, then no further explanation was required; if it forms a continuous manifold, then the explanation for its metric relations must be sought in the binding forces acting on it.

> The solution to these questions can only be found by starting from the previous conception of phenomena, proven by experience, for which Newton laid the foundation, and gradually reworking it, driven by facts that cannot be explained from it. Investigations, like the one conducted here, which start from general concepts, can only serve to ensure that this work is not hindered by the limitations of the concepts and that progress in recognizing the connection of things is not hindered by traditional prejudices.
>
> This leads into the area of another science, into the the area of physics, which the nature of today's proceeding probably does not allow us to enter. [Riemann 1868/1876, 286]

Riemann's final statement sounded very much like an invitation to join in a discussion, and in all likelihood some remarks were offered by those present. Unfortunately, we have no way of knowing what was said, in particular by Gauss. He was in failing health, and may not have wanted to prolong this formality beyond what was absolutely necessary. As for Riemann, he never once called attention to the ideas in his lecture, which disappeared until 1868, the year Dedekind put it into print as [Riemann 1868/1876] (see Section 3.2).

## 6.3 Threads of Reception

When Riemann's lecture first appeared in volume 13 of the *Göttinger Abhandlungen*, Dedekind attached a note to it indicating that he planned to publish mathematical commentary on the text. He surely would have included such a commentary at the time it was printed, except for the fact that Riemann's hints about his approach to curvature were very sketchy. In keeping with the general audience to whom he spoke, Riemann suppressed all the supporting mathematical apparatus. Dedekind noticed this immediately, of course, but he also realized that he faced a serious challenge in trying to explicate things like

Riemann's curvature tensor.[7] The long history of mathematical formalisms (vectors, tensors, etc.) that accompanied physical theories had little to do with the broader interest in Riemann's essay [Reich 1994]. Its third section, pertaining to the geometry of space, was for most readers the real topic of interest, and many misunderstood Riemannian curvature as a property that implied the existence of a fourth dimension.[8]

In any case, Dedekind's announced publication never appeared, for reasons he later explained in a letter to Heinrich Weber from 14 March 1875:

> In the following years (mainly in 1867, I believe) I devoted myself to this subject for a long time, but later I completely abandoned publication, partly because others (Christoffel, Lipschitz, Beltrami) had taken up this subject, and partly because in 1869, the indispensable preliminary work for the second edition of Dirichlet's number theory forced me to devote myself to an entirely different field, namely the creation of a general, but completely comprehensive theory of ideal numbers. [Scheel 2015, 57]

From Dedekind's letter, written on 18 July 1867 to Jakob Henle (quoted in 12.1), we know that he was, indeed, deeply immersed in Riemann's unpublished papers at that time. He therein expressed his "hope to complete [this commentary] before the end of the semester." Putting these two letters together, it is evident that Dedekind felt overwhelmed by the task of promoting Riemann's legacy alongside that of his teacher Dirichlet, whose number-theoretic studies were far closer to Dedekind's own mathematical interests. Back in 1867, though, he felt a tremendous sense of obligation to make Riemann's work more widely accessible, not least because his widow had entrusted him with this task. At some point, Dedekind realized he had to shift gears and begin preparing the second edition of Dirichlet's lectures on number theory [Dirichlet 1871], which contains an early version of Dedekind's famous theory of ideals.

Dedekind may have still hoped to finish this task in 1868, but he was in large part constrained by the fact that he wanted not only to explicate the text of [Riemann 1868/1876], but also to reconcile its contents with the brief explanations Riemann had given in his unpublished "Commentatio," the 1861 Paris prize paper Section 8.3). The three authors Dedekind mentioned in his letter to Heinrich Weber were, of course, writing under no such constraint, and their responses to Riemann's varied a good deal [Klein 1927, 188–199]. As a differential geometer with strong interests in physics, Beltrami saw Riemann's methods as opening the door to higher-dimensional spaces based on different possible line elements. The two papers by E.B. Christoffel ([Christoffel 1869]) and Rudolf Lipschitz ([Lipschitz 1869]) appeared next to one another in volume 70 of Crelle. In fact, their submission dates in January 1869 were only a day apart, though neither author seems to have in contact with the other either

[7] This terminology only arose much later in the wake of general relativity. Einstein introduced tensors as generalized vectors in his famous survey paper [Einstein 1916], but Marcel Grossmann had already done so in [Einstein/Grossmann 1913], which closely followed [Ricci/Levi-Civita 1901], though the concept of tensors appears nowhere in their text.

[8] On the general fascination with the fourth dimension, see [Volkert 2023]. In Germany, this ran parallel with but largely independent from the reception on non-Euclidean geometry, for which see [Volkert 2013].

before or after. Were they inspired to write these papers by reading Riemann's lecture? The manner in which Christoffel cited that recent publication at the very end of his paper [Christoffel 1869, 70] would seem to suggest otherwise.[9] As Klein emphasized, Christoffel's paper was by far the more influential, whereas Lipschitz contributions (which included [Lipschitz 1870], [Lipschitz 1876]) were far closer to the spirit off Riemann's work [Rowe 2022].

Only months after Dedekind put Riemann's lecture into print, the story of its fame began. In that same year, Hermann Helmholtz and Eugenio Beltrami tied together Riemann's ideas with their own reflections on the empirical grounds of spatial conceptualization and non-Euclidean geometry. Thus, the timing for the posthumous publication of Riemann's lecture surely could not have been better, as the subsequent discussions and debates over the geometry of space had an entirely different character from those that dominated the preceding era.[10] Helmholtz was at this time professor of physiology in Heidelberg, where he had been investigating optical phenomena, in which the concept of a visual field played a key role. Riemann had some acquaintance with this area of research, which was pursued by two Göttingen professors, Listing and Ruete, whom he knew.[11] One might well wonder whether this field of inquiry had some influence on Riemann's general approach to metrics on higher-dimensional manifolds.

Helmholtz learned about Riemann's work from Ernst Schering, who in April 1868 sent him a copy of [Schering 1867]. In all likelihood, Schering had communicated earlier with Helmholtz in connection with Riemann's posthumously published article on acoustical mechanics [Riemann 1867b/1876]. This paper disputed certain views advanced by Helmholtz, who therefore responded in [Helmholtz 1869]. Helmholtz responded to Schering's letter on 21 April by thanking him "for sending the two small essays concerning Riemann," though it remains unclear what the second piece may have been. Regarding Schering's account of Riemann's life, Helmholtz was struck by what he read regarding the Habilitation lecture on the hypotheses underlying geometry. Schering's text strongly hinted that Riemann's lecture was little more than a generalization of Gauss's work on surface theory, the only part of his geometrical investigations that Gauss had chosen to publish [Schering 1867, 311].

Helmholtz wondered whether Riemann's text might still be available, while anticipating (and hoping) Riemann's findings would fully vindicate his own work:

> In the last two years I have been working on the same subject in connection with my investigations into physiological optics, but have not yet completed and published the work because I still hoped to be able to generalize individual points. In particular, I cannot yet make everything as general for three dimensions as I can for two. Now I see

[9] The claim in [Farwell/Knee 1990, 239] that Riemann's Habilitation lecture was clearly triggered Christoffel's investigation lacks any supporting evidence.

[10] The philosophical discussions in Germany were dominated by those who defended or attacked Kantian epistemology. Going back further, the debates from the early eighteenth century were directly tied to theological issues, as in the famous Leibniz-Clarke exchange [Koyré 1957].

[11] See the final note on [Riemann 1876/1892/1902, 1902: 113].

> from the few hints you give about the results of the work that Riemann came to exactly the same results as I did. [Koenigsberger 1902, 138]

Offprints of [Riemann 1868/1876] were by now available, so Schering sent Helmholtz a copy. He continued to pound home that Riemann's ideas were part of a Gaussian program, as Schering surely hoped to publicize through the Gauss *Werke.* He added the comment that "the most essential moment in Riemann's investigation is the theorem that the quantity defined by Gauss as a measure of curvature means a differential invariant for homogeneous second-degree differential expressions of first order with two variables" Helmholtz replied on 18 May by sending a short note for the Royal Society, to which he belonged as a corresponding member. This seems not to have been published, probably because Helmholtz had already indicated that he wished to prepare a lengthier paper on the same subject.

Only a few days later, on 22 May, Helmholtz delivered a short lecture in Heidelberg, in which he described the connection between his investigations and those of Riemann. He noted that his work was:

> ...largely completed when Riemann's habilitation lecture "On the Hypotheses Underlying Geometry" appeared, in which the same investigation is carried out with a slightly different approach. On this occasion, we learned that Gauss had also worked on the same topic, and that his famous treatise on the curvature of surfaces is the only published part of this investigation. [Helmholtz 1868/1869, 198]

This, of course, was the opinion Ernst Schering had conveyed to him. In any event, on Helmholtz's reading, Riemann's lecture lent additional authority to his own work. Indeed, he underscored the importance of Riemann's concept of an $n$-dimensional manifold as this related to his concept of fields of vision. Even though this example does not appear in Riemann's text, Helmholtz pointed out that Riemann did mention the manifold representing the system of colors.

> Riemann begins by explaining how the general characteristics of space – its continuity and the multiplicity of its dimensions – can be expressed analytically by stating that every particular element within the manifold it presents, that is, every point, can be determined by measuring $n$ continuously and independently varying quantities (coordinates). If $n$ such things are necessary, then space is, as he calls it, an $n$-fold extended manifold, and we ascribe $n$ dimensions to it. (*ibid.*)

Here, in his Heidelberg lecture, Helmholtz overlooked the class of discrete manifolds, which played a major part in Riemann's microphysical speculations. Moreover, Helmholtz called a Riemannian metric a generalized Pythagorean metric, thereby emphasizing that the Euclidean case retained its special status.

In Riemann's approach, the given line element was the only invariant, whereas Helmholtz sought to derive its special form under the assumption that space presumes the free mobility of congruent 3-dimensional figures. This led him to adopt a system of four axioms, from which he then attempted to prove that these led to spaces with constant vanishing curvature. Helmholtz's first three axioms were uncontroversial, but the fourth axiom demanded what he called monodromy, a term Riemann had introduced in [Riemann 1857a, 68]. For an $n$-dimensional space, Helmholtz required that, whenever $n - 1$ points of a body remain fixed, any other point of the body will be constrained to move on a closed curve. He was evidently motivated to add this axiom because in the plane one encountered spiral curves that were invariant under a subgroup of rotations, but failed to close. By the late 1880s, Sophus Lie re-examined Helmholtz's work on the Riemann-Helmholtz space problem from the vantage point of his theory of continuous groups, showing among other things that the monodromy axiom was superfluous. This terrain later became a major area of research after David Hilbert highlighted its importance in his fifth Paris problem, which eventually led to the modern theory of topological groups (for details, see [Freudenthal 1964]).

Helmholtz reiterated his indebtedness to Riemann in the famous paper "Ueber die Thatsachen, die der Geometrie zum Grunde liegen" [Helmholtz 1868], writing

> ...I must confess that even though the publication of Riemann's investigations deprived me of priority for a number of my own work results, it was of no small importance to me that in dealing with this unusual and rather discredited subject ... to see that such an excellent mathematician had considered the same questions worthy of his interest, and that having encountered him as a companion was an important guarantee for the correctness of the path I had chosen. [Helmholtz 1868, 195]

The similarity with Riemann's title was surely meant to be striking, but even more so the emphasis on facts (*Tatsachen*) as opposed to hypotheses. Helmholtz's entire undertaking aimed to show that our conceptual understanding of the properties of space derived from the experience of moving rigid bodies from place to place without their changing size or shape. In idealized form, this unarticulated principle served as the foundation for Euclid's theory of congruent figures. In Riemann's lecture, he treated the corresponding manifolds of positive constant curvature as a special case of his theory, whereas Helmholtz sought to derive this type of metric from his axioms. As Lie later showed, his argument assumed local properties that Helmholtz had ignored. His aim, in any case, was to identify a minimal set of axioms based on empirical principles from which the properties of 3-dimensional Euclidean space could be derived. Beyond the four axioms set forth earlier in his Heidelberg lecture, he now added two more: 5) space is 3-dimensional, and 6) space is infinite, which he (incorrectly) assumed must imply that its curvature was zero. He ended by asserting:

> ...the entire possibility of our system of spatial measurements ... depends on the existence of natural bodies that nearly correspond to our conception of rigid bodies. The factual basis for our measurement of space is the independence of congruence from place, from the direction of two superimposed bodies, and from the path that leads from one to the other. [Helmholtz 1868, 221]

Helmholtz saw his reflections on the foundations of geometry as, in effect, a special case that conformed with Riemannian principles. This might seem a fair assessment, except for one twist. It soon transpired that most commentators interpreted these two essays as part of the then rising interest in the works of Bolyai and Lobachevsky on non-Euclidean geometry. Moreover, by this time, Gauss's longstanding interest in this topic – including his familiarity with the works of these two pioneering figures – was well established. If Riemann ever knew about this, no solid evidence indicating that he did has ever been brought to light.

Helmholtz probably had little or no idea of non-Euclidean geometry up until the day he received a letter from Eugenio Beltrami, written on 24 April 1869. That letter is amusing to read today, as Beltrami went on at some length about the apparent boundary curves of a pseudo-sphere, which cannot be completely embedded in Euclidean 3-space (as Hilbert later showed). The thrust of his argument was that a pseudo-sphere can always be extended, an indication that it lives in an infinite non-Euclidean space of negative curvature. Helmholtz readily agreed, and thus sent a small note amending his earlier claim in his first Heidelberg lecture [Helmholtz 1868/1869, V: 31–32]. In a later Heidelberg lecture, "On the Origin and Meaning of Geometric Axioms" [Helmholtz 1870/1883], Helmholtz gave an intuitive account of geometry on a Beltrami pseudo-sphere that was intelligible for non-mathematicians.

Beltrami had struggled over the differences between non-Euclidean geometry in dimension two – which he studied directly on the surface of a pseudo-sphere [Beltrami 1868a] – and the notion of a 3-dimensional geometry with corresponding negative curvature [Beltrami 1868b]. In the case of surfaces, Beltrami had proved that those of constant curvature have the property that their geodesics can be expressed in terms of linear equations. Only after reading Riemann's essay, though, did he convince himself of the possibility that non-Euclidean metrics were also valid in dimension three, as he announced at the outset of his "Fundamental Theory of Spaces of Constant Curvature" (see [Stillwell 1996, 41] and [Scholz 1980, 108–112]). Beltrami's work also met with considerable resistance, even from no less a figure than Luigi Cremona [Volkert 2013, 60–61]. Although Beltrami had opportunities to speak with Riemann in Pisa, the latter never mentioned the ideas in his Habilitation lecture, which Beltrami only learned about after its publication. Thus it was Beltrami – not Riemann or Helmholtz – who signaled the connection between their two studies and non-Euclidean geometry.

This point was easily lost, however, as it was mainly Helmholtz's semi-popular publications that served to enliven the subsequent controversies. Helmholtz had no doubt foreseen that he would be attacked by those who saw his work as undermining the precepts of Kantian epistemology [Laßwitz 1883]. Some twenty years later, he wrote:

> The Kantians of strictest observance emphasize above all the points where, in my opinion, Kant suffered from the imperfect development of the special sciences of his time and became involved in errors. The core of these errors are the axioms of geometry, which he regards as forms of intuition given a priori, but which are in fact propositions that could be tested through observation and, if they were incorrect, possibly refuted. I have tried to prove this last point. But this eliminates the possibility of being able to provide metaphysical foundations for natural science, in which Kant actually believed. [Koenigsberger 1902, 141]

Riemann, who never set foot in any comparable public arenas, harbored similar thoughts regarding the weaknesses of Kant's views on space and time. That he never mentioned the name Kant in his Habilitation lecture should come as no surprise. All he needed to say were the names he dropped at the outset: Gauss and Herbart, both of whom had distanced themselves publicly from Kantian orthodoxy. Just as his aim had little to do with Helmholtz's effort to refute Kant's epistemological views, so was his interest framed by an analysis of the topological and metrical properties of manifolds – a concept only implicit in his other works. Whether or not he had ever heard of the work of Bolyai and Lobachevsky, their approach to the space problem had virtually nothing in common with his. Not until his unpublished works came out in the *Werke* [Riemann 1876/1892/1902] could readers begin to appreciate the broader context of physical interests that originally motivated his lecture. By then, however, its reception belonged to the wider frame of discourse concerned with non-Euclidean geometry, the fourth dimension, and the overturning of conventional views of space.

Heinrich Weber was struck by how quickly the Leipzig astrophysicist Friedrich Zöllner drew not only on Riemann's lecture but on his philosophical writings in the long and highly polemical preface to his *Principien einer elektrodynamischen Theorie der Materie* [Zöllner 1876, lxxv–lxxvi]. This work appeared only months after the publication of Riemann's *Werke*. Commenting on this in a letter to Dedekind from 10 December 1876, Weber had this to say:

> Have you perhaps seen that in Zöllner's new book on electrodynamics, which is essentially a reprint of Weber's treatises, he refers several times in the long and very peculiar preface to the philosophical part of Riemann's works, on the whole very appreciatively, although he has a fundamentally different opinion on one main point. Incidentally, I believe that the good Zöllner is gradually going astray in his philosophical speculations. [Scheel 2015, 160]

Weber's last remark proved to be right on the mark, as not long afterward Zöllner gained widespread fame and notoriety for his seances with the American medium Henry Slade. In [Zöllner 1876], he cited Riemann's views as support for his claims regarding the reality of a four-dimensional space, which soon hereafter became a topic of lively discussions (see [Volkert 2023]). One of ironies that leaps out in Zöllner's preface is that his *bête noire* in Germany was none other than Hermann Helmholtz, in large part because the latter sided with the advocates of British field physics against the electrodynamic theory advocated by Wilhelm Weber and his followers. Thus, Riemann's ideas regarding the fundamental principles underlying our understanding of space were taken up by both Helmholtz and Zöllner, despite the fact that they held highly antithetical views on closely related matters.

# Chapter 7
# Elliptic and Abelian Functions

## 7.1 Elliptic Functions

Although Riemann never published directly on elliptic functions, they played a major role in his lecture courses, serving as a bridge to his general theory of Abelian functions. Although Abel and Jacobi developed the theory of elliptic functions in terms of a complex variable, their approach was formal and non-geometrical.[1] Riemann's entire approach, on the other hand, took the Gaussian plane as foundational. Moreover, his geometric representation made clear why the integrals corresponding to functions of the type $y^2 = P_3(x)$ or $y^2 = P_4(x)$ lead to doubly periodic functions when the variables are complex. Indeed, the corresponding Riemann surface has genus $p = 1$. This is the case of a torus, which arises from four simple branch points. As Riemann often emphasized, these geometric methods – hinging, of course, on the existence theorems he derived using Dirichlet's principle – made it possible to circumvent the complicated analytic machinery that made earlier work so impenetrable. This hardly meant, however, that Riemann himself avoided complicated calculations when these were called for; in fact, he was deeply steeped in Jacobi's analytic formalisms.

Historically, the theory of elliptic functions arose from longstanding efforts to understand the properties of elliptic integrals, which have the form

$$\int R(x, \sqrt{P(x)})dx,$$

where $R$ is a rational function and $P(x)$ is a polynomial of degree three or four with distinct roots. Unlike the integral of a rational function $R(P_m(x), Q_n(x))$, where $P_m(x)$, $Q_n(x)$ are polynomials of degree $m$ resp. $n$, an elliptic integral cannot be expressed as a sum of elementary functions. These more complicated integrals tended to arise quite often, however, in particular as exact solutions to physical problems. Adrien-Marie Legendre famously devoted twenty years, from

[1] On these developments in general, see [Houzel 1978]; for Abel's contributions, see [Houzel 2004].



D. E. Rowe, *Bernhard Riemann: His Life and Wondrous Mathematical Legacy*,
https://doi.org/10.1007/978-3-032-25457-3_8

1811 to 1830, to studying the transformation properties of elliptic integrals. He proved the fundamental result that every elliptic integral can be expressed as a linear combination of three special types, known today as the three Legendre forms. Toward the end of his life, he was amazed to see this theory turned on its head, as it were, by Niels Henrik Abel and C.G.J. Jacobi, who showed that instead of viewing an elliptic integral as a function of its upper endpoint $x$ in

$$u(x) = \int_0^x R(t, \sqrt{P(t)})dt,$$

one could study the inverse function $x(u)$. Apparently, Legendre had never considered this possibility.

Thus, by the mid-1820s, Abel and Jacobi independently created the theory of elliptic functions by inverting elliptic integrals and showing this leads to doubly periodic functions. At most one of the periods can be real, however, which explains why the theory of elliptic functions pointed mathematicians in the direction of complex analysis. As emphasized by Bottazzini and Gray, however, one should not leap to the conclusion that Abel and Jacobi developed their theories as special types of complex functions [Bottazzini & Gray 2013, 49–50]. Indeed, they never cited the work of Cauchy, who along with Gauss was the first to consider the properties of genuine functions of a complex variable. It was Riemann, who first built on their work, which motivated his entire approach to elliptic functions and higher transcendentals. As a student of Jacobi, he also had the opportunity to absorb the latter's unique contribution to the theory, namely, the theory of $\theta$-functions. In Riemann's hands, this would become the tool needed to construct his general theory of Abelian functions, inspired by Jacobi's inversion problem.

Before describing Riemann's approach to elliptic functions, we should first take note of an earlier result that had great significance for his general theory, namely Abel's addition theorem. Abel had treated this as a result applying to real integrals, but for Riemann the proper context was the complex domain. In this setting, one begins with an irreducible algebraic equation $F(z, w) = 0$ and corresponding Abelian integrals $\int R(w, z)dz$, where $R$ is a rational function of $z, w$ lying on the curve $F(z, w) = 0$. Roughly stated, Abel's theorem asserts that any sum of Abelian integrals with arbitrary upper endpoints can be reduced to another (smaller) sum of integrals whose upper end points are algebraically related to those of the original sum. The maximum number of integrals required is what came to be known as the genus of the underlying curve $F(z, w) = 0$.

To illustrate this by way of a simple example, consider the hyperelliptic case given by $w^2 = P_6(z)$, where $P_6$ is a sixth-degree polynomial with distinct roots. Here the genus of the curve is $p = 2$. Abel recognized that any given polynomial $f(z, w)$ will determine $p$, the maximum number of linearly independent Abelian integrals irrespective of the form of $R$. His theorem in this present case then asserts that an arbitrary finite sum of such integrals can be expressed as a sum of just two of them, thus:

$$\sum_{m=1}^{M} \int_{0}^{z_m} R(w,z)dz = \int_{0}^{a} R(w,z)dz + \int_{0}^{b} R(w,z)dz + EF,$$

where $a$ and $b$ are algebraic functions of the $z_m$ and $w_m$, and $EF$ (for elementary functions) do not affect the result.

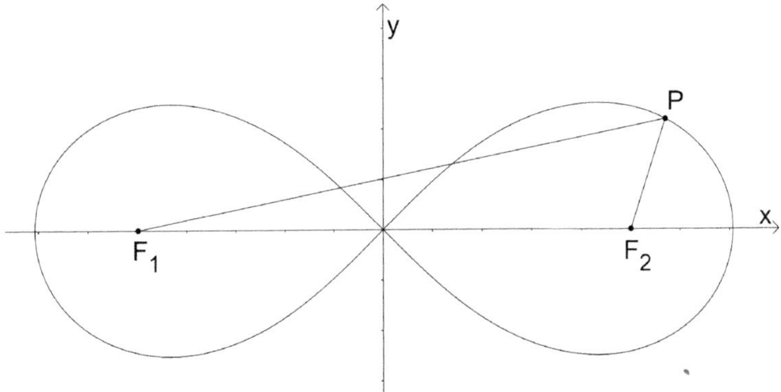


Fig. 7.1: A lemniscate curve with its two foci.

One of the earliest examples of this had already been investigated by Giulio Fagnano and Leonhard Euler in the mid-eighteenth century. Their interest centered on Jakob Bernoulli's lemniscate curve, which in polar coordinates corresponds to the equation $r^2 = \cos 2\theta$. Synthetically, a lemniscate consists of the locus of points in the plane for which the product of the distances from two fixed foci is constant (see Fig. 7.1). This definition makes evident its close relationship with the ellipse and hyperbola, which satisfy corresponding properties for the sum and difference of the two distances. Calculations of the arc length of the lemniscate curve in Cartesian coordinates lead to the elliptic integral

$$\int_0^u \frac{dx}{\sqrt{1-x^4}}.$$

Euler found the relation

$$\int_0^u \frac{dx}{\sqrt{1-x^4}} + \int_0^v \frac{dx}{\sqrt{1-x^4}} = \int_0^{f(u,v)} \frac{dx}{\sqrt{1-x^4}},$$

where

$$f(u,v) = \frac{u\sqrt{1-v^4} + v\sqrt{1-u^4}}{u^2v^2},$$

which is a symmetric algebraic function of the two upper endpoints $u, v$. Gauss later studied the associated elliptic function, thus the function obtained as the inverse of this integral. Although his results only appeared posthumously, Riemann was probably the first to study the related papers in his estate, as he initially took charge of that topic for volume 3 of Gauss's *Werke.*[2]

Gauss noted the analogy with the fundamental trigonometric functions $\sin z$ and $\cos z$, for $z \in \mathbb{C}$. These can be introduced by inverting the integrals:

$$z = \int_0^{\sin z} \frac{dx}{\sqrt{1-x^2}} = \int_{\cos z}^1 \frac{dx}{\sqrt{1-x^2}},$$

where the integrals satisfy the simplest case of an addition rule, namely,

$$\int_0^{\sin u} \frac{dx}{\sqrt{1-x^2}} + \int_0^{\sin v} \frac{dx}{\sqrt{1-x^2}} = \int_0^{\sin(u+v)} \frac{dx}{\sqrt{1-x^2}}.$$

Analogously, these two lemniscatic integrals lead to the special elliptic functions lemniscatic sine and cosine, usually abbreviated sl, cl:

$$z = \int_0^{\mathrm{sl}\, z} \frac{dx}{\sqrt{1-x^4}} = \int_{\mathrm{cl}\, z}^1 \frac{dx}{\sqrt{1-x^4}}.$$

This connection was noted in *Disquisitiones Arithmeticae* [Gauss 1801/1966], where Gauss dealt with the theory of cyclotomic equations. In his posthumous papers, he worked out the theory of sl and cl as elliptic functions with periods that depend on the fundamental constant $\varpi$, which plays the same role as $\pi$ for the trigonometric functions. It can be found directly from tables of integrals for:

$$\varpi = 2 \int_0^1 \frac{dx}{\sqrt{1-x^4}} = 2.62205\ldots,$$

whereas Gauss showed how to find it by means of $\pi$ and the algorithm for the arithmetic-geometric mean (AGM):

$$\varpi = \frac{\pi}{\mathrm{AGM}(1, \sqrt{2})}.$$

Like their trigonometric counterparts, sl and cl are related by a phase shift: $\mathrm{cl}\, z = \mathrm{sl}(\frac{1}{2}\varpi - z)$, corresponding to $\cos z = \sin(\frac{1}{2}\pi - z)$. Their two periods are real and pure imaginary numbers of equal length: $2\varpi, 2\varpi i$.

Riemann's approach to elliptic functions was heavily indebted to Jacobi's work, especially his monograph *Fundamenta nova theoriae functionum ellipticarum* [Jacobi 1829]. Jacobi introduced an amplitude function $\phi(u)$ by inverting the integral

$$u = \int_0^{\phi} \frac{d\theta}{\sqrt{1 - m \sin^2 \theta}},$$

[2] Ernst Schering pointed this out in [Gauss 1863–1933, 3: 492], though in a misleading statement that led Karl Hattendorff to rebuke him in [Hattendorff 1869]; see Section 12.1.

where the modulus $m$ was taken as a constant, normally with $0 < m < 1$. Writing $\text{am}(u, m)$ for $\phi$, he introduced the three fundamental elliptic functions: sin $\text{am}(u, m)$, cos $\text{am}(u, m)$, del $\text{am}(u, m)$ (abbreviated sn, cn, dn), where del $\text{am}(u, m) = \frac{d\,(\text{am}(u,m))}{du}$. From these, he defined nine more related elliptic functions. All have essentially the same periods: one real equal to $2K$ (or $4K$) and the other pure imaginary $2K'i$ (or $4K'i$).

Reading about Jacobi's theory today, one should not imagine that he treated $m$ as a variable; the study of elliptic modular functions only came about after his death. Riemann, in fact, was one of the first to work on modular functions: see the fragment edited by Dedekind, which he published posthumously in [Riemann 1876/1892/1902, 1876: 427–437]. Nor should the reader jump to the conclusion that Jacobi's transcendental functions were meant to induce a lattice of rectangles in $\mathbb{C}$, even if that appears completely self-evident today. In fact, Jacobi made no reference to the complex plane, which was of course the setting for Riemann's entire theory.

Riemann's approach to elliptic functions cannot be found in his published works, but he took up and developed that topic in two lecture courses on complex functions, taught over two semesters. The first of these took place during the academic year 1855–56; he taught the second, one of his last courses, in 1861–62. Based on these, Hermann Stahl published [Riemann 1899], a textbook on elliptic functions to which he added additional explanatory notes.[3] Stahl also included a section dealing with Weierstrass's theory, which by the time his book appeared had completely displaced Jacobi's approach, still another reason why Riemann's theory of elliptic functions never gained a wide audience.

Hattendorff's text provides direct confirmation that Riemann studied Gauss's unpublished works related to elliptic functions.[4] In fact, when he first presented Abel's theorem, Riemann remarked that his proof followed the unpublished argument that Gauss had given for a special case.[5] In his lecture courses, Riemann often made comments about earlier achievements of others, in this case the three most prominent names were Gauss, Abel, and Jacobi.

Following Hattendorff's *Ausarbeitung*, we have the opportunity to see how Riemann dealt with the surface $T$ for a standard elliptic integral, given by

$$w = \int \frac{a\,dz}{\sqrt{(z-\alpha)(z-\beta)(z-\gamma)(z-\delta)}}.$$

This corresponds to a genus one surface with two leaves and four branch points. One can obtain $T$ by taking two cuts, one joining the branch points $z = \alpha$, $z = \beta$,

[3] For Sections I-IV, he used Hattendorff's elaboration of the winter lecture of 1861/2 (Cod. Ms. B. Riemann 32 : 1); for Section V, a transcription of the winter lecture 1855/6 by E. Schering. In addition, he also examined a number of loose sheets on which Riemann had made short notes for the lecture [Riemann 1899, iv-v].

[4] On Riemann's role as a member of the Gauss Commission, see Section 12.1.

[5] If $F(s, z) = 0$ defines an algebraic function $s(z)$, Gauss introduced a new variable $\zeta = G(s, z)$, where $G$ is an entire function. Riemann showed that if $\zeta$ is branched like $s(z)$ then it need only be a rational function; see (Cod. Ms. B. Riemann 32 : 1, pp. 167–170).

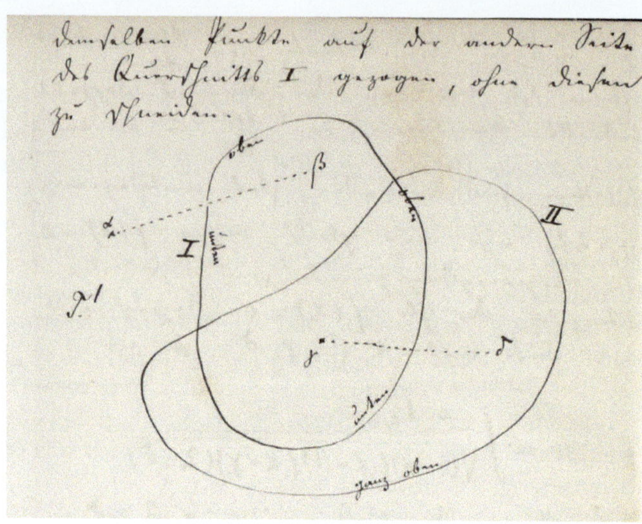


Fig. 7.2: A drawing by Hattendorff from Riemann's lecture course 1861/62.

the other between $z = \gamma$, $z = \delta$, and reattaching the leaves so that a closed curves will pass from one to the other before returning to its starting position (see Fig. 7.2). Riemann then showed how two standard cross cuts will transform $T$ into a simply connected surface $T'$ (Fig. 7.3). The first cross cut I passes between the two pairs of branch points, where the diagram indicates how I passes from one leaf to the other (*oben* and *unten*) as it crosses the virtual cuts joining the pairs of branch points. The second cross cut II now starts at a point $P$ on I and encircles the branch points $z = \gamma$, $z = \delta$, but without intersecting I a second time; it thus remains completely above (*ganz oben*, see Fig. 7.2). The integral $w$ has a periodicity module determined by $K$ and $K_1$, which for any point $w_o$ determine a parallelogram.

Riemann's approach was not only novel but also in some ways influential, even though Stahl's book appeared far too late to affect ongoing research.[6] As Stahl himself noted in the foreword, careful study of elliptic functions (genus $p = 1$) offered the best possible preparation for taking up Riemann's general theory [Riemann 1899, iii]. When mathematicians today study elliptic functions via the lattice of parallelograms determined by their two periods and by constructing the complex torus obtained by identifying their opposite sides, they take up ideas first set forth by Riemann. This aspect of his legacy has been heralded over and again, but in the present context one should also emphasize the very large debt Riemann owed to Jacobi, whose technically daunting results and methods completely permeate the lectures as edited by Karl Hattendorff. This strong influence of Jacobi can easily be overlooked, especially since the modern theory

[6] Felix Klein's influential booklet [Klein 1882/1923] took up the case of elliptic functions in the spirit of Riemann; see Section 11.6.

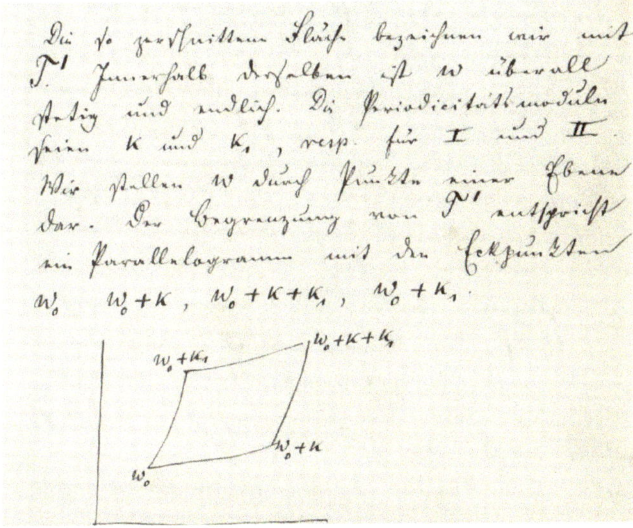
Die so zerschnittene Fläche bezeichnen wir mit T'. Innerhalb derselben ist w überall stetig und endlich. Die Periodicitätsmoduln seien k und $k_1$, resp. für I und II. Wir stellen w durch Punkte einer Ebene dar. Die Begrenzung von T' entspricht ein Parallelogramm mit den Eckpunkten $w_0$, $w_0+k$, $w_0+k+k_1$, $w_0+k_1$.


Fig. 7.3: Hattendorff's drawing of the simply connected surface $T'$.

of elliptic functions discourages mathematicians from looking so far backward. Even Stahl's version of Riemann's lectures contains numerous additions that cannot be found in Hattendorff's text, the following being one such case.[7]

Rather than starting from Legendre's theory of elliptic integrals, a major topic for Riemann, Stahl began by describing the simplest possible single-valued doubly periodic functions in the complex plane. In the spirit of Riemann, he showed one could deduce several of their properties without appealing to any special analytic forms. Thus, if $\phi(v)$ has periods $k_1$ and $k_2$, then

$$\phi(v) = \phi(v + m\,k_1 + n\,k_2), \qquad \text{for} \quad m, n \in \mathbb{Z}.$$

Taking any $v_o \in \mathbb{C}$, the four points $v_o, v_o + k_1, v_o + k_2, v_o + k_1 + k_2$ form the vertices of a parallelogram $P$. Assuming $\phi(v)$ has no zeroes or singularities on $\partial P$, it must have poles on the interior; otherwise $\phi$ would be a bounded holomorphic function, which by Liouville's theorem must be constant. Furthermore, the contour integral $\int_P \phi(v)dv = 0$, so the sum of the residues vanishes by Cauchy's theorem.

This means, that $\phi$ must have at least two poles in $\mathrm{int}(P)$. Furthermore, $\phi'(v)$ is also doubly periodic with periods $k_1$ and $k_2$, which is also the case for $\phi'(v)/\phi(v)$. From this, Stahl concluded that if $\phi$ has $m$ first-order zeroes and $\mu$ simple poles in $P$, then

[7] Laugwitz presented the example below [Laugwitz 1996/1999, 93–94], but without mentioning that it appears in [Riemann 1899, 5–7].

$$\int_{\partial P} \frac{\phi'(v)}{\phi(v)} dv = \int_{\partial P} d\,log\phi(v) = 2\pi i(m - \mu) = 0,$$

hence the number of poles equals the number of zeroes in $P$.[8]

Thus, in the simplest case, $\phi$ has just two first-order poles with opposite residues $A, -A$ at two points $v_1, v_2$ in the interior of $P$, and the derivative $\phi'(v)$ then has second-order poles at those same points, so it must have zeroes at four points in the interior of $P$. One of these, $\alpha_1$, will be the midpoint between the two poles, thus $\alpha_1 = (v_1 + v_2)/2$; the other three are $\alpha_2 = \alpha_1 \pm k_1/2, \quad \alpha_3 = \alpha_1 \pm k_2/2, \quad \alpha_4 = \alpha_1 \pm (k_1 + k_2)/2$; (of the two options $\pm$ only one corresponds to a point in $P$). Thus, $\phi'(\alpha_i) = 0$ and the function $(\phi'(v))^2$ has double zeroes in those four points and poles of order four at $v_1, v_2$. Let the four function values $\phi(\alpha_i) = a_i$ and consider the fourth-degree function $w = (z - a_1)(z - a_2)(z - a_3)(z - a_4)$. For each of the points $v_1, v_2$, one has $z = \infty$, so these are fourth-order poles for $w$; likewise at $\alpha_i$ the factor $z - a_i$ yields a second-order zero. Accordingly, Stahl concluded that the algebraic functions $w$ and $(\phi'(v))^2$ were identical up to a constant $C$, which leads directly to the differential equation

$$C \cdot \frac{dz}{dv} = \sqrt{(z - a_1)(z - a_2)(z - a_3)(z - a_4)}$$

and an elliptic integral of the first kind:

$$v = \int_{z_o}^{z} \frac{dz}{\sqrt{(z - a_1)(z - a_2)(z - a_3)(z - a_4)}},$$

with corresponding elliptic function $z(v)$.

## 7.2 Jacobi's Inversion Problem

Whereas elliptic functions arose by inverting single integrals of rational functions $R(z, w)$, where $w^2 = P_n(z)$ for $n = 3, 4$, the next higher cases ($n = 5, 6$) required inversion of two integrals, since here $p = 2$ by Abel's theorem. Already in 1832, Jacobi hit upon a way to invert two such integrals at once by taking ordered pairs of the arguments, thus $(z_1, z_2)$ and the corresponding sums $(w_1, w_2)$ of two integrals:

$$\int_0^{z_1} \frac{dz}{\sqrt{P(z)}} + \int_0^{z_2} \frac{dz}{\sqrt{P(z)}} = w_1, \int_0^{z_1} \frac{z\,dz}{\sqrt{P(z)}} + \int_0^{z_2} \frac{z\,dz}{\sqrt{P(z)}} = w_2.$$

Here $P(z)$ is a polynomial of degree $n = 5, 6$ without multiple roots.

In the general case $P_n$, these integrals lead to two-sheeted Riemann surfaces of genus $p = (n - 2)/2$ for $n$ even and $p = (n - 1)/2$ when $n$ is odd; thus, for

---

8 The same proof shows that a principal divisor of an elliptic function always vanishes [Brieskorn/Knörrer 1986, 314–315].

hyperelliptic integrals $p \geq 2$. Analogous to the case above, one now requires the inversion of $p$ integrals:

$$\int \frac{z^{k-1}dz}{P_n(z)}, \quad k = 1, 2, \ldots p.$$

For $p = 2$, Jacobi was able to show that $z_1 + z_2$ and $z_1 z_2$ are single-valued functions of $w_1$, $w_2$ and also to establish a corresponding addition theorem. These functions of $(w_1, w_2)$ turn out to have four periods. Explicit formulas for these special hyperelliptic functions were published independently by Adolph Göpfel in 1847 and by Georg Rosenhain four years later. Rosenhain's paper was written for a prize competition announced by the Paris Academy, which sought to promote the theory of Abelian transcendentals. At one point in his text, Rosenhain apologized for not having had the time to complete the calculations needed for a full solution [Rosenhain 1851/1895, 1895: 65–66]. Nevertheless, in 1849 Joseph Liouville reported that Rosenhain's lengthy contribution was worthy of the prize.

Riemann may have known the paper by Adolph Göpfel, which appeared in Crelle, but he definitely studied [Rosenhain 1851/1895], a work sometimes credited with having introduced theta characteristics to solve this problem.[9] It remains an open question whether this technically demanding work influenced Riemann in one way or another. Heinrich Weber pointed to a particular argument in Rosenhain's text that dealt with the quotients of his theta functions. This brought to mind Riemann's derivation of the bilinear relations satisfied by the periods of Abelian integrals, except that their multiplicity only becomes transparent when operating over the complex numbers. Weber considered Rosenhain's complex argument as reflecting only an understanding of certain strands of the corresponding Riemann surface rather than the full object itself [Rosenhain 1851/1895, 92].

In modern language, mathematicians would later describe the case developed by Göpfel and Rosenhain as a Riemann surface branched over six points. If $n = 5$, then the sixth point is $z = \infty$, which is another good reason to begin with the so-called Riemann sphere $\mathbb{C} \cup \infty$ rather than the complex plane $\mathbb{C}$. Carl Neumann made use of the former in his textbook [Neumann 1865/1884], but Riemann's published work referred only to $\mathbb{C}$ (the notion of one-point compactification came, of course, much later). It is worth noting the importance of continuity arguments in this early work as well. Rosenhain appealed to what he called the "principle of continuity," a standard idea among geometers in the nineteenth century. In his lecture course from 1855/56, Riemann cited works by Cauchy and Puiseux for their discussions of the continuity of the roots of algebraic equations as functions of their coefficients as well as their permutations (Cod. Ms. B. Riemann 37, 42–43).

One should keep in mind that Riemann was equally concerned with methods for constructing complex functions in several variables. This was, after all, the

[9] The role of theta characteristics in Riemann's work is discussed in Section 11.4.

purpose of his theta series, which as a practical tool were known to converge very quickly. He and Karl Weierstrass thus shared this interest, even though the latter was strongly opposed to arguments based on geometrical principles or intuitive physical ideas such as Dirichlet's principle. When Weierstrass solved Jacobi's inversion problem for general hyperelliptic integrals in [Weierstrass 1854], he did so by appealing to Abel's theorem and purely algebraic arguments. If a mathematician had told him that these methods sufficed for handling a special class of genus $p$ Riemann surfaces, namely those with two leaves and $n = 2p$ branch points, we can probably imagine Weierstrass's reaction: he would have just shrugged.

Nevertheless, he was more than willing to grant Riemann's his due. In his report for the Berlin Academy, nominating Riemann as a corresponding member, he began by praising the ideas in his dissertation (see the introduction to Chapter 5) and went on to underscore the importance of his more recent work:

> Through his later work, Mr. Riemann has factually proven that his new ideas are not the result of idle speculation, but the healthy fruit of a natural development of science. Using the example of the hypergeometric series – in a treatise that is included in the writings of the Göttingen Society – he showed that even in the case of a transcendent, which has already been dealt with as often as the one mentioned, his methods are still capable of revealing new properties of it. Of even greater importance are the three papers that appeared in the Journal for Mathematics at the end of 1857. In them it is demonstrated that the algebraic functions and the functions arising from them through integration belong to the same size domain, which is characterized by very specific features and is self-contained. A necessary consequence of this is that even the most general periodic functions, whose basic properties have their origin in Abel's theorem about the integrals of algebraic differentials, can all be traced back to the transcendent, which is as simple as it is distinguished by the most remarkable properties, which Göpel, Rosenhain and one of the undersigned had independently arrived at in their investigations into Abel's functions. A theory that leads to such a result needs no further proof of its fruitfulness. [Biermann 2022, 28]

## 7.3 Abelian Functions

Riemann's paper on Abelian functions [Riemann 1857b] has an unusual structure, reflected in the circumstance that it was originally published as three short notes followed by a long paper containing its many new results.[10] One should keep in mind that these four papers constitute Riemann's first publication in *Journal für die reine und angewandte Mathematik*, Germany's only journal exclusively devoted to mathematics. Since very few of its readers knew his dissertation [Riemann 1851/1876], one can easily imagine their shock in trying to understand not only the methods, but also many of the results that he unveiled in this difficult paper. Actually, it appeared as four separate papers: the main body of the text contains two parts, which are preceded by three introductory papers, in the last of which Riemann summarized the topics and results of the

[10] For a detailed technical account of its contents, see [Houzel 1978].

fourth part. He characterized the results in Part I of part four as belonging to the theory of systems of algebraic functions and their integrals that share the same branching structure. Sections 1 to 5 deal with determinations of such systems of functions by means of their discontinuities and branching behavior, whereas 6 to 10 discuss their properties in terms of rational expressions in two variables linked by an algebraic equation. Riemann took up the birational transformations of these functions in Sections 11 to 13, where he underscored his concept of algebraic equations of the same class, including his results on those equations of lowest degree.

The final topic in Part I (Sections 14 to 16) concerns Abel's theorem for everywhere finite integrals of equally branched algebraic functions as this applies to the integration of a system of differential equations. This topic prepares the groundwork for Part II, which introduces $p$-fold infinite $\theta$-series to express Jacobi inversion functions in $p$ variables. The great highlight that follows gives Riemann's general solution of the Jacobi inversion problem for systems of finite integrals derived from algebraic functions represented by a $2p+1$-fold connected closed Riemann surface. As described in Section 3.7, Riemann not only took pains to credit Weierstrass for his solution of this problem in the case of hyperelliptic integrals, he also alluded to the latter's even more recent work, which he assumed would soon be published:

> To what extent the later parts agree with the work presented here, not only with regard to results but also the methods leading to them, will to a large extent only be revealed by the promised detailed presentation of the same. [Riemann 1857b, 102]

Far from marking the end of this story, Riemann's findings and methods launched several waves of new investigations (see [Bottazzini & Gray 2013]). In several respects, he anticipated this, even if many responses went off in directions he and his followers considered less fruitful (see Chapter 11). In his summary, Riemann especially noted that the theory of $\theta$-functions he set forth in Part II did not concern the general case, since one of the parameters required for their definition was subject to an algebraic condition. This meant that for $p > 3$, instead of having $p(p+1)/2$ parameters, there were only $3p-3$. About this, Riemann commented:

> This part of the treatise also constitutes a theory of this special genus of $\theta$-functions; the general $\theta$-functions are excluded here, but can be treated according to a very similar method. [Riemann 1857b, 94]

In the first three sections, which precede Part I, Riemann reviewed the main methods he introduced earlier in his dissertation, though with some important modifications. Earlier he had only touched on surfaces that were not simply connected, whereas in the second note he gave a new definition of connectivity that applied to these cases. Thus, a closed curve that does not bound a region of the surface leads to an integral with periodicity. Riemann's definition reads:

> If $n$ closed curves $a_1, a_2, \ldots, a_n$ can be drawn on a surface $F$, so that they neither individually nor together completely bound any part of $F$, but which, with the addition any other closed curve, do form the complete boundary of part of $F$, then this surface is called $(n+1)$-fold connected. [Riemann 1857b, 85–86]

As noted by Erhard Scholz, Riemann operated with such systems of curves much in the same spirit as in basic homology theory [Scholz 1980, 64–67]. Thus, to show that the connectivity number $n+1$ is well defined, he considered a second system of curves $b_1, b_2, \ldots, b_n$ that do not bound. Taking the system $a_1, a_2, \ldots, a_n$ as above, Riemann systematically interchanged the $b_i$ with $a_i$, showing that at each stage they satisfy they same bounding conditions. Hence, the original system $b_1, b_2, \ldots, b_n$ meets the definition above, namely that adding another curve will produce a region bounded by the system.

In the third note, Riemann brought up Dirichlet's name for the first time in connection with the method Riemann used for proving the existence of complex functions with given conditions:

> As a basis for investigating transcendental functions, it is above all necessary to establish a system of sufficient and independent conditions for their determination. In many cases, particularly with integrals of algebraic functions and their inverse functions, a principle can be used which Dirichlet employed to solve this problem for a function of three variables satisfying Laplace's partial differential equation. He was probably prompted by a similar idea of Gauss, who used this for a number of years in his lectures on forces acting proportionally to the inverse square of the distance. [Riemann 1857b, 1876: 90]

Did Riemann make use of Dirichlet's principle merely as a heuristic device, as Felix Klein later claimed [Klein 1926, 264]? Klein wrote that Weierstrass informed Riemann the principle was flawed [Klein 1895, 492]. This seems unlikely, but Riemann probably realized one needed to employ this argument with care. Immediately after the above passage, he added this qualifying remark:

> For the application to the theory of transcendentals, however, there is one particularly important case for which this principle in its simplest form is not applicable .... This case arises when the function has prescribed discontinuities at points in the region where it is to be determined. ... I will present the principle here in the form required for the intended application and allow myself to refer to the presentation of the same given in my doctoral dissertation. (*ibid.*)

This and other references to [Riemann 1851/1876] led several mathematicians to study that text for the first time. He made another reference to it in the introduction summarizing the contents of what followed. There he called attention to the significance of his notion of a (birational) class of algebraic equations and his concrete results in Section 13 on the equations of lowest degree in a given class. In his dissertation, which focused on characterizing functions by means of their values on the boundary of a region, Riemann had little to say about functions defined over an unbounded surface. He thus began Part I by addressing this situation:

> If $s$ is the root of an irreducible equation of degree $n$, whose coefficients are entire functions of degree $m$ of $z$, then each value of $z$ corresponds to $n$ values of $s$, which change continuously with $z$ wherever they do not become infinite. Therefore, if the branching mode of this function is represented by an unbounded surface $T$ spread out in the $z$-plane, then this surface is $n$-fold in each part of the plane, and $s$ is then a single-valued function of position in this surface.

A bit later, Riemann introduced the notation $F(\overset{n}{s}, \overset{m}{z}) = 0$ to describe this situation. Here one might have expected him to introduce the Riemann number sphere $\mathbb{C} \cup \infty$, but throughout this text he carefully distinguished between $z = \infty$ and finite points in $\mathbb{C}$. Thus, we are to think of the closed surface $T$ with $z = \infty$ lying on each of the $n$ sheets of the surface, unless it happens to be a branch point.

Riemann then alluded to a key theme:

> Every rational function of $s$ and $z$ is evidently also a single-valued function of position in the surface $T$ and thus possesses the same branching pattern as the function $s$, and it will be shown below that the converse is also true.

This cryptic claim seems to say that algebraic functions can be characterized by those Riemann surfaces which are unbounded. The converse statement claim is taken up in Section 5, where he derives an algebraic function based on the properties of such a surface. One should take care to avoid identifying this theorem with the modern formulation, i.e., that the Riemann surfaces of algebraic functions are compact and orientable; those topological concepts only emerged much later.[11]

Part I thus deals with systems of uniformly branched algebraic functions and their integrals, defined by general properties rather than concrete expressions, and introduced by applying Dirichlet's principle. In Section 2 Riemann discussed the behavior of such functions at a simple pole:

> ...for a point on the surface $T$, let a function be said to be *infinitely small of first order* if its logarithm increases by $2\pi i$ during a positive revolution around a region surrounding that point where the function remains finite and non-zero.

At points of higher singularity, one can apply the principle of superposition. Thus, if at a point $P \in T$ a function becomes infinitely small or infinitely large of $n$th order, one can view this as equivalent to having $n$ points of first order that coincide with or come infinitely close to $P$. Citing Cauchy, Riemann wrote down a local Laurent series with finitely many terms for such points: $A \log r + Br^{-1} + Cr^{-2} + \ldots$.

In Section 3, he showed how Dirichlet's principle can be used to describe a class of functions $\omega$ by means of the Laurent series for their poles and their module of periodicity. To begin, Riemann obtained a simply connected surface $T'$ by taking $2p$ cross cuts on the closed surface $T$. The decomposition of $T$ follows the case discussed above in Section 7.1, where here each successive cross cut begins by taking a point on the previous cut. Riemann noted that $\partial T'$, the boundary curve of $T'$, is connected. For the initial data, he sets $z = x + yi$ on $T$ and lets $\alpha + \beta i$ be a function of $x$ and $y$ with poles at a given set of points $\epsilon_v \in T'$. At each of these points $\epsilon_v$, Riemann described the singularity by writing:

$$\phi_v(r_v) = A_v \log r_v + B_v r_v^{-1} + C_v r_v^{-2} + \ldots.$$

[11] Laugwitz evidently thought that Riemann had proved the essential part of this theorem, even though he lacked a great deal of the machinery that is needed for a solid proof, e.g., Camille Jordan's classification of topological surfaces. Rather oddly, Laugwitz cited the proof Riemann gave in his later lecture course from the summer semester of 1861, even though Riemann presented the same argument already in Section 5 [Laugwitz 1996/1999, 90–91]. For a modern proof of this key theorem, see [Springer 2001, 286–289].

These $\epsilon_v$ are then joined by non-intersecting lines $\ell_v$ to a fixed point $O$ whenever the residue $A_v \neq 0$. The function $\alpha + \beta i$ is then so defined on the remainder of $T$ to be continuous outside the lines $\ell_v$ and the $2p$ cross cuts, where there are jumps of $-2\pi A_v$ across the former and given real constants $h^1, h^2, \ldots h^{2p}$ for the latter. Riemann called the latter constants the *periodicity module* of the function. To ensure this periodicity requires that the sum of the residues $A_v$ vanishes. Finally, he noted that the integral

$$\Omega(\alpha) = \int \left( \left( \frac{\partial \alpha}{\partial x} - \frac{\partial \beta}{\partial y} \right)^2 + \left( \frac{\partial \alpha}{\partial y} + \frac{\partial \beta}{\partial x} \right)^2 \right) dT$$

must be finite, a requirement for applying Dirichlet's principle.

Riemann now used this indirectly by defining an everywhere continuous function that agreed with $\alpha + \beta i$ on the simply connected surface $T'$, but had a purely imaginary periodicity module. He then took $\omega$ to be the difference between these two functions, while noting that $\omega$ not only has the same singularities as $\alpha + \beta i$ but also the real parts of their periodicity modules agree, so the latter can be chosen arbitrarily. Once these are fixed, this determines the imaginary part of the periodicity module for $\omega$ as well, so it is defined everywhere on $T$ up to a constant. Many readers surely found this argument either problematic or difficult to follow, so this existence proof proved to be the crucial threshold that separated followers of Riemann from all those who viewed his methods as doubtful.

In Section 4, Riemann turned to a discussion of the three basic types of functions $\omega$ that can arise. Those that are everywhere finite he called integrals of the first kind, in keeping with Legendre's classification of elliptic integrals. These arise by integrating holomorphic functions, and Riemann typically wrote them as $w_i$, where $i = 1, \ldots, p$, since there are $p$ linearly independent integrals of this type on a surface $T$ of genus $p$.[12] Integrals of the second kind have one simple pole, so all terms in the Laurent series vanish at that point except for $Br^{-1}$; Riemann wrote these as $t_j$. The integrals of the third kind have only logarithmic infinities, which appear in pairs whose residues are $A$ and $-A$, thus whose sums vanish.

In Section 5, he carried this analysis over to a consideration of single-valued functions defined on a closed surface $T$ of genus $p$ with $m$ simple poles. Taking $w_i, \quad i = 1, \ldots, p$ to be a basis for $p$ functions of the first kind and $t_j$ for the functions describing the $m$ poles, Riemann wrote these functions in the form:

$$s = \beta_1 t_1 + \beta_2 t_2 + \cdots + \beta_m t_m + \alpha_1 w_1 + \alpha_2 w_2 + \cdots + \alpha_p w_p + C.$$

If $s$ is single-valued, then the coefficients $\alpha, \beta$ must be such that the periodicity module for $s$ vanishes. These $2p$ moduli depend linearly on the $m + p + 1$ coefficients, so this leads to a homogeneous system of $2p$ linear equations, which can

---

[12] As is well known, Clebsch was the first to use the term genus (*Geschlecht*) in connection with Riemann's theory; see [Lê 2020]. Riemann always referred instead to the connectivity of a surface, which has the disadvantage that not everyone followed his definition. Despite the anachronism, I will use the term genus freely here.

always be solved when $m \geq p+1$. This condition, usually written $m > p$, came to be known as the Riemann inequality. It gives a lower bound $h = m - p + 1$ for the dimension of the linear space of meromorphic functions on a Riemann surface. The corresponding equality, due to Gustav Roch, takes into account that the $2p$ linear equations for the moduli may not be linearly independent, as Riemann himself noted (see Section 11.4). This can happen when one or more of the $\beta_i$ vanish, as when some of the poles lie on an algebraic curve of lower order than would be the case were these points in general position.

If $m \geq p+1$, the dimension of the linear space of meromorphic functions $h \geq 2$, so it must contain nonconstant functions. Since the $m$ poles can lie anywhere on the surface, each represents an additional degree of freedom, so the totality of single-valued meromorphic functions with $m$ simple poles on a closed surface of genus $p$ depends on $2m-p+1$ constants. As Riemann expressed this:

> The number of arbitrary constants contained in a function $s$ that becomes infinite of the first order only for $m$ points of the surface $T$ and otherwise remains continuous is in all cases $2m - p + 1$.[13] [Riemann 1857b, 108]

Riemann next turned in Section 6 to a function $s$ defined implicitly by a polynomial equation $F(\overset{n}{s}, \overset{m}{z}) = 0$. The rather obscure treatment that follows aimed to illuminate the branching behavior of the corresponding Riemann surface, though the discussion reveals why these surfaces are not easy to describe. An algebraic geometer will immediately recognize that $F = 0$ defines a complex one-dimensional subset of $\mathbb{C}^2$, but this is not the surface Riemann means, nor is it the projective closure of this complex curve. What Riemann seems to have meant was the surface obtained by desingularizing the curve, which amounted to separating out its double points and then removing these. He does not state this explicitly, but this would appear to be the upshot of his discussion.

He refers to $s$ as the root of an equation of degree $n$ whose coefficients are entire functions of degree $m$ in $z$, only afterward introducing the notation above. To show this, he assumes the given surface has $n$ leaves, so when $z$ is not a branch point there are $n$ values $s_1, \dots, s_n$. Taking $\sigma$ to be an arbitrary quantity, he forms the product $(\sigma - s_1(z))(\sigma - s_2(z)) \dots (\sigma - s_n(z))$, which is single-valued when $z$ is not a branch point. Assuming the $m$ poles $\zeta_1, \dots, \zeta_m$ are all finite, Riemann forms the product

$$(\sigma - s_1(z)) \dots (\sigma - s_n(z)(z - \zeta_1) \dots, (z - \zeta_m),$$

which is finite for finite $z$ and infinite of order $m$ when $z = \infty$. Thus, $s$ satisfies an equation of the form $F(\overset{n}{s}, \overset{m}{z}) = 0$, which Riemann showed to be equivalent to the power of an irreducible algebraic equation.

He noted further that these same conclusions apply for the derivatives of the functions $\omega$ introduced by means of Dirchlet's principle. By construction, $\frac{d\omega}{dz}$ will be single-valued because along a cross section $\ell$ the difference between the two

---

[13] Klein emphasized that this space of functions forms a connected manifold [Klein 1882/1923, 542].

values of $\omega$ on either side remains constant. From this, Riemann concluded that a given Riemann surface $T$ determines a system of algebraic functions branched like $T$ and the integrals of such functions. The properties of any $\omega$ thus depend on the connectivity of $T$, whereas this system of functions depends only on the location of its branch points.

For an algebraic function $s$ given by $F(\overset{n}{s}, \overset{m}{z}) = 0$, Riemann calculated the number of branch points $\mu$, treating those of higher order as multiple simple branch points. For these, $s$ takes on fewer than $n$ values, which means that

$$F(s) = a_0 s^n + a_1 s^{n-1} + \cdots + a_n = 0$$

and $F'(s) = 0$ have common roots. The maximal number of branch points is then $w = 2m(n-1)$, though this depends on their position on $T$. One must also take the possibility of double points into account, and for $r$ of these one gets $w = 2m(n-1) - 2r$. In Section 7, Riemann related this number to the genus $p$ of a closed surface $T$. Although his result was purely topological, he obtained it by analytic means.[14]

After taking $2p$ cross sections to decompose $T$, he applied Dirichlet's principle to the simply connected surface $T'$ by defining a suitable branch of $\log z$ on an equivalent circle in the $\zeta$-plane, while making use of the Riemann mapping theorem. Integrating around its boundary, he obtained:

$$\int d \log \frac{dz}{d\zeta} = 2\pi i(w - 2n).$$

Taking $s$ to be a segment on the boundary of $T'$ and $\sigma$ to be the corresponding segment on the circle, one has:

$$\log \frac{dz}{d\zeta} = \log \frac{dz}{ds} + \log \frac{ds}{d\sigma} - \log \frac{d\zeta}{d\sigma}.$$

Integrating this relation over the entire boundary then yields:

$$\int d \log \frac{dz}{d\zeta} = (2p-1)2\pi i + 0 - 2\pi i = (2p-2) \cdot 2\pi i,$$

thus

$$2\pi i(w - 2n) = (2p-2)2\pi i,$$

and therefore $w - 2n = 2(p-1)$. Comparing this with the result derived in Section 6 ($w = 2m(n-1) - 2r$), Riemann obtained the genus of $T$ as

$$p = (n-1)(m-1) - r.$$

[14] In [Riemann 1876/1892/1902, 1892: 143], Heinrich Weber noted that Carl Neumann obtained the formula $w - 2n = 2(p-1)$ without appealing to Dirichlet's principle in the second edition of [Neumann 1865/1884].

In Section 8 Riemann again considered a function $s$ with $m$ simple poles derived from the relation $F(\overset{n}{s}, \overset{m}{z}) = 0$. Such an $s$ depends on $m-p+1$ parameters and he showed how it can be represented as a rational function of $s$ and $z$. The same also applies for $\frac{d\omega}{dz}$, and in Section 9 Riemann showed how to obtain a suitable system of such rational functions.

If $\omega$ is everywhere finite, then $\frac{d\omega}{dz}$ will be infinite at its branch points and otherwise always finite. The denominator of $\frac{d\omega}{dz}$ must have the property that it vanishes at the branch points and double points, which is the case for

$$\frac{\partial F}{\partial s} = a_0 n s^{n-1} + a_1(n-1)s^{n-2} + \cdots + a_{n-1}.$$

To ensure that $\frac{d\omega}{dz}$ remains finite except at branch points and infinitesimally small of second order for $z = \infty$, the numerator $\phi$ must be an entire function that vanishes at the $r$ double points and has degree less than that of $F(\overset{n}{s}, \overset{m}{z})$. Riemann therefore introduced

$$w = \int \frac{\phi(\overset{n-2}{s}, \overset{m-2}{z})dz}{\frac{\partial F}{\partial s}} = -\int \frac{\phi(\overset{n-2}{s}, \overset{m-2}{z})ds}{\frac{\partial F}{\partial z}}.$$

Furthermore, these $\phi$ depend on $(n-1)(m-1) - r = p$ arbitrary constants, corresponding to the number of linearly independent integrals of the first kind. One can thus find a basis for these $\phi$ functions, from which Riemann derived the general form:

$$w = \alpha_1 w_1 + \alpha_2 w_2 + \cdots + \alpha_p w_p + C.$$

He remarked further that one can find similar construction for the other $\omega$ functions corresponding to integrals of the second and third kinds, which he omitted to do.

In Section 10, Riemann made some observations regarding quotients of such functions. Each of the $\phi_i$ vanishes at $m(n-2) + n(m-2) - 2r = 2(p-1)$ points of $F = 0$, which can be used to determine quotients that become infinite at $p$ or fewer points. In the following Section 11 Riemann discussed the relationship between two algebraic functions, given by $F(\overset{n}{s}, \overset{m}{z}) = 0$ and $F(\overset{n_1}{s_1}, \overset{m_1}{z_1}) = 0$, with the same branching behavior as that of a given surface $T$. He used his mapping theorem to compare these functions on another surface $T_1$ to conclude that each could be transformed into the other by means of a birational mapping. Taking their corresponding data into account, he deduced that

$$(n_1 - 1)(m_1 - 1) - r_1 = (n-1)(m-1) - r = p,$$

thus two functions that are birationally equivalent have the same genus.

As Riemann then noted in the following Section 12, the converse does not hold. He began with this definition:

> Consider all irreducible algebraic equations in two variables, such as $F(s,z) = 0$ and $F_1(s_1, z_1) = 0$, as belonging to the same class if there are rational substitutions of $s_1$ and $z_1$ for $s$ and $z$ such that $F(s,z) = 0$ transforms into $F_1(s_1, z_1) = 0$ and at the same time $s_1$ and $z_1$ are rational functions of $s$ and $z$.

In Section 13, Riemann presented a first classification for birationally equivalent functions based on those of lowest degree and given genus $p$. If one takes any rational function $\zeta$ of $s$ and $z$, then this determines

> a system of equally branched algebraic functions. In this way, every equation evidently leads to a class of systems of equally branched algebraic functions. These can be transformed into one another by introducing a function of the system as an independently variable quantity, and indeed all equations of a class lead to the same class of systems of algebraic functions, and conversely every class of such systems leads to a class of equations.

Riemann considered the general case where $\zeta$ has $\mu$ leaves on a $(2p+1)$-connected surface. The number of branch points is then $2(\mu+p-1)$, whereas $\zeta$ depends on $2\mu-p+1$ complex constants, leaving $2\mu+2p-2-(2\mu-p+1) = 3p-3$ undetermined. Since these remaining parameters can assume arbitrary values, Riemann concluded that the class of systems of equally branched $(2p+1)$-connected $\mu$-valued functions and the class of algebraic equations belonging to it depends on $3p-3$ continuously changing quantities, which he called the moduli of this class. He observed that this formula holds for $p \geq 2$; for $p = 1$ it equals 1. Today there is a vast literature dealing with moduli spaces, a term Riemann never used.[15]

Starting with Section 14, Riemann's text becomes increasingly complex, making it pointless to continue describing it section by section. His arguments in closing out Part I were devoted to his particular version of Abel's theorem, for which he gave necessary and sufficient conditions that two finite sets of points of the same size on a Riemann surface could be the zeros and poles of a meromorphic function. This result played a key role in his approach to solving Jacobi's inversion problem, the main highlight in Part II to be discussed below. The key technical tool Riemann needed to accomplish this was his theta function for any finite number of complex variables.

As already noted in Section 1.3, Riemann was familiar with Jacobi's theory of theta functions even before he took the latter's course on elliptic functions in Berlin. For he had been studying the latter's formidable monograph, *Fundamenta nova theoriae functionum ellipticarum* [Jacobi 1829], before he arrived in the Prussian capital. This classic study introduced the three fundamental Jacobi elliptic functions $(\mathrm{sn}, \mathrm{cn}, \mathrm{dn})$ from which nine others are derived; later, in Section 52, Jacobi introduced a first example of a theta function [Bottazzini & Gray 2013, 46–48].

Jacobi explored the properties of sn, cn, and dn by representing them by means of infinite sums and products. To accomplish this, he first introduced two new functions $s$ and $q$, which are defined in terms of ratios of $u, K, K'$:

15 Klein was probably the first to emphasize that the moduli form a higher-dimensional connected manifold; see [Klein 1882/1923, 558–559].

$$s = e^{\frac{i\pi u}{K}}, \qquad q = e^{-\frac{\pi K'}{K}}.$$

Replacing $K'/K$ with the complex number $\tau$ having positive imaginary part, one can define $q = e^{i\pi\tau}$, leading to the Jacobi theta function

$$\theta(z) = \sum_{n=-\infty}^{+\infty} (-1)^n q^{n^2} e^{2\pi i n z}.$$

Jacobi then showed one can construct a function with periods 1 and $\tau$ by forming the quotient $\theta(z+\frac{1}{2})/\theta(z)$. Riemann adopted this same approach for elliptic functions, but more significantly he extended it to define $\theta$-functions in $p$ variables, one of his principal tools for solving Jacobi's general inversion problem for Abelian integrals.

An important difference between them is that Riemann's $\theta(v_1, v_2, \ldots v_p; B_\mu^\nu)$ depends on a $p\times p$ symmetric matrix $B_\mu^\nu = (a_{\mu\nu})$, which he could utilize as input data taken from $p$ Abelian integrals of the first kind. Without taking account of its precise form, we merely note the fact that a finite-valued $\theta$-function is determined up to a constant factor by two properties:

1) $\theta(v_\mu) = \theta(v_\mu + \pi i)$, and

2) $\theta(v_1, v_2, \ldots v_p) = (e^{2v_\mu + a_{\mu\mu}})\, \theta(v_1 + a_{1\mu}, v_2 + a_{2\mu}, \ldots v_p + a_{p\mu})$, where the terms $a_{n\mu}$ refer to the $n$th column in the matrix $(a_{\nu\mu})$. These columns form a basis for the periods of $p$ Abelian integrals of the first kind, $u_1, \ldots, u_p$. Riemann chose a careful system of cross cuts of $T$ leading to $p$ pairs of curves $a_\nu$, $b_\nu$ and a special binary relation between the periods for the integrals $u_\mu$, $\mu = 1, \ldots, p$. He chose these so that the period matrices were $A_\mu^\nu = i\pi\delta_{\mu\nu}$ and $B_\mu^\nu = a_{\mu\nu}$, where the latter is symmetric and its real part is negative definite. Only the matrix $B_\mu^\nu = a_{\mu\nu}$ enters into $\theta$. Conditions 1) and 2) show that $\theta$ satisfies certain periodicity relations, which in modern terminology determines a lattice $\Lambda$ on $\mathbb{C}^p$. Riemann referred instead to systems with congruency relations (Section 15).

In Section 18, he explained how to evaluate $\theta$ in the present setting. In place of the $v_\mu$ one sets $u_\mu$ evaluated at a fixed lower endpoint $z_o$. This $\theta(u_\mu; B_\mu^\nu)$ can then be viewed as a mapping $T \longrightarrow \mathbb{C}^p/\Lambda$ defined by $z \mapsto (u_\mu(z))$. Riemann focused attention on its zeros (the $\theta$ divisor in modern terms). Taking fixed arbitrary points $e = (e_1, \ldots, e_p)$, he showed that if $\theta(u_\mu - e_\mu)$ does not vanish identically for all $z \in T$, then $\theta(u_\mu - e_\mu) = 0$ for exactly $p$ points $\eta_1, \ldots, \eta_p \in T$, which vary for each choice of $e = (e_1, \ldots, e_p)$. To prove this, Riemann made use of contour integrals for $d \log \theta$ on the decomposed surface obtained from $T$, an argument similar to the one he used to derive the relationship between the number of branch points and the genus of a closed surface (see above).

If $E \subset T$ consists of those points $e = (e_1, \ldots, e_p)$ for which $\theta(u_\mu - e_\mu) = 0$ for all $z \in T$, then let $X = T - E$. Riemann's argument establishes a bijection between $X$ and the congruence classes for sums of $p$ integrals $u_\mu$ via the points $\eta_\nu$ for which $\theta(u_\mu - e_\mu) = 0$, where each $\eta_\nu$ is transformed in $\sum u_\mu(\eta_\nu)$. Conversely, he argued that if $\theta(r) = 0$, then $r \equiv \sum_1^{p-1} \eta_\nu$, where the points

$\eta_1, \eta_2, \ldots, \eta_{p-1}$ are uniquely determined for each such $r$. These results drew on Riemann's version of Abel's theorem, which he proved in Sections 14 and 15. His analysis of the zeros for this theta function was somewhat incomplete, causing him to return to this topic in [Riemann 1866/1876].

As noted in Section 3.7, Riemann was well aware that this work presented a major challenge for Berlin's leading analyst, Karl Weierstrass. Through Borchardt, he also knew that Weierstrass was working on a sequel to his earlier paper in which he solved the Jacobi inversion problem for general hyperelliptic integrals (Section 7.2). Riemann later learned that he had already submitted a paper on the general case, but decided to withdraw after reading what Riemann had accomplished. Two years later, in 1859, when he nominated Riemann to become a corresponding member of the Berlin Academy, Weierstrass sang praises for the originality of his work in the field of complex functions (see the quote in the introduction to Chapter 5).

As time passed, Weierstrass's views gradually hardened; he later expressed these in an oft-cited letter to Hermann Amandus Schwarz, written on October 3, 1875:

> The more I think about the principles of function theory – and I do so incessantly – the the firmer becomes my conviction that this must be based on the foundation of algebraic truths and that, consequently, it is not the proper way if, instead of building on simple and fundamental algebraic theorems, one appeals to the "transcendental," to express myself briefly – however seductive on first glance ... that manner of thinking might be, by which Riemann has discovered so many of the most important properties of algebraic functions. (That any route must be permitted to the researcher during the period of seeking is self-evident; what is at issue here is merely the systematic foundation.) [Neuenschwander 1981c, 96–97]

In his posthumously published *Werke*, one can also read Weierstrass's own explanation for his decision to withdraw the paper he wrote in 1857 from publication:

> I had already presented a direct solution to this problem in a detailed treatise to the Berlin Academy in the summer of 1857. However, I withdrew the manuscript, which had already been submitted to the printer, because a few weeks later Riemann published a paper on the same problem but based on completely different principles than mine, and which did not immediately reveal that its results completely agreed with mine. Proving this required several investigations, mainly of an algebraic nature, which were not entirely easy for me to carry out and took a great deal of time. However, after this difficulty had been resolved, a thorough revision of my treatise seemed necessary. ... not until the end of 1869 was I able to give the solution to the general inversion problem the form in which I have presented it in my lectures since then. [Weierstrass 1902, 9–10]

# Chapter 8
# Zeta Function and Paris Prize

Soon after Riemann was appointed as Dirichlet's successor in 1859, he took the opportunity to visit with mathematicians in the capital cities of Prussia and France (Sections 4.2 and 4.3). He was well received in both cities, experiences that inspired him to write two shorter papers that drew relatively little attention at the time. Today these are among his most famous works, though for very different reasons. In the first, he introduced the Riemann zeta function, obtained by extending Euler's zeta function to the complex numbers. The title accurately conveys his larger aim in doing so, namely to estimate "the number of prime numbers below a given quantity" (Ueber die Anzahl der Primzahlen unter einer gegebenen Grösse) [Riemann 1859/1876]. Riemann here attempted to refine earlier approximations of the distribution of primes, producing an intriguing string of ideas, but no definitive findings. As Harold M. Edwards put it: "The real contribution ... lay not in its results but in its methods" Although the text is only eight pages long, the arguments in it are exceedingly difficult to follow.[1]

The background that led to Riemann's other paper, [Riemann 1876b], remains far less clear, nor can anyone say with assurance what motivated him to write it. Given the fact that he had long been living on the brink of poverty, fame and fortune may have been two important factors. During his trip to Paris in the spring of 1860, he met a number of distinguished mathematicians, which makes it likely that one or more of them informed Riemann that the French Academy had announced a prize problem involving the conduction of heat in a homogeneous body. As an expert in this area, he may have found the problem itself attractive, particularly since solving it would bring a handsome reward: a gold medal valued at 3000 francs. Some commentators, however, have implied that Riemann had an entirely different reason for writing this paper, namely to showcase his unpublished transformation theory for quadratic differential forms, a theme he only hinted at in his Habilitation lecture from 1854. This aspect of the Paris prize paper, which like his lecture on the foundations of geometry

[1] Those who wish to try should take advantage of the expert guidance offered in [Mazur/Stein 2016]; whereas Edwards1974 is a technically more demanding guidebook.



D. E. Rowe, *Bernhard Riemann: His Life and Wondrous Mathematical Legacy*,
https://doi.org/10.1007/978-3-032-25457-3_9

remained unknown at the time of Riemann's death, was what attracted the interest of later mathematicians. Lack of relevant source material makes it simply impossible to determine whether this was a motivating factor, which only adds to the mystery. In any event, Riemann submitted this paper to the Academy at the last minute, namely on the closing date of 1 July 1861. He also apologized for writing in Latin rather than in French. Since the arguments he presented were incomplete, his submission failed to win the prize.

## 8.1 Euler's Zeta Function

Riemann had undoubtedly already encountered Euler's version of the zeta function at an early age, as Euler put it on full display in Volume 1 of his *Introductio in analysin infinitorum* [Euler 1748]. In this celebrated "introductory" monograph, Euler defined special transcendental functions – logarithm, exponential, and trigonometric functions – without any recourse to differentiation or integration. In short, this was a truly high-brow pre-calculus textbook. In particular, he derived the famous relation

$$e^{i\phi} = \cos\phi + i\sin\phi\,,$$

from which follows $e^{i\pi} + 1 = 0$, sometimes called the most beautiful formula in mathematics.

In Chapter XV of his *Introductio*, Euler gave several examples of results derived from his product formula for the real zeta function:

$$\zeta(s) = \sum_{n=1}^{\infty} \frac{1}{n^s} = \prod_p \frac{1}{1-p^{-s}},$$

where the product is over all primes $p$ [Ayoub 1974]. The proof is another example of Euler's eye for uncanny formal manipulations. Here, he peeled off each term in $\zeta(s)$ starting with the second, generating the desired product on the right in the end. Thus, he started with

$$\frac{1}{2^s}\zeta(s) = \frac{1}{2^s} + \frac{1}{4^s} + \frac{1}{6^s} + \ldots,$$

which subtracted from the series for $\zeta(s)$ yields:

$$(1 - \frac{1}{2^s})\zeta(s) = 1 + \frac{1}{3^s} + \frac{1}{5^s} + \frac{1}{7^s} + \ldots.$$

Repeating the same procedure using the second term on the right, he then removed all the terms with multiples of 3 in the denominator:

$$(1 - \frac{1}{3^s})(1 - \frac{1}{2^s})\zeta(s) = 1 + + \frac{1}{5^s} + \frac{1}{7^s} + \frac{1}{11^s} + \ldots.$$

One easily sees how this algorithm mimics the famous sieve of Eratosthenes, except that Euler used it to eliminate all the terms on the right, save the number 1:

$$\cdots + (1 - \frac{1}{11^s})(1 - \frac{1}{7^s})(1 - \frac{1}{5^s})(1 - \frac{1}{3^s})(1 - \frac{1}{2^s})\zeta(s) = 1,$$

hence

$$\zeta(s) = \prod_p \frac{1}{1 - p^{-s}}.$$

Ancient Greek mathematicians already knew one could never count all the prime numbers. By way of comparison, Archimedes might have expressed this by alluding to the easily bounded number of sand grains in an imaginary sphere with radius equal to the distance from the Earth to the Sun. Before Georg Cantor brought forth the notion of completed infinities, mathematicians thought in terms of objects without bounds. Thus, following Gauss or Kronecker, we cannot conceive of all prime numbers (or any other infinite set), but we can know that there are infinitely many. Euclid presented the classical proof, practically a one-liner: if $p_1, p_2, \ldots, p_n$ were all the primes, then $P = (p_1 \cdot p_2 \cdot \cdots \cdot p_n) + 1$ would not be divisible by any prime, q.e.d. Euler showed the divergence of the zeta function for $s = 1$, which yields the harmonic series

$$\sum_{n=1}^{\infty} \frac{1}{n} = 1 + \frac{1}{2} + \frac{1}{3} + \cdots + \frac{1}{n} + \ldots.$$

In beginning courses in calculus, students encounter this as the critical case for the $p$-series

$$\sum_{n=1}^{\infty} \frac{1}{n^p},$$

which converges for $p > 1$ and diverges for $p \leq 1$.

Ordinary intuition might suggest that if we deleted a large infinite number of terms from the harmonic series the result ought to converge, as the $p$-series does for $p = 1.0000000001$. So what happens if we toss out all the terms where $n$ is not a prime number? It turns out, as Euler showed using his product formula, that $1/2 + 1/3 + 1/5 + 1/7 + 1/11 + \ldots$ diverges. We might be tempted to interpret Euler's finding as showing that, in some sense, the primes are densely distributed among the natural numbers. But here again our intuition tells us very little, since we are dealing with two divergent series, one highly regular and the other sporadic. The more usual situation would involve two convergent series with known limits, in which case one can investigate how quickly they converge. Some converge very slowly, as does the famous Leibniz series

$$1 - \frac{1}{3} + \frac{1}{5} - \frac{1}{7} + \cdots = \frac{\pi}{4}.$$

(Here, in accordance with Riemann's rearrangement theorem (Section 6.1), the ordering of the terms is essential.) A series can diverge very slowly, too. For example, were we to sum the first 50 million terms of the Euler series the result would be a number less than four! [Zagier 1977, 13]. Empirical findings suggest the likely existence of arbitrarily large pairs of twin primes, though this remains an open problem. On the one hand, one can find stretches of composite numbers exceeding any given length, since there no primes lying between $n!+2$ and $n!+n$. In 1915, the Norwegian mathematician Viggo Brun proved that no matter how many twin primes might exist, the sum of their reciprocals must converge.

Riemann's mathematical interests centered on analysis and physics. Unlike Gauss, Dirichlet, and Dedekind, number theory held no great appeal for him, making his paper on the zeta function exceptional.[2] Yet this singular work was hardly a conventional contribution to number theory; on the contrary, Riemann's paper was a virtuoso display of his new methods in complex analysis. He thus set himself the task of finding a new analytic expression for approximating the number of primes below a given large number $x$. With such an asymptotic representation he hoped to reveal a global pattern underlying this distribution. The preceding remarks can be taken as hints pointing to the peculiar difficulties this type of problem poses.

We can only guess as to when Riemann first learned about questions involving the distribution of primes. All indications, however, point to a date long before he finally wrote his famous paper "On the Number of Primes less than a given Quantity" [Riemann 1859/1876]. Today that paper's fame has little to do with Riemann's principal aim and results; instead, mathematicians know about it in connection with Hilbert's eighth Paris problem [Hilbert 1900/1935], a conjecture Riemann formulated, but was unable to prove. Riemann's hypothesis states that all the "non-trivial zeros" of his zeta function have their real part lying on the line $s = 1/2$. When Hilbert announced this challenge in 1900, it was virtually the only remaining mystery in Riemann's paper, and he himself took it to be correct. In fact, he suggested that several other problems were likely to be within reach once Riemann's claims regarding the distribution of the primes had been fully proved, including Goldbach's conjecture and the infinitude of twin primes (both remain open questions today). In 1998 Steve Smale placed Riemann's hypothesis at the top of his list of eighteen important unsolved problems, and the Clay Institute later made it one of the seven millennium problems for the mathematics of the twenty-first century. Since this book restricts its focus to the events of Riemann's lifetime and the decade following his death in 1866, I will only attempt to show how the Riemann hypothesis first arose.

---

[2] André Weil went so far as to maintain that Riemann showed "a complete lack of interest for number theory and algebra," despite his exposure to the work of Dirichlet, Eisenstein, and Dedekind [Dugac 1976, 210].

## 8.2 Context and Text

In 1859 Riemann became a corresponding member of the Berlin Academy, an honor that led to his memorable trip to Berlin in the company of Dedekind (see Section 4.2). As an act of gratitude, he sent Weierstrass his cryptic paper on the zeta function. The mysteries surrounding this short paper are many, beginning with what and when he learned about the conjectures that pointed in the direction of the prime number theorem, which was eventually proved by Jacques Hadamard in 1896. In his introductory remarks, Riemann alluded to the interest taken by both Gauss and Dirichlet in this problem, though neither of them published directly on it. One could well imagine that Riemann's own interest in the global distribution of primes arose from reading *Théorie des nombres* [Legendre 1830], if he indeed devoured both volumes of this classic as a pupil in the Lüneburg Gymnasium. His mathematics teacher loaned him this work, but probably only the first volume.[3]

As described in Section 4.2, Riemann's paper was prompted by discussions with Kronecker when he and Dedekind visited Berlin in September 1859. This emerges from the letter Riemann sent to Weierstrass not long after he had returned to Göttingen (see below). He began his text with a note of thanks:

> I believe I can best express my gratitude for the honor which the Academy has bestowed on me in naming me as one of its correspondents by immediately availing myself of the privilege this entails to communicate an investigation of the frequency of prime numbers, a subject which because of the interest taken by Gauss and Dirichlet in it over many years seems not wholly unworthy of such a communication. ... [Riemann 1859/1876, 145]

Riemann's primary goal was to shed light on an older idea that had long interested Gauss: the prime number conjecture. Don Zagier gave a simple empirical argument as motivation for this idea [Zagier 1977, 9]. If one calculates $\pi(x)$ – the number of primes less than or equal to $x$ – for powers of 10, thus $\pi(10) = 4, \pi(100) = 25, \pi(1000) = 168$, etc., a clear pattern emerges among the ratios $x/\pi(x)$. For each increase in order of magnitude for $x$, this ratio increases by a constant factor, $x/\pi(x) \approx 2.3$, which coincides with $\log_e(10) \approx 2.3$. Thus, one might plausibly guess that

$$\pi(x) \approx \frac{x}{\log x}, \quad \text{or} \quad \lim_{x\longrightarrow\infty} \frac{\pi(x)}{\frac{x}{\log x}} = 1,$$

the conjecture Hadamard eventually proved.

Gauss did not, of course, know $\pi(10^n)$ for powers greater than $n = 6$, but he liked to calculate logarithms. When he was 14, his patron Carl Wilhelm Fer-

[3] Laugwitz noted that André Weil found evidence suggesting that Riemann's interest may have been sparked during his two years in Berlin when he often conversed with Eisenstein [Laugwitz 1996/1999, 173]. Eisenstein wrote down a proof of the functional equation for the Dirichlet series $L(s)$, though this was dated 7 April 1849, when Riemann was about to leave Berlin [Weil 1989]. He had, in any case, few if any contacts with Eisenstein after the events of 1848; see also Section 1.5.

dinand, the Duke of Brunswick, gave him a collection of mathematics books, which included a book listing natural logarithms for both natural numbers and for primes. This source may well have led Gauss to formulate his original hunch, which he recalled in a letter to Franz Encke from December 24, 1849 (see Section 4.2). Since Riemann had already returned to Göttingen by this time, he must have learned about the content of this letter while studying there, though probably not from Gauss himself. Far more likely, he heard about Gauss's long-standing interest in the distribution of primes through Gauss's junior colleague, Benjamin Goldschmidt, who long before took over the project of finding large primes from his former mentor.

Another obvious source of information for Riemann would have been through Dirichlet, who in 1837 published seminal work that helped launch analytic number theory [Merzbach 2018, 105–108]. In that year, he sketched a proof that if $a$ and $d$ are coprime, the arithmetic progression $a + n\,d$, must contain infinitely many primes; the full proof by means of his $L$ series appeared in [Dirichlet 1839]. Unlike power series, whose regions of convergence are circles, Dirichlet series converge in half-planes of the form $\mathrm{Re}(s) > c$. The most famous Dirichlet series is the classical zeta function, which converges in the half-plane $\mathrm{Re}(s) > 1$. A general Dirichlet series depends on an odd prime $p$, a second prime $q$ not equal to $p$, and the roots of unity $\omega$ obtained from $\omega^{p-1} = 1$. Letting $\gamma$ be a fixed natural number that depends on $q$, one can introduce the infinite geometric series:

$$1 + \omega^{\gamma}\frac{1}{q^s} + \omega^{2\gamma}\frac{1}{q^{2s}} + \omega^{3\gamma}\frac{1}{q^{3s}} + \cdots = \frac{1}{1 - \omega^{\gamma}\frac{1}{q^s}}.$$

Using this, Dirichlet introduced the product formula

$$\prod \frac{1}{1 - \omega^{\gamma}\frac{1}{q^s}} = \sum \omega^{\gamma}\frac{1}{n^s} = L,$$

where the sum is taken over primes except for $p$, and $n$ is any number not divisible by $p$. This expression represents $p-1$ distinct equations, which Dirichlet wrote as $L_0, L_1, \ldots, L_{p-2}$. He used these series, which bear his name, in proving asymptotic properties of prime numbers in arithmetic progressions. For example, he could show why prime numbers of the form $4n + 1$ and $4n + 3$ appear in approximately the same ratios on a large scale. Even though Riemann's research interests were generally far removed from classical number theory, these analytic results surely would have caught his attention at some point during his student years. Exactly when, however, remains a mystery.[4]

As never before, Riemann revealed his incredible virtuosity as an analyst in [Riemann 1859/1876]. He drew on methods in Fourier series, theta-series, complex integration, and an array of formal manipulations reminiscent of Jacobi. Riemann realized, of course, that he could not simply pass to the complex domain by treating the exponent in

[4] For other possible sources of information, which may have influenced Riemann's work on the zeta function, see [Laugwitz 1996/1999, 172–173].

$$\zeta(s) = \sum_{n=1}^{\infty} \frac{1}{n^s}$$

as a complex number, since this expression only converges when $\mathrm{Re}(s) > 1$. Ultimately, he hoped to make use of the zeros of $\zeta$, none of which lie in that region of the complex plane. So his first task was to find a way to extend the domain of definition of $\zeta(s)$ to all of $\mathbb{C}$. By using the gamma function, Riemann showed that $\zeta$ can, indeed, be extended holomorphically to the entire complex plane, except for a single pole at $s = 1$.

Already here some confusion may arise. To begin with, Riemann wrote $\Pi(s)$ in place of the usual $\Gamma(s+1)$, which was first introduced by Legendre and eventually universally adopted. Gauss had instead written $\Pi(s)$ for Euler's factorial function in his study of the hypergeometric series [Gauss 1812], and Edwards was pleased to reintroduce that notation in his book. Thus, following Riemann, for $s > -1$

$$\Pi(s) = \int_0^{\infty} e^{-x} x^s \, dx,$$

can be defined for complex numbers $s$ in the half-plane $\mathrm{Re}\, s > -1$, and for all natural numbers $s$, $\Pi(s) = s!$. Edwards also gave the formula for the extended zeta function in Riemann's notation, writing

$$\zeta(s) = \frac{\Pi(-s)}{2\pi i} \int_{+\infty}^{+\infty} \frac{-x^s}{e^x - 1} \frac{dx}{x},$$

where the contour integral follows a path from $+\infty$ down the positive real axis, circles the origin in the positive direction and returns in the opposite direction. This formula, however, was *not* given by Riemann; instead, Weber derived it in his commentary for the second edition of the *Werke* [Riemann 1876/1892/1902, 1892: 154]. Unfortunately, it was given incorrectly there [Edwards 1974, 10]. Not that such things mattered very much; Riemann's paper was extremely sketchy and very difficult to follow. On one point, though, his argument was very clear: his extension of Euler's zeta function was based on a relationship between $\zeta(s)$ and $\Pi(s)$:

$$\Pi\left(\frac{s}{2} - 1\right) \pi^{-\frac{s}{2}} \zeta(s),$$

which remains invariant when $s$ and $1 - s$ are interchanged. Riemann here remarked that this property prompted him to look for an expression of $\zeta(s)$ in terms of $\Pi(\frac{s}{2} - 1)$ rather than using $\Pi(s-1)$ [Riemann 1859/1876, 146].

The expression he found established a functional equation for $\zeta(s)$ valid for all $s \in \mathbb{C}$ and with symmetry about the line $s = 1/2$:

$$\zeta(1-s) = \pi^{-s} 2^{s-2} \Pi(s-1) \cos \frac{\pi s}{2} \zeta(s).$$

Based on this, Riemann introduced a new function $\xi(t)$ by shifting over to the symmetry line with $s = 1/2 + it$:

$$\xi(t) = \Pi(\frac{s}{2})(s-1)\pi^{-s/2}\zeta(s).$$

He calculated an integral expression for $\xi(t)$ and proceeded to discuss its zeros. Since these are identical with the "critical zeros" of $\zeta(s)$, Riemann barely mentioned the latter and wrote instead about the zeros $t = \alpha$ of the function $\xi(t)$.

These first appear in the formula ([Riemann 1859/1876, 148]):

$$\log \xi(t) = \sum \log \left(1 - \frac{t^2}{\alpha^2}\right) + \log \xi(0).$$

This follows from Riemann's general principles because $\log \xi(t)$ has logarithmic singularities at $t = \alpha$ and no other singularities; hence $\log \xi(t)$ is determined by this data up to a constant, which can be set to $\log \xi(0)$. Following Edwards, this second result leads to a product representation, namely

$$\xi(t) = \xi(0) \prod_{\alpha} \left(1 - \frac{t^2}{\alpha^2}\right),$$

which combined with the above product for $\xi(t)$ leads to an expression for $\log \zeta$ and ultimately to the key analytic expression $f(x)$ given below.[5]

Riemann knew that the non-trivial zeros of the zeta function were located symmetrically about the line $s = 1/2 + it$ and within the range $0 \leq \mathrm{Re}(s) \leq 1$. His text, however, becomes increasingly murky as he approached the famous Riemann hypothesis:

> The number of roots of $\xi(t) = 0$, whose real part lies between 0 and $T$, is approximately
>
> $$\frac{T}{2\pi} \log \frac{T}{2\pi} - \frac{T}{2\pi},$$
>
> because the integral $\int d \log \xi(t)$ equals $\ldots (T \log \frac{T}{2\pi})i$, and this integral equals the number of roots of $\xi(t) = 0$ in this domain multiplied by $2\pi i$.

Edwards noted that this obscure argument was only proven to be correct in 1905, but what followed was even more baffling:

> One finds, in fact, about this many real roots within these bounds, and it is very likely that all the roots are real. Of course a stringent proof of this would be desirable, but after a few fleeting unsuccessful attempts I have put this search aside, since it is not needed for the purpose of my investigation. [Riemann 1859/1876, 147–148]

Riemann's purpose was clearly to improve on Gauss's conjecture for an asymptotic law for $\pi(x)$. In his letter to Encke, written in 1849, he indicated that $\pi(x)/x$ was approximately given by $\int dn/\log_e n$. Edwards referred to a table published after Gauss's death that was based on the formula $dn/\log n$, which Edwards took to mean

[5] Edwards wrote this as $J(x)$ instead of $f(x)$, and he rejected Riemann's change of variable $s = 1/2 + it$, which simplifies some of the calculations [Edwards 1974, 16–21, 25–33].

$$\pi(x) \sim \int_2^x \frac{dt}{\log t} = \mathrm{Li}(x) - \mathrm{Li}(2).$$

This accords with what Riemann wrote at the end of his paper:

> In fact the comparison of $\mathrm{Li}(x)$ with the number of primes less than $x$, which was undertaken by Gauss and Goldschmidt and pursued up to $x =$ three million, shows that the number of primes is already less than $\mathrm{Li}(x)$ in the first hundred thousand and that the difference, with minor fluctuations, increases gradually as $x$ increases. The thickening and thinning of primes which is represented by the periodic terms in the formula has also been observed in the counts of primes, without, however, any possibility of establishing a law for it based on these fluctuations. [Riemann 1859/1876, 153]

This passage reveals very clearly that Riemann knew – probably already for a long time – about the work Gauss described in his letter to Encke (see Section 4.2). He was also well aware of the fact that $\mathrm{Li}(x)$ seemed to consistently overestimate the number of primes, though it nevertheless gave a very accurate measure of their relative frequency. At some point Riemann began thinking about a straightforward way to refine this estimate, namely, by taking powers of primes into account. The idea was to count not only the prime numbers but also their powers, thus he treated the square of a prime as half a prime, the cube of a prime as a third, etc. This led Riemann to the approximation

$$f(x) = \pi(x) + \frac{1}{2}\pi(x^{1/2}) + \frac{1}{3}\pi(x^{1/3}) + \cdots + \frac{1}{n}\pi(x^{1/n}) + \ldots,$$

which for any given $x$ will have only finitely many terms since for $n$ sufficiently large $x^{1/n} < 2$, which implies $\pi(x^{1/n}) = 0$. He then inverted this by means of a method due to August Ferdinand Möbius, obtaining:

$$\pi(x) = f(x) - \frac{1}{2}f(x^{1/2}) - \frac{1}{3}f(x^{1/3}) - \frac{1}{5}f(x^{1/5}) + \frac{1}{6}f(x^{1/6}) + \cdots + \frac{\mu(n)}{n}f(x^{1/n}) + \ldots,$$

where $\mu(n)$ is 0 for $n$ a prime square, 1 when $n$ is an even product of primes, and -1 for an odd product.

The truly astounding result Riemann derived in his paper was an exact analytic expression for $f(x)$ that led to a theoretical estimation for $\pi(x)$ complete with an error term. This was Riemann's formula

$$f(x) = \mathrm{Li}(x) - \sum_{\alpha}[\mathrm{Li}(x^{\alpha}) + \mathrm{Li}(x^{1-\alpha})] + \int_x^{\infty} \frac{dt}{t(t^2-1)\log t} + \log \xi(0),$$

where the roots $\alpha, 1-\alpha$ appear in pairs.[6] This gives Riemann's more elaborate estimate with finitely many terms for

$$\pi(x) \sim \sum_{n=1}^{\infty} \frac{\mu(n)}{n}\mathrm{Li}(x^{1/n}).$$

[6] The constant $\log \eta(0) = -\log 2$ was given incorrectly by Riemann, which caused some confusion; see [Edwards 1974, 31].

Assuming $N$ terms are required, the error in this estimate is given by

$$\sum_{n=1}^{N} \sum_{\alpha} \mathrm{Li}(x^{\alpha/n}) + \text{smaller terms},$$

which thus depends on knowing the zeros of the zeta function. Riemann noted that the series $\sum[\mathrm{Li}(x^{\alpha}) + \mathrm{Li}(x^{1-\alpha})]$ converges only conditionally, so he ordered the roots according to $|\mathrm{Im}\,\alpha|$.

One sometimes encounters Riemann's approximation formula written as

$$\pi(x) = R(x) + \sum_{\rho} R(x^{\rho}),$$

where the sum is over the critical zeros of the zeta function. This formulation utilizes the power series

$$R(x) = 1 + \sum_{n=1}^{\infty} \frac{(\log x)^n}{n\, n!\zeta(n+1)},$$

which was first published by the Danish mathematician Joran Gram in 1884. His name is better known from the Gram-Schmidt method for orthogonalizing a basis of vectors, a technique already used by Laplace.

Riemann's results were, of course, far ahead of the time when mathematicians could actually make use of his formula. Not until the era of high-speed computers would one even know whether this elaborate machinery led to results that went beyond Gauss's original approximation for $\pi(x)$. During the 1950s, one of the pioneers in electronic computational methods, D.H. Lehmer, began comparing Riemann's predictions with those based simply on $\mathrm{Li}(x)$ by tabulating $\pi(x)$ for increments of $x$ between one and ten million. His findings not only confirmed what Riemann had already observed, namely that Gauss's formula consistently gave a number greater than $\pi(x)$, but also that those deviations exceeded in all cases the results predicted by Riemann, often by a factor of ten. Moreover, Riemann's formula undercounted just as often as it overcounted [Edwards 1974, 35]. Subsequent tests have shown that the zeros of the zeta function reflect oscillations of the primes around their predicted positions.

Returning to the events of 1859, on 26 October Riemann sent a letter to Weierstrass, in which he offered some interesting comments about his paper, including those parts he had not yet proved to his own satisfaction. Another part of this letter became known after Borchardt published it in [Riemann 1869/1876] (see Section 4.2), but the original letter was afterward lost. A draft turned up later among Riemann's posthumous papers, though this only appeared in print in the 1990 edition of the *Werke* [Riemann 1859/1990, 823–825], despite the fact that its existence had already been announced a century earlier. This letter was written shortly before Riemann submitted his paper, but he wanted to give a brief summary of the main result, assuring Weierstrass that he would "easily convince himself," despite the brevity of the presentation, that everything was correct, except for two assertions he had not been able to fully prove. For those who wish to struggle with the original paper, this summary offers a perhaps useful guide, so let me try to convey the gist of this.

The synopsis above attempts to evade all of the many technical difficulties, but Riemann's letter highlights an integral expression for $\xi(t)$ that plays a central part in his argument. This integral involves $\psi'(x)$, the derivative of an infinite exponential series, both introduced on [Riemann 1859/1876, 147], at which point Riemann entered into the discussion of the zeros. In his letter, he writes $t = 1/2 + \alpha i$ for the roots of $\xi(t)$, stating that they are probably all real, but in any case their imaginary parts lie between $\frac{1}{2}i$ and $-\frac{1}{2}i$. Another point is largely notational, as Riemann never used $\pi(x)$, which became standard after his time. Instead, he wrote $F(x)$, a function which took the discontinuities into account when $x$ happened to be a prime number. He thus defined

$$F(x) = \frac{\pi(x+0) + \pi(x-0)}{2},$$

and then related this to $f$:

$$f : \sum \frac{1}{n} F(x^{1/n}).$$

Applying Möbius inversion leads to

$$F(x) = \sum (-1)^{\mu} \frac{1}{m} f(x^{1/m}).$$

With this minor difference in notation, his derivation is the same as that given above. In his letter and on [Riemann 1859/1876, 147], Riemann key formula for $f(x)$ appears just as above.

The editors of the original 1876 edition simply reprinted Riemann's paper, possibly because it had not yet attracted much attention. By the early 1890s, though, the situation suddenly began to change. Around this time, Heinrich Weber was preparing the second edition of Riemann's *Werke*, to which he made relatively few changes, though he did add an appendix to the paper on the zeta function. There he began by citing a key passage from Riemann's till then unknown letter to Weierstrass, where Riemann described the two results he had not been able to prove to his own satisfaction. Evidently he thought they were within his reach, but he needed to use a complicated expression for $\xi$, one "which he had not yet simplified enough to be able to communicate" [Riemann 1876/1892/1902, 1892: 154]. This remark led to another story, familiar to experts through Carl Ludwig Siegel's work on what today is known as the Riemann-Siegel formula.[7]

[7] For an idea of just how complicated Riemann's expression was, see [Edwards 1974, 155–159].

## 8.3 Riemann's Paris Prize Paper

Among Riemann's many posthumously published papers, [Riemann 1876b], the manuscript he submitted to the Paris Academy in 1861 counts among the most controversial. As described in Section 4.3, he wrote it in hopes of winning a gold medal offered by the Academy for a solution to a problem in heat conduction. Since Riemann apparently never spoke with anyone other than his wife about this paper, the circumstances that led him to write it remain a matter of speculation. The subsequent controversies surrounding it, however, have far more to do with the content of Riemann's text than his motives for writing it.

A draft of Riemann's manuscript was among the several texts his widow, Elise Riemann, placed in the hands of Richard Dedekind shortly after her husband's death (see Section 12.1). Dedekind decided to publish the two texts Riemann had written in 1854 in connection with his Habilitation, but not the "Commentatio." He nevertheless made a careful study of its second section, which contained important clues about the mathematical technicalities Riemann had excluded from his lecture text "On the Hypotheses underlying the Foundations of Geometry" [Riemann 1868/1876]. As described in Section 6.3, Dedekind originally planned to furnish these missing pieces in a separate commentary, a project he decided to abandon.[8] Riemann's "Commentatio" [Riemann 1876b] was later studied by Heinrich Weber took over the editorship of his Collected Works (see Section 12.3).

By the spring of 1876, when work on that edition was coming to a close, Weber wrote a revealing letter to Dedekind on the 18th of May. Weber's interest in the mathematical methods employed in Riemann's "Commentatio" led him to wonder whether the text might provide all that was needed to solve the Paris problem. He informed Dedekind that he would ask Frau Riemann's permission to send the contents of the Paris manuscript to Karl von der Mühll in Leipzig. A native of Basel with expertise in mathematical physics, von der Mühll took courses taught by Riemann and Wilhelm Weber before going on to Königsberg, where he took his doctorate under Franz Neumann in 1866. Heinrich Weber likely heard about him after he joined the Königsberg faculty in 1875. He reported to Dedekind that

> von der Mühll ... is, or at least used to be, a very clever calculator, so I hope he might be able to carry out the calculations. I have the idea that it might be possible, once the matter has been sorted out, to obtain part of the prize from the Paris Academy for Frau Riemann, if one can show that the problem had already been solved back then.

[8] See [Scheel 2015, 57]. Much later, his notes for that commentary fell into the hands of Emmy Noether when she was editing Dedekind's Collected Works [Dedekind 1930–1932]. She then sought the advice of Ludwig Bieberbach, who years later sent the manuscript back to the SUB in Göttingen, where it was incorporated into Dedekind's estate in 1966. The text was then published with a French translation and commentary in [Sinaceur 1990]. Darrigol suggested that this document was later used by Heinrich Weber for his commentary on the Paris prize manuscript [Riemann 1861a/1876]. Evidence for this seems to be lacking, though, as the title of Dedekind's text refers only to the 1854 lecture (see the reproduction in [Sinaceur 1990, 236], which was incorrectly transcribed in the plural form "Abhandlungen").

> However, I am not yet clear about how or in which way this could be achieved, and I ask you, of course, not to talk about it for the time being. Above all, the problem must first be solved. [Scheel 2015, 141]

Frau Riemann then gave Weber permission to send the text to von der Mühll, but nothing is known about how the latter responded. The letter cited above, however, makes apparent that Weber approached this text (and several others as well) as presenting a mathematical challenge, whereas the usual considerations related to the editing of the text itself were of secondary importance for him.[9]

On 10 December 1876, thus roughly a half-year after the *Werke* had appeared, Weber wrote these lines about his note on the "Commentatio":

> I like to believe that Lipschitz's works, and perhaps also those of Christoffel, may contain a large part of the content of that note. However, I believe that the purpose of making Riemann's formulas understandable for readers far removed from the subject matter through the briefest possible explanations would not have been achieved by simply 'citing' these extensive works. Incidentally, I must firmly reject any hint of ownership of this note; I did nothing more than compile from your papers what seemed to me to have a direct reference to Riemann's works, and I only made this compilation myself because I did not at the time dare to ask you to do this work. As far as citing other works, I made it my principle to only mention those which have a direct reference to Riemann's publications. [Scheel 2015, 159–160]

Dedekind and Weber were of one mind that the second section of the prize paper was of great importance for understanding Riemann's Habilitation lecture. As Darrigol wrote, they "inaugurated a tradition of regarding the *Commentatio* as an opportunity for Riemann to summarize some of the calculations he must have done to prepare his habilitation lecture" [Darrigol 2015, 70]. No one disputed, of course that these two works differed in many respects, but for most commentators the importance of the former was merely the light it shed on the latter. That was long the standard viewpoint up until it was contested by Ruth Farwell and Christopher Knee in [Farwell/Knee 1990]. Unlike nearly all other commentators, they took the position that Riemann was genuinely interested in the physical problem posed by the Academy. He had, after all, spent much of his career working on various topics and problems in mathematical physics.

Their paper begins with a lengthy critique of the historical literature, citing authors who have echoed the orthodox position first advocated by Dedekind and Weber, though treating Felix Klein's lengthy analysis in [Klein 1927] as an exceptional case. According to this new interpretation, Riemann's two works should not be linked because they differed not only in subject matter but also by virtue of their motivation. More specifically, they argued that the second part of Riemann's "Commentatio" was concerned with a problem in differential invariants (or in more modern language tensor analysis) and had nothing to do with differential geometry. As they put this, the Paris prize paper

[9] Olivier Darrigol, who critically examined Riemann's notes for the "Commentatio," concluded that what Dedekind and Weber wrote as commentary deviated from the line of argument Riemann likely followed. But he also added that since "[they] were not pretending to be doing history, they cannot be blamed for having creatively mixed ideas that Riemann conceived separately" [Darrigol 2015, 90].

> ...can justifiably be cited as a significant contribution to tensor analysis and was thus eventually relevant to Einstein's theory of general relativity just as Riemann's conception of space was. However, we hold that Riemann's geometry and tensor analysis developed relatively independently. [Farwell/Knee 1990, 224]

Karin Reich, in her detailed history of the tensor concept and the analytical works that culminated with the absolute differential calculus of Ricci and Levi-Civita [Reich 1994], took a sympathetic view of the standpoint advocated in [Farwell/Knee 1990].

Alberto Cogliati, on the other hand, argued that the attempt to draw a sharp line between works devoted to differential geometry, on the one hand, and the analytical methods of invariant theory, on the other, was not only historically misleading but also an unlikely way to gauge an author's motivation, particularly in a case like Riemann's. In Cogliati's view:

> ...every attempt to confine Riemann's Commentatio to a specific area of mathematical research runs the risk [of providing] an incomplete picture of it, since, as is well known, a broader look at Riemann's entire mathematical production reveals a complex net of reciprocal influences among most diverse research interests such as analysis, mathematical physics and geometry. [Cogliati 2014, 92]

Clearly, any hope of uncovering Riemann's original motives for writing the Paris prize paper is chimerical, but we can still at least try to understand the thrust of his argument.

As was noted in [Farwell/Knee 1990, 225], this story actually began in 1855, when the French Academy proposed a related question, but since this drew no responses the problem was reformulated in 1858 to read:

> To determine the caloric state of any solid homogeneous body such that a system of isothermal curves, at a given instant, remains isothermal after a given time, in such a way that the temperature at a point can be expressed as a function of time and two other independent variables. [Farwell/Knee 1990, 240]

For his prize entry, Riemann submitted the motto: "Et his principiis uia sternitur ad majora" (And from these beginnings a way is opened to greater things). Some readers have surely taken this to mean that the text itself was meant to portend something much greater. The full title of Riemann's manuscript read: "Commentatio mathematica, qua respondere tentatur quaestioni ab Ill$^{ma}$ Academia Parisiensi propositae" (A mathematical treatise in which an attempt is made to answer the question proposed by the most illustrious Academy of Paris). An English translation of the Latin text of Riemann's "Commentatio" was presented in [Farwell/Knee 1990], and this will be cited accordingly below.

At the very outset, Riemann explained the strategy he would follow for solving this problem:

> We shall consider the question proposed by the Academy in such a way that we shall first solve a more general question: [to find] the properties of a body which determine the conduction of heat and the distribution of heat within it such that there exists a system of lines which remain isothermal. Then: from the general solution of this problem we shall distinguish those cases in which the properties vary from those in which the properties remain constant, that is where the body is homogeneous. [Farwell/Knee 1990, 240]

Riemann thus dealt at first with the general case where the body is not assumed to be homogeneous, but nevertheless possesses an invariant system of isothermal curves. Riemann thus proposes to find conditions that allow for the elimination of one spatial coordinate in an appropriate coordinate system, thereby determining a system of isothermal curves by varying the other two coordinates. He then discussed under which conditions such a coordinate system can be transformed into an equivalent system for a homogeneous thermal body. Riemann thus divided his paper into two independent parts, the second of which is far shorter and was of far greater interest at the time it appeared (see Section 12.3). Part one, on the other hand, led him to consider a number of special cases that were both difficult to follow and tedious to read. He broke these considerations off at one point, simply writing "we shall not linger on this investigation which is more long-winded than difficult" [Farwell/Knee 1990, 244].

Riemann began by introducing $u$, which measures the temperature at time $t$ and point $(x_1, x_2, x_3)$ in Cartesian coordinates. The heat equation for a homogeneous body is then

$$\Delta u = \frac{\partial^2 u}{\partial x_1^2} + \frac{\partial^2 u}{\partial x_2^2} + \frac{\partial^2 u}{\partial x_3^2} = h\frac{\partial u}{\partial t},$$

where $h$ is the specific heat for the entire volume. For an inhomogeneous thermic body, Riemann introduced a more general heat equation for $u$ based on a $3 \times 3$ symmetric matrix $(a_{\iota,\iota'})$, where the $a_{\iota,\iota'}$ vary with time. Were the body thermally homogeneous, then the $a_{\iota,\iota'}$ would be constant, which Riemann had shown was equivalent to a coordinate system $x_i$ with metric $\sum dx_i^2$, the Euclidean case. His first task, then, was to transform the equation for $u$ to curvilinear coordinates $s_1, s_2, s_3$ and determine conditions for expressing $u$ as a function of $s_1, s_2$, and $t$. Once this had been achieved, he could turn to the problem of determining when this coordinate system was equivalent to another with constant coefficients. The argument is not entirely straightforward, however, as the coefficients in the transformed heat equation do not transform with the variables directly.

Riemann first assumed that the change in temperature will be the same in opposite directions, so the matrix $a_{\iota,\iota'}$ will be symmetric. Furthermore, since heat moves from a warmer to a cooler region, it must be positive definite. He then invoked a variational argument to deduce the corresponding coefficients $b_{\iota,\iota'}$ for the heat equation in the new coordinates $s_1, s_2, s_3$. Letting $A$ be the determinant of $a_{\iota,\iota'}$ and $B$ the determinant of $b_{\iota,\iota'}$, he noted that $A = B \cdot \det(J)$, where $J$ is the Jacobian matrix for the coordinate transformation. This $J$ then leads to the transformation equation

$$b_{\mu,\nu} = \sum_{\iota,\iota'} a_{\iota,\iota'} \frac{\partial s_\mu}{\partial x_\iota}\frac{\partial s_\nu}{\partial x_{\iota'}} \Big/ \det(J), \quad k = h\Big/\det(J),$$

where $k$ is the specific heat in the new coordinate system.

Riemann realized, though, that he could avoid dealing with det$(J)$ by passing over to the adjoint matrices instead. Thus, if $\alpha_{\iota,\iota'}$, resp. $\beta_{\iota,\iota'}$, is the adjoint of $a_{\iota,\iota'}$, resp. $b_{\iota,\iota'}$, then

$$\beta_{\mu,\nu} = \sum_{\iota,\iota'} \alpha_{\iota,\iota'} \frac{\partial x_i}{\partial s_\mu} \frac{\partial x_{i'}}{\partial s_\nu},$$

and therefore

$$\sum_{\iota,\iota'} \alpha_{\iota,\iota'} dx_\iota \, dx_{\iota'} = \sum_{\iota,\iota'} \beta_{\iota,\iota'} ds_\iota \, ds_{\iota'},$$

whereas

$$h/A = k/B.$$

At this stage, Riemann summarized again the strategy for the first part of his paper. Writing $(\alpha, \beta, \gamma)$ instead of $s_1, s_2, s_3$, he considers different cases for the differential equations for $u$ that lead to solutions independent of the third coordinate $\gamma$. When this can be achieved for $b_{\iota,\iota'}$ and $k$, he can then construct the corresponding differential form $\sum_{\iota,\iota'} \beta_{\iota,\iota'} ds_\iota \, ds_{\iota'}$. The problem then reduces to deciding, for given $a_{\iota,\iota'}$ and $h$ with associated $\alpha_{\iota,\iota'}$, whether $\beta_{\iota,\iota'}$ and $k$ can be transformed into $\alpha_{\iota,\iota'}$ and $h$. In the case of a homogeneous thermal body, the differential form $\sum_{\iota,\iota'} \alpha_{\iota,\iota'} dx_\iota \, dx_{\iota'}$ will have constant coefficients, which means it can be written as $\sum dx_i^2$. The problem thus becomes one of deciding whether the differential form $\sum_{\iota,\iota'} \beta_{\iota,\iota'} ds_\iota \, ds_{\iota'}$ is equivalent to a flat metric, a type of problem Riemann had raised earlier in his Habilitation lecture.

In Part 2, Riemann derived a four-index differential expression $(\iota\iota', \iota'', \iota''')$, which he showed vanishes in the case of a flat metric. His derivation involves straightforward calculations starting from

$$b_{\iota,\iota'} = \sum_{\nu} \frac{\partial x_\nu}{\partial s_\iota} \frac{\partial x_\nu}{\partial s_{\iota'}},$$

and differentiating with respect to $s_{\iota''}$. Combining the result with the derivatives of $b_{\iota,\iota'}$ with respect to $s_\iota$ and $s_{\iota'}$, Riemann obtained three-index expressions $p_{\iota,\iota',\iota''}$ essentially equivalent to Christoffel symbols. Taking derivatives of these, he then obtained an invariant fourth-order expression, later identified as the Riemann curvature tensor, which vanishes when the metric is flat. This central part of Riemann's argument was technically demanding, but it has never been a source of controversy. He had addressed this special case in his Habilitation lecture, where he wrote:

> In a flat $n$-fold extended manifold, the measure of curvature is zero at every point and in every direction; however, ... to determine the metric relations it suffices to know that at every point it vanishes in $n(n-1)/2$ surface directions whose measures of curvature are mutually independent. [Riemann 1868/1876, 263–264]

In other words, if the sectional curvature vanishes for any independent set of $n$ vectors at every point of a manifold, then the metric is flat.

What Riemann wrote next in the "Commentatio," however, has aroused much interest as well as controversy. After introducing this four-index symbol

$(\iota\iota'\,,\,\iota'',\iota''')$, Riemann went on to interpret this analytic entity geometrically. At this point, to avoid the awkward accents Riemann used, it will be convenient to adopt Cogliati's notation and write: $(\iota\iota'\,,\,\iota'',\iota''') = (ij,\, lt)$. He first formed the differential invariant:

$$\sum_{ijlt}(ij,\, lt)(ds_i\delta s_j - ds_j\delta s_i)(ds_l\delta s_t - ds_t\delta s_l),$$

an expression reminiscent of the one Riemann described in his Habilitation lecture. From this he constructed the quotient:

$$\frac{1}{2}\frac{\sum(ij,\, lt)(ds_i\delta s_j - ds_j\delta s_i)(ds_l\delta s_t - ds_t\delta s_l)}{\sum b_{ij}ds_i ds_j \sum b_{ij}\delta s_i\delta s_j - \left(\sum b_{ij}ds_i\delta s_j\right)^2},$$

which he referred to as (III). Riemann asserted that (III) remains invariant if any independent linear expressions are substituted for $ds_i$ and $\delta s_i$. Moreover, the maximum and minimum values of (III) are not affected by the form of $\sum b_{ij}ds_i ds_j$ or the values of $ds_i$ and $\delta s_i$. These values therefore "can be used to ascertain whether two forms can be transformed into each other" [Farwell/Knee 1990, 252]. This thus suggests a method for addressing Christoffel's equivalence problem, which the latter had solved using algebraic invariants.

At this point, Riemann went on to make these telling remarks:

> These investigations can be illustrated by a geometrical example which, although unusual, will be a useful addition.[10] The expression $\sqrt{\left(\sum b_{\iota,\iota'} ds_\iota\, ds_{\iota'}\right)}$ $[\sqrt{(\sum b_{ij} ds_i\, ds_j)}]$ can be regarded as a line element in a more general space of $n$ dimensions extending beyond the bounds of our intuition. But if in this space all possible shortest lines are drawn from the point $(s_l,\, s_2, \ldots, s_n)$ with first-order variations of $s$ dots, then these lines will form a surface which can be visualized in the space of our common intuition. Thus expression (III) will then be a measure of the curvature of this surface at the point $(s_l,\, s_2, \ldots, s_n)$. [Farwell/Knee 1990, 252]

We cannot know whether Riemann wrote this passage in order to elucidate his ideas from seven years earlier, but even so the parallels between the two are obvious. Nevertheless, a variety of different opinions have been expressed as to how Riemann arrived at his fragmentary results. In recent times, Olivier Darrigol made a valiant effort to unlock the secrets of Riemann's "Commentatio" in [Darrigol 2015], which has some of the flavor of a scientific detective story. He first framed a hypothesis about how Riemann found his key results, which he hoped to confirm by studying the extant notes in Riemann's estate. He must have experienced a true Eureka moment when he discovered a note that began with $\delta\int\sqrt{p_1^2 s_1^2 + p_2^2 s_2^2 + p_3^2 s_3^2} = 0$, the variational equation for geodesics of a diagonal metric [Darrigol 2015, 64]. Alongside this, he set forth a new interpretation for the presentation of curvature in Riemann's Habilitation lecture as a

[10] Cogliati considered this an inaccurate and misleading translation, so he wrote instead: "These investigations can be illustrated by means of a certain geometrical interpretation, which, though based upon unusual concepts, is nonetheless worth pointing out" [Cogliati 2014, 84].

direct generalization of Gaussian curvature in surface theory. Darrigol's technical mastery of the mathematics involved is truly impressive, and his findings lead to interesting new conclusions regarding the long-term reception of the "Commentatio." These are all difficult matters that, thankfully, lie far beyond the scope of the present book.

# Chapter 9
# Mathematical Physics

During his lifetime, Riemann published four papers on topics in physics, two of which have come to be regarded as of major importance.[1] These publications, however, provide only a glimpse of the larger role physical ideas played in his overall conception of mathematical research. Riemann was hardly a traditionalist, of course, but his approach to mathematical physics was nevertheless strongly influenced by the works of his predecessors, Gauss and Dirichlet. Gauss published pioneering work on potential theory, whereas Dirichlet taught his standard lecture course on this subject no fewer than eight times between 1839 and 1858 [Merzbach 2018, 248]. Section 3.3 presented an overview of remarks from Riemann's very first lecture course on partial differential equations and their applications to physics, which he taught during the 1854/55 semester. At that time, he took the position that advances in mathematical analysis since the time of Newton had greatly outpaced those in experimental physics. Not surprisingly, this standpoint is reflected in his own physical works, which were generally more impressive for their mathematical virtuosity than for the physical insights they uncovered.

Riemann's early interests in physics were also briefly mentioned in Section 5.1, which took note of five tentative theses he considered defending as part of the ritual of attaining a doctoral degree. These theses only came to light amidst the new source material Max Noether and Wilhelm Wirtinger unveiled in 1902, but their existence makes evident that already in 1851 Riemann was well aware of ongoing debates over physical issues [Riemann 1876/1892/1902, 1902: 112–113]. The parallelism between electrical and magnetic phenomena raised questions about whether both could be viewed as imponderable fluids, a view Riemann rejected with regard to magnetism. He was certainly well aware that Michael Faraday generally opposed Wilhelm Weber's approach to electromagnetic phenomena [Darrigol 2000, 101–102], though why Riemann thought that Faraday's "induction in curved lines" was untenable remains a mystery. Faraday considered this as lending strong argument for his view that electromagnetic

[1] These are his paper on acoustic shock waves [Riemann 1860] and his study of motions of fluid bodies that maintain their form as an ellipsoid [Riemann 1861a/1876].



D. E. Rowe, *Bernhard Riemann: His Life and Wondrous Mathematical Legacy*,
https://doi.org/10.1007/978-3-032-25457-3_10

induction was transmitted via contiguous particles rather than by action-at-a-distance [Darrigol 2000, 101–102]. Faraday's position was certainly compatible with the viewpoint Riemann would later adopt.

For Riemann's teaching, the importance of physical ideas was more than readily apparent. With that in mind, the present chapter aims to describe his general physical views and methodological approaches without, however, attempting to analyze Riemann's published works on topics in mathematical physics in any detail. This chapter along with the one that follows aim mainly to show how Riemann's work on experimental as well as theoretical physics reflect important currents within the Göttingen scientific milieu. While surveying the broader terrain of interests that informed his work, our attention will also turn to parallel currents of research which may have influenced Riemann's thinking. As background for what follows, it will be expedient to begin with a brief account of potential theory.

## 9.1 Potential Theory

The historical roots of potential theory date back to the eighteenth century, an era when Newtonian mechanics reigned supreme. Many of the leading mathematicians from that period developed new methods that aimed to bolster that paradigm. In the background stood Newton's inverse square law for universal gravitation, which opened the door to modern celestial mechanics. As noted in Section 3.2, Riemann's approach to physics owed much to Newtonian ideas, though he hoped to deepen their foundations by probing the mechanisms underlying gravitation, electromagnetism, and light, a problem Newton surely considered but never addressed. On the other hand, Newton made an important first contribution to a fundamental problem in potential theory. In the *Principia*, he famously proved that the gravitational force associated with a homogeneous spherical body was equivalent to the field generated by a single point mass, namely one in which the entire mass of the body lies at the center of the sphere.

In the meantime, Charles-Augustin de Coulomb established an inverse square law for electrostatics, later extended to magnetism by André-Marie Ampère. Coulomb's law for charged particles was identical in form with Newton's gravitational law for mass points, except that the force was repulsive for charges with the same sign. For charged bodies, one needed to study arbitrary continuous distributions given by a density function $\rho$. This problem was pursued in France by Laplace and Poisson, who introduced the famous partial differential equations

$$\left(\frac{\partial V}{\partial x}\right)^2+\left(\frac{\partial V}{\partial y}\right)^2+\left(\frac{\partial V}{\partial z}\right)^2=0 \quad \text{or} \quad \Delta V=0,$$

and for a matter density $\rho$,

$$\left(\frac{\partial V}{\partial x}\right)^2 + \left(\frac{\partial V}{\partial y}\right)^2 + \left(\frac{\partial V}{\partial z}\right)^2 = -4\pi\rho \quad \text{or} \quad \Delta V = -4\pi\rho,$$

where $\Delta$ is the so-called Laplace operator.

Potential theory eventually became a full-fledged mathematical enterprise following the appearance of Gauss's "Allgemeine Lehrsätze" ("General Theorems relating to the Forces of Attraction and Repulsion acting in Inverse Proportion to the Square of the Distance") [Gauss 1840a].[2] This classic study laid the groundwork for subsequent works by Dirichlet and Riemann, from which arose the longstanding Göttingen tradition in mathematical physics [Merzbach 2018, 186–190, 245–246]. Gauss's general theorems applied to the force laws for gravity, magnetism, and electricity, but like Newton he offered no hypothesis about the underlying physical causes that supported these. Riemann announced courses on this topic four times, leading to the publication of [Riemann/Hattendorff 1876], which was based on the lectures he delivered in the summer semester of 1861.

As described in Section 1.3, Riemann took Benjamin Goldschmidt's course on geomagnetism during his first semester in Göttingen. Goldschmidt taught a variety of topics, including courses on theoretical and popular astronomy. In his second semester, Riemann planned to sign up for Goldschmidt's course on probability theory, partly to ensure the necessary quorum. Whether or not he had by this time already gained a clear understanding of Gauss's approach to potential theory, he continued to immerse himself in Gauss's work during the following academic year in Berlin. There he encountered another leading expert on potential theory: Gustav Lejeune Dirichlet.

Given our scant knowledge about what Riemann learned in the courses he took as a student, it is probably impossible to narrow down precisely when he first learned the fundamentals of potential theory. My earlier guess that he picked up this knowledge during his first semester, when he took Goldschmidt's course on geomagnetism, is merely a conjecture. What we do know, though, is that starting with the winter semester of 1849/50 and for several semesters thereafter, Riemann borrowed the reports published by the Magnetic Union [Neuenschwander 2022, 44–45]. Among those reports was the one for the year 1839, which contained Gauss's "Allgemeine Lehrsätze" [Gauss 1840a]. Since Riemann never cited that important work, let it be noted that he surely read it well before he wrote his dissertation (5.2). That later famous work was, in fact, strongly influenced by Riemann's prior knowledge of potential theory, as can easily be explained.

The earlier works by Laplace and Poisson set the stage for Gauss, who studied a general problem pertaining to any potential function $V$ corresponding to an inverse square force. This involved finding a density function $\rho$ that produces a constant potential on a given surface $S$. Such a potential can be expressed as a

[2] The pioneering study by George Green, *Essay on the Application of Mathematical Analysis to the Theories of Electricity and Magnetism* (Nottingham, 1828), only became known after 1850. William Thomson republished it with a brief biography of Green in three issues of *Journal für die reine und angewandte Mathematik*: 39 (1850), 73–89; 44 (1852), 356–374; and 47 (1854), 161–221.

surface integral

$$V = \int_S \frac{\rho ds}{r},$$

where $r$ is the distance between points on $S$. Gauss's intuitive argument was to consider all distributions on $S$ with the same total mass $M$ and from these find the one that minimizes the energy, thereby leading to a state of equilibrium. He thus considered the energy integral

$$I = \int_S V\rho ds,$$

which can be transformed, up to a constant factor, into the volume integral

$$\iiint \left( \left(\frac{\partial V}{\partial x}\right)^2 + \left(\frac{\partial V}{\partial y}\right)^2 + \left(\frac{\partial V}{\partial z}\right)^2 \right) dxdydz.$$

If the surface $S$ is bounded, then let $r_o$ be the supremum for all distances between the points on $S$, and let $M$ be the total mass for the density function $\rho$. Then the integral $I$ is bounded below by $M^2/r_o$, from which Gauss concluded that there exists a distribution $\rho$ that minimizes $I$.[3]

Gauss's "argument" for minimizing this integral was later taken up by Dirichlet, as Riemann apparently only learned *after* completing his dissertation. There Riemann utilized a similar argument to establish the existence of a harmonic function satisfying given boundary conditions. Riemann dubbed this Dirichlet's principle, while noting that Gauss had already employed essentially the same idea [Riemann 1857b, 90]. Since Riemann was familiar with the argument in [Gauss 1840a] well *before* he wrote his dissertation, these circumstances appear quite strange and certainly worthy of attention. For the validity of "Dirichlet's Principle" was not only questioned during Riemann's lifetime, it was later exposed by Weierstrass as the Achilles' heel in Riemann's entire approach to complex functions [Monna 1975].[4]

The significance of potential theory for Riemann's function theory is easy to appreciate when one considers the importance he attached to the Cauchy-Riemann differential equations as fundamental for the concept of analytic functions (see Section 5.2). Thus, for a function

$$f(z) = f(x + iy) = u(x + iy) + iv(x + iy)$$

to be complex-differentiable, the real-valued functions $u$ and $v$ must satisfy

[3] Gauss produced a similar argument in an unpublished manuscript from 1827 [Gauss 1863–1933, 3: 479–480]. Schering commented that his sketch, which dealt with Laplace's equation in two dimensions, anticipated Riemann's approach to complex analysis [Gauss 1863–1933, 3: 494–495].

[4] On the evolution of the mathematical theory of potential after 1860, see [Archibald 1996].

$$\frac{\partial u}{\partial x} = \frac{\partial v}{\partial y} \quad \text{and} \quad \frac{\partial u}{\partial y} = -\frac{\partial v}{\partial x}.$$

This condition then implies that $u$ and $v$ satisfy the 2-dimensional Laplacian equations:

$$\Delta u = \Delta v = 0,$$

i.e., $u$ and $v$ are harmonic functions. From this starting point one can derive the usual properties of analytic functions via their local power series representations, but for Riemann this was of secondary importance, unlike for Weierstrass. The Riemannian approach avoided explicit formulas and instead used the methods devised by Gauss and Dirichlet in potential theory. For a given region $U$ in the complex plane, he applied Dirichlet's principle to derive the existence of uniquely given harmonic functions $u$ and $v$ defined on $U$ and taking given values on $\partial U$, its boundary. The latter came to be known as the Dirichlet problem in complex analysis. Its analogue in $\mathbb{R}^3$ is known as the Dirichlet problem in potential theory.

## 9.2 Terrestrial Magnetism

Riemann's intense initial interest in terrestrial magnetism was clearly influenced by the ongoing project first launched by Alexander von Humboldt and which came to fruition in Göttingen through the collaboration of Gauss and Weber. This activity later took on a legendary quality, making it an ideal subject for the novelist Daniel Kehlmann, author of *Measuring the World*. One might even say that during this era the Göttingen astronomical observatory became the hub for a new trend in scientific research based on measurements taken with precise new instruments. During the 1830s, the Göttingen Magnetic Union coordinated the synchronized activities undertaken at over fifteen magnetic observatories. Readings were normally taken six times each year using special devices designed by Gauss and Weber. These were conducted continuously every five minutes for a period of 24 hours. Humboldt had dreamed of undertaking this project on an international scale, but to do so meant engaging the British scientific community with its far-flung Empire.

In 1838 Weber spent some time in England, where he met the astronomer John Herschel, who had recently returned from South Africa. Herschel voiced his support for the Magnetic Union, but the British Association for the Advancement of Science had its own ideas about how best to organize this research and they eventually gained government support for it. Thus, by the 1850s, geomagnetic research had greatly expanded under British leadership with some thirty-five observatories strewn throughout Europe, another six in Asia, two in Africa, three in North America, and four in the South Seas. Göttingen was no longer the hub; it was merely another node in this enlarged network.

After Weber departed from Göttingen, research on geomagnetism was mainly in the hands of Gauss's assistant, Benjamin Goldschmidt, whom Riemann met

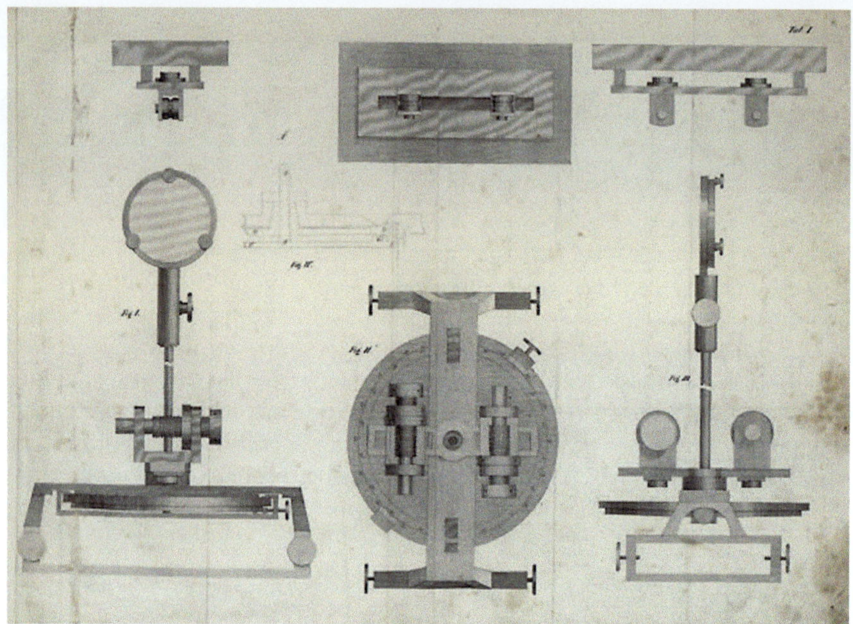

Fig. 9.1: Parts of a Bifilar-Magnetometer.

during his first semester of study. Weber had hoped to continue working on terrestrial magnetism in cooperation with Gauss, but the latter signaled that due to the separation he no longer had any real interest in pursuing this work further. Weber continued nevertheless to assist Gauss for a number of years, before accepting a professorship in Leipzig in 1843. There he undertook a major revision of Ampère's theory of electrical forces in closed circuits (see Section 2.4). At the same time, Weber obtained funds from the Ministry in Saxony for building an isolated, iron-free, magnetic observatory like the one in Göttingen. He also planned to publish a seventh volume of the *Resultate*, though this never materialized. In the meantime, the number of magnetic observatories had increased dramatically, making it impossible for a single scientist with a small group of assistants to deal with all the data.

Gauss's "General Theory of Terrestrial Magnetism" [Gauss 1839] employed potential theory to describe the quantitative features of magnetic force (Section 9.1). Rather than developing a physical theory, however, he used spherical harmonics to describe magnetic phenomena. One year later, he published the first general account of potential theory, his "Allgemeine Lehrsätze" [Gauss 1840a], a work that effectively launched this field [Archibald 1989a]. Drawing on these theoretical works, Gauss and Weber published their *Atlas of Terrestrial Magnetism* [Gauss/Weber 1840], a compilation based on calculations made from a great deal of observational data. These led to good approximations for the poles, axes, and moments of the earth's magnetic field as well as the ranges of magnetic intensity, etc. These results were summarized in the text and then displayed visually by a series of maps.

The Atlas presented eighteen pairs of maps, one using a stereographic projection, the other a Mercator projection. These showed various contour lines, such as isogonics connecting points with equal magnetic declination, isoclinics showing places with equal magnetic dip, and isodynamics indicating equal magnetic intensity. On two of these charts, equipotential lines were presented for the first time in history. When Riemann first learned about the Gauss-Weber Atlas remains a mystery, but in later years he borrowed it from the Göttingen University Library on several occasions [Neuenschwander 2022]. In Gauss's preface to [Gauss/Weber 1840], he described in detail who was responsible for which parts of this Atlas, a project he merely oversaw. The actual work was carried out by Weber and Goldschmidt, though Gauss mainly credited the former:

> My esteemed friend, Professor Weber, who spares no sacrifice when it comes to serving science, undertook to create such a visualization through a series of maps that graphically represent, in the greatest possible detail, all the magnetic conditions for the entire surface of the Earth, as that theory yields them. ... This explanation exhausts everything necessary for understanding the maps and for assessing the usefulness they can provide, so completely that nothing remains for me to add except the wish that this laborious and meritorious work may find just recognition among the friends of natural science [Gauss 1863–1933, 377–378]

Over several years, Riemann studied the volumes from the series co-edited by Gauss and Weber, *Resultate aus den Beobachtungen des magnetischen Vereins*, which he regularly withdrew from the Göttingen University Library. Moreover, he later gave a report on the determination of magnetic constants using a bifilar magnetometer, the device invented by Gauss and Weber for measuring magnetic fluctuations.[5] Riemann's report was based on Gauss's article [Gauss 1840b], reprinted in [Gauss 1863–1933, 2: 404–426]. This describes in detail how four particular constants can be determined from measurements taken with his bifilar-magnetometer (Fig. 9.1). He was surely also familiar with the earlier unifilar magnetometer, which was described in detail in the *Resultate* (for an account of both instruments, see 10.1).

## 9.3 Weberian Electrodynamics

Only fragmentary evidence survives relating to Riemann's activities in experimental physics during the early 1850s. Even so, this confirms that his physical theorizing went hand-in-hand with careful laboratory experiments undertaken

[5] The contents of Riemann's report have apparently since vanished. Dedekind dismissed it as unimportant when he first advised Heinrich Weber on potential documents for the Riemann edition. On 14 March 1875, he wrote: "The papers relating to bifilar magnetometers probably only contain preparations for a lesson Riemann gave (probably in 1852 or 1853 as an assistant to W. Weber) in the mathematical-physical seminar" [Scheel 2015, 58].

as a member of Weber's small team of researchers. From a theoretical standpoint, Riemann's general methodological approach aligned perfectly with the standards advocated by Gauss and Weber in all their physical research. During the years 1851–53, Riemann was the junior member in Weber's small group of three, which also included Gustav von Quintus-Icilius, who habilitated in 1849, the year Weber returned to Göttingen. Weber had been conducting experiments in an effort to detect diamagnetic polarization, but his findings had been greeted with skepticism by Michael Faraday [Darrigol 2000, 101–108]. In 1852, Weber undertook another series of experiments, this time enlisting the assistance of his two staff members as well as two close colleagues, Professors Listing and Sartorius. Each of these four independently replicated Weber's experiment; his findings and theirs were then presented together in [W. Weber 1852]. Two years later, Riemann again did experimental work on electrical phenomena alongside Weber and his colleague Rudolf Kohlrausch (see Section 2.7).

It was In Leipzig during the mid-1840s that Wilhelm Weber found his groundbreaking results in electrodynamics. The professorship he assumed there in 1843 had previously been held by Gustav Fechner, who soon shifted his interests to the field of psychophysics. Fechner nevertheless continued to develop a theory of electrical forces that was compatible with Ampère's work as well as with Faraday's discovery of electromagnetic induction [Wise 1981]. This led to Fechner's two-stream theory of electricity [Fechner 1845], which Weber would adopt one year later in positing his new force law [Darrigol 2000, 61–64]. The latter extended Coulomb's electrostatic law to the case of moving currents, where Weber assumed the electrical force depended on the first two derivatives of the distance $r$ separating two "particles" $e$ and $e'$.[6]

Weber reasoned that one needed the first derivative to account for the repulsive force between two collinear current elements and the second derivative to capture the attractive force between current elements that move parallel to one another [W. Weber 1846]. His force law thus took the form:

$$F = \frac{e\,e'}{r^2}\left[1 - \frac{1}{C^2}\left(\frac{dr}{dt}\right)^2 + \frac{2r}{C^2}\left(\frac{d^2r}{dt^2}\right)\right],$$

where $C$, known today as Weber's constant, is given by the ratio between electrostatic and electrodynamic units of charge. Weber and Kohlrausch conducted experiments with a Leyden jar in 1855 to determine its value experimentally and relate it to other testable forces. They found it to be roughly equal to $C = \sqrt{2}\,c$, though it was only later that Maxwell underscored the connection between electromagnetic propagation and the velocity of light. Weber's law is thus often written as

$$F = \frac{e\,e'}{r^2}\left[1 - \frac{1}{2c^2}\left(\frac{dr}{dt}\right)^2 + \frac{r}{c^2}\left(\frac{d^2r}{dt^2}\right)\right].$$

[6] Weber's original theory involved two types of electrical fluids, so it was not truly particulate.

In [W. Weber 1848], he showed that this can be derived from the potential

$$V = \frac{e\,e'}{r}\left(1 - \frac{r^2}{2c^2}\right).$$

From this, Weber was able to derive Ampère's force law and other related findings (see [Darrigol 2000, 402–403]).

An important part of Weber's electrodynamical research resulted from his use of sophisticated instrumentation and precision measurements. This was especially apparent in Leipzig, but the groundwork had already been laid in Göttingen during the years 1831 to 1837 when he collaborated with Gauss in refining the instruments they used to measure terrestrial magnetism. As Darrigol emphasized, Weber's invention of the electrodynanometer was directly inspired by the idea behind Gauss's bifilar magnetometer [Darrigol 2000, 49–65]. In broader terms, one should not overlook the important role played by technicians like Moritz Meyerstein who built made-to-order scientific tools for Gauss and Weber (see Section 10.1).

During Riemann's lifetime, Weber's law remained the paradigmatic result for Continental electrodynamics, as it was also supported by Franz Neumann's phenomenological theory of induction [Archibald 1989b]. This utilized a potential function adapted from Ampère's force law for closed currents [Darrigol 2000, 47]. In the meantime, a very different approach was slowly unfolding in Great Britain, where the experimental physicist Michael Faraday pursued an early form of field physics. Faraday's electric and magnetic lines of force eventually gained cogency in the hands of James Clerk Maxwell, whereas William Thomson pursued a program that remained rooted in mechanical principles, ultimately rejecting Maxwell's new theory of electromagnetism. By the 1870's Helmholtz had become an adherent of the more physically grounded British approach, while subjecting Weber's theory to the fundamental criticism that it contradicted conservation of energy [Caneva 2021]. These developments culminated in the 1880s with the famous experiments of Heinrich Hertz, which firmly established Maxwellian electrodynamics as the new paradigm [Darrigol 2000, 209–264].

As described in Section 2.7, not long after Weber introduced his new inverse-square law for electrodynamics Riemann had begun to develop an ether theory to account for both gravity and light. This theory aimed to create mathematical foundations that would link the laws of motion with their underlying causes rather than merely stipulating a formula for action at a distance. Later, in 1858, Riemann found a way to derive Weber's law in electrodynamics by modifying Poisson's static potential function so that it propagated over finite time intervals [Riemann 1867a/1876] (for details, see [Archibald 1989a, 54–56]). His note began:

> To the Royal Society I take the liberty of communicating a remark which brings the theory of electricity and magnetism into close connection with that of light and radiant heat. I have found that the electrodynamic effects of galvanic currents can be explained if one assumes that the effect of an electric mass on the others does not occur instantaneously, but propagates to them at a constant speed (equal to the speed of light within

> the limits of observational error). Under this assumption, the differential equation for the propagation of the electric force becomes the same as that for the propagation of light and radiant heat. [Riemann 1867a/1876, 270]

Riemann's basic idea was to introduce a retarded potential function, which he showed would propagate with a velocity $\alpha$ comparable to the speed of light. This potential $U$ was thus defined via the equation

$$\frac{\partial^2 U}{\partial t^2} - \alpha^2 \left( \frac{\partial^2 U}{\partial x^2} + \frac{\partial^2 U}{\partial y^2} + \frac{\partial^2 U}{\partial z^2} \right) + \alpha^2\, 4\pi\rho = 0,$$

which can be viewed at the non-homogeneous wave equation for $U$:

$$\Delta U - \frac{1}{\alpha^2} \frac{\partial^2 U}{\partial t^2} = 4\pi\rho.$$

Riemann withdrew this paper, so it never appeared during his lifetime. When Dedekind received the paper Riemann's widow had collected, this was one of the three he decided could be published without further emendation. He therefore submitted it to Poggendorff's *Annalen*, which prompted Rudolf Clausius to write the editor, pointing out a mathematical error. Heinrich Weber later wrote that he found Clausius's analysis convincing, but he also conjectured that Riemann may have withdrawn this note after discovering the same mistake. Weber nevertheless decided to include it the first edition of Riemann's *Werke*, in part because of interest in the methodology (see also [Reiff/Sommerfeld 1902]). Carl Neumann, son of the Königsberg physicist Franz Neumann, later took up retarded potential in electrodynamics himself [Archibald 1989a, 57–64]. These then came to play a role in the debates between followers of Weber's theory and its principal German critic, Hermann von Helmholtz.

In his unpublished memorial lecture, Ernst Schering had this to say about the motivation behind Riemann's original paper:

> Riemann sought to explain Weber's fundamental law, which accounted for the effects of galvanic currents in terms of interactions between moving electrical particles, by assuming a gradual spread of the forces emanating from a point, while showing that all phenomena are explained to within sufficient accuracy based on the observations then currently available. He submitted a draft of his investigation to the Royal Society of Sciences, but later expressed his satisfaction that it went unpublished because he had since found a clarification of his law that showed it satisfies certain general principles as do other fundamental laws of forces. He based this on Cauchy's theory of motion for the luminiferous aether, which he derived based on a minimizing condition for the value of a certain integral. By adding some simple terms to the function occurring in this integral, he gave it a form such that this minimizing condition includes the laws of all the relevant forces, its form being analogous to that of the Gaussian principle of least constraint. (*ibid.*)

These remarks make clear that Schering knew about the contents of "Beitrag zur Elektrodynamik" [Riemann 1867a/1876] before it appeared in *Annalen der Physik und Chemie* soon after Riemann's death. Interestingly, Schering's explanation as to why Riemann withdrew it from publication differs from the one

Weber offered in [Riemann 1876/1892/1902, 1892: 293]. The integral expression Schering described above is evidently that found in the fragment published posthumously under the title "Gravitation und Licht" [Riemann 1876/1892/1902, 1892: 538], which has usually been dated to the period 1853–54. On the other hand, the paper Riemann withdrew was submitted to the Göttingen Scientific Society in February 1858, which would mean that if Schering's assertion was correct, then the undated fragment must have been composed some five years later than had been presumed.[7]

After he took on principal responsibility for preparing Gauss's *Werke*, Schering likely had little time to pursue these ideas further. On the other hand, he later actively blocked Klein's efforts to obtain support from the Göttingen Scientific Society for publishing additional works by Riemann [Gierl 2004, 180]. Schering's memorial lecture for Riemann from December 1866 only appeared long after his death in 1897, whereas his brief obituary article [Schering 1867] had an immediate impact, because he sent an offprint to Hermann von Helmholtz, thereby sparking his paper on what today is called the Riemann-Helmholtz-Lie space problem [Koenigsberger 1902, 138] (for details, see Section 6.3).

## 9.4 Research on Open Currents

With the above background in mind, it becomes easier to appreciate Riemann's early interest in problems involving open currents. Section 2.7 touched on his research on residual charges in a Leyden jar, the topic of a lecture he delivered in September 1854 at the annual meeting of natural scientists held that year in Göttingen. The upshot of Riemann's talk was an argument in support of Franklin's single fluid theory of electricity, as opposed to the dualistic theory then favored by many German scientists [Riemann 1876/1892/1902, 1876: 48–53]. Rudolph Kohlrausch, a close collaborator of Wilhelm Weber, had initially encouraged Riemann to write up his results for publication in Poggendorff's *Annalen der Physik und Chemie*, but for unclear reasons Riemann later decided to withdraw his paper, which only appeared posthumously in [Riemann 1876/1892/1902, 1876: 345–356].

Ernst Schering was a young student in Göttingen at that time, one of the few who regularly attended Riemann's lectures. His memorial lecture for Riemann was described earlier in Section 12.1. Schering's feigned intimacy with Riemann was harshly exposed by Karl Hattendorff, but even so his memorial lecture suggests he was well acquainted with Riemann's physical views. Schering described his approach as one that began by eliminating the effects of reciprocal forces that acted at a distance, as these did not involve spatial properties as such. He then transformed the laws governing masses, bound electrical charges, and

[7] I know of no other traces of this unified aether theory in the extant sources, including Riemann's lectures on the mathematical theory of gravity, electricity, and magnetism from the summer of 1861 (https://gdz.sub.uni-goettingen.de/id/PPN631015833).

closed galvanic currents by viewing the corresponding forces in terms of movements of a homogeneous space-filling medium. The points in space from which these forces emanate he took to be places where this medium became infinitely dense, thereby corresponding to spatial singularities that lie outside the medium proper. Mathematically this Ansatz involved a Pfaffian transformation of the expression for the virtual moment of the forces. Physically, Riemann drew on the observations of Kohlrausch for the reciprocal action between dielectric bodies as well as the electrical action within them. He combined these findings with Franklin's hypothesis, according to which an electrified body will tend to stabilize in a definite state [Schering 1909/1990, 1909: 377–378]. Schering pointed to these theoretical and experimental findings as providing the motivation behind Riemann's lecture at the annual meeting of German natural scientists, held in Göttingen on 21 September 1854 [Riemann 1854].

Riemann's posthumously published paper contains a third section that he crossed out, though for unclear reasons [Riemann 1876/1892/1902, 1876: 348–349]. Heinrich Weber speculated that Riemann did not want to publish views that might arouse protest among the physicists of that era, though he also surmised that the paper was withdrawn because he did not wish to make certain editorial changes. Whatever the reasons may have been, he soon found another topic that perhaps suited him better. Thus, soon after dropping his paper on electrical residues in Leyden jars, he offered Poggendorff another contribution to physics: his theoretical treatment of the phenomenon known as Nobili's color rings [Riemann 1876/1892/1902, 54–61]. This paper and its prehistory were carefully analyzed in [Archibald 1991], which serves as the basis for what follows.

Leopoldo Nobili discovered this phenomenon in 1825, when he first noticed multicolored concentric bands formed on a plate. In carrying out various experiments, Nobili had the idea of replacing a thin wire electrodes with an extended conductor, such as a plate or disc made of silver, platinum, gold or brass. As a of byproduct of electrolysis, the plate was coated with a chemical substance, which Nobili studied using different materials. Roughly twenty years later, Edmond Becquerel undertook the first mathematical description of these phenomena. Analyzing the surface of the plate, he found it to be coated with a thin layer of lead peroxide, which steadily decreased in thickness in proportion to the distance from the cathode. Becquerel also noted the similarity with earlier observations of Newton's rings.

Riemann had studied this work as well as the subsequent investigations published by Emil du Bois-Reymond and Wilhelm Beetz. Regarding Becquerel's work he noted that it was based on the assumption that the current enters through a point on the surface and spreads out along straight lines according to Ohm's laws. Riemann's findings, however, undercut the claim that the flow lines could be straight, which again raises the question why he objected to Faraday's curved magnetic lines. In any event, Riemann's methodological approach differed sharply from those taken by his predecessors, as Archibald noted:

> Here it is the mathematics that is of primary interest, and indeed a great deal of specialized knowledge was required to understand the details of Riemann's treatment, let alone to produce it. Riemann's work follows clearly in the footsteps of Dirichlet ... via partial differential equations and boundary- value problems. Riemann's aim ... [was the] passage from the infinitesimal law to the description of phenomena by rigorous mathematical methods, without simplifying assumptions. One seeks to know whether the physical law in question is actually the case, not merely whether it describes experience approximately under certain simplifying assumptions. [Archibald 1991, 268–269]

These methodological concerns can be found again in several of Riemann's later papers on topics in mathematical physics, two of which deserve mention here. Given the breadth of his interests, it should come as no great surprise that Riemann took an intense interest in Helmholtz's works. In Section 4.6, a brief account was given of Riemann's investigation of the physiology of the oral apparatus, an interest directly related to his reading of *Lehre von den Tonempfindungen* (Sensations of Tone) [Helmholtz 1863]. This same work also inspired his paper on acoustics [Riemann 1860], one of his many pioneering studies. At the outset, he wrote:

> Although the differential equations that hold for the movement of gases have long been known, their integration has almost always only been carried out for the case when the pressure differences can be viewed as infinitely small fractions of the total pressure, and until recently investigators have been content to take only the first powers of these fractions into account. [Riemann 1860, 145]

That situation had now changed in Riemann's view as a result of Helmholtz's investigations of overtones in music, for which he had to take second-order terms into account. If the restriction to linear differential equations had in the past sufficed for the study of acoustics

> ... given the great progress that has recently been made by Helmholtz in the experimental treatment of acoustical questions, the results of more precise calculations could perhaps provide experimental research with some clues in the not too distant future. This, apart from the theoretical interest that the treatment of nonlinear partial differential equations offers, may justify the present communication (*ibid.*).

## 9.5 On the Physics of Space

Riemann never made any effort to publicize his probationary lecture after he delivered it in 1854, which explains why references to it during his lifetime have yet to be found. One contemporary of Riemann, however, may have been aware of the import of his ideas and discussed these with him. Ernst Schering claimed as much in his obituary for Riemann, where he wrote that his former teacher continued to ponder the implications of microphysical studies for Newton's laws of mechanics and the geometry of space. He added that on numerous occasions, Riemann shared his latest reflections on these matters with him [Schering 1867, 311]. Not long before this obituary appeared, Schering delivered a memorial lecture before the Göttingen Scientific Society, in which he gave more details regarding the motivation behind Riemann's reflections.

After describing the key role played by Gauss's intrinsic theory of surfaces and their differential invariants, he went on to say:

> [Riemann] particularly emphasized how our empirical knowledge of space based on observable bodies does not allow us to draw any conclusions regarding conditions that might pertain to geometric structures that are immeasurably large or small, and he drew attention to the distinction between the concepts of unlimited and infinite structures. He indicated that our ignorance of the immeasurably small leaves open the question whether space has a continuous or a discrete construction, but he also noted how this issue bears an influence on our views of natural laws based on Newton's natural philosophy. Riemann engaged deeply with these problems and he shared his ongoing results with me. [Schering 1909/1990, 1909: 377]

These remarks suggest that Schering might have already been familiar with Riemann's text from 1854 long before Dedekind finally published it in [Riemann 1868/1876]. Whether or not that was the case, however, he apparently never engaged with these ideas publicly until after Riemann's death. Schering's relations with Riemann and his legacy were, indeed, mysterious and difficult to understand (see 12.1). Nevertheless, he understood the thrust of Riemann's argument, having possibly discussed these ideas with him over an extended period of time. He was thus aware of the strategy Riemann had adopted for pursuing an ether physics that aimed to unify the fundamental forces of nature.[8]

Riemann's approach to the foundations of geometry clearly reflected his broader interests in the physical properties of spatial phenomena. Given his enormous talents as a mathematician, one might have thought that his philosophical orientation had no room for German *Naturphilosophie* in the tradition of Schelling and Goethe. This, however, would be to overlook Riemann's religious background and training, which channeled not only his personality but also much of his thought. Moreover, he lived during a time when science had not yet thrown Christian apologists into a panic, as Darwinism did soon afterward. Even the heated exchanges between physiologists in the materialism controversy of the 1850s apparently made little impression on Riemann (see 3.3).

A kindred spirit, who happened to have a deep appreciation for Riemann's speculative ideas, was the English philosopher-mathematician William Kingdon Clifford. In 1873, Clifford published an English translation of Riemann's 1854 lecture on the foundations of geometry in *Nature*. Three years later, he offered these noteworthy thoughts about the import of these ideas in "On the Space-Theory of Matter":

> Riemann has shown that just as there are different kinds of lines and surfaces, so there are different kinds of space of three dimensions; and that we can only find out by experience to which of these kinds the space in which we live belongs. In particular, the axioms of plane geometry are true within the limits of experiment on the surface of a sheet of paper, and yet we know that the sheet is really covered with a number of small

---

[8] Important documentary evidence relating to Schering's interest in these matters can be found in the manuscript Aetherbewegungen [und] Hypothesen der Geometrie in Cod. Ms. B. Riemann 36, which can be accessed online at https://gdz.sub.uni-goettingen.de.

> ridges and furrows, upon which (the total curvature not being zero) these axioms are not true. Similarly, he says, although the axioms of solid geometry are true within the limits of experiment for finite portions of our space, yet we have no reason to conclude that they are true for very small portions; and if any help can be got thereby for the explanation of physical phenomena, we may have reason to conclude that they are not true for very small portions of space.

The affinity Clifford felt for Riemann's ideas clearly derived from their shared interest in microphysics, i.e., physical phenomena in the small. In Riemann's case, he almost never revealed his more speculative ideas publicly, so the publication and translation of his Habilitation lecture was practically the first glimpse of his philosophical position. Nevertheless, his lectures on partial differential equations and potential theory clearly complemented the ideas he espoused in connection with ether physics. Clifford, on the other hand, had no inhibitions when it came to speculating about the ultimate nature of things, so that the publication of [Riemann 1868/1876] provided him with a welcome opportunity to expand on Riemann's ideas.

> I wish here to indicate a manner in which these speculations may be applied to the investigation of physical phenomena. I hold in fact
>
> (1) That small portions of space are in fact of a nature analogous to little hills on a surface which is on the average flat; namely, that the ordinary laws of geometry are not valid in them.
>
> (2) That this property of being curved or distorted is continually being passed on from one portion of space to another after the manner of a wave.
>
> (3) That this variation of the curvature of space is what really happens in that phenomenon which we call the motion of matter, whether ponderable or etherial.
>
> (4) That in the physical world nothing else takes place but this variation, subject (possibly) to the law of continuity.
>
> I am endeavouring in a general way to explain the laws of double refraction on this hypothesis, but have not yet arrived at any results sufficiently decisive to be communicated. [Clifford 1876]

Clifford was only 35 at the time of his death in 1879; he, too, died of tuberculosis. A number of physicists afterward set forth various ether theories, including the influential electron theory of Hendrik Antoon Lorentz. Aside from Clifford, however, few others took inspiration from Riemann's physical ideas.[9]

Finally, just a last word about what later made this forgotten lecture so famous (see [Jost 2025]). One should tread lightly here, rather than simply leaping from Riemann to Einstein and space-time curvature. It has occasionally been noted that [Riemann 1868/1876] was one of the several texts that Einstein and two friends in Bern (the "Olympia Academy") read and discussed a few years before Einstein hit on the ideas that led to (special) relativity [Einstein 1989, xxiv–xxv]. It seems, though, that this encounter with Riemann's ideas left no impression on him at the time, nor did he later do more than pay lip service to Riemann, while helping to make the mathematician's name become famous.

---

[9] An exception might be made for the Italians: Beltrami, Betti, and Ricci; see [Tazzioli 2025, 30, 36].

Einstein's path to his gravitational field equations is a much-studied topic that need not detain us here (see [Janssen/Renn 2022]). As is well known, he framed his new theory of gravitation within the context of 4-dimensional spacetime physics, which made it seem natural to view Riemann's lecture as an anticipation of central ideas in Einstein's general theory of relativity. Hermann Weyl even invoked the magic word "divination," though he made clear that Riemann's physical speculations differed fundamentally from Einstein's [Weyl 1923, 46]. This stands in contrast to the position taken by Weyl's former mentor, David Hilbert, who imagined he had unlocked the secret behind Riemann's bold attempt to unite light and gravitation.[10] Circumstances such as these helped to create tremendous interest in Riemann's ideas, though his lecture still remains one of the more mysterious chapters in his life.

[10] See his assertion at the outset of [Hilbert 1915, 32], a claim he later dropped in his revised text [Hilbert 1924].

# Chapter 10
# Gaussian Influences

Gauss was well past his prime when Riemann first met him, which partly explains why they had few if any substantive discussions concerning scientific matters. What Riemann seems to have known about Gauss's ongoing interests he heard from Wilhelm Weber, with whom he worked closely. Nevertheless, practically from the moment he first set foot in Göttingen in April 1846, Riemann began studying various works of the legendary Carl Friedrich Gauss. On April 30, he borrowed his *Disquisitiones Arithmeticae* and Volume 6 of the *Commentationes*, which contains the fundamental treatises on biquadratic residues [Gauss 1828a] and differential geometry of surfaces [Gauss 1828b]. Later in the semester, he took out Gauss's fundamental study on terrestrial magnetism [Gauss 1833/1995], which undoubtedly supplemented what he was then learning in the course taught by Gauss's protégé Benjamin Goldschmidt, who had recently acquired the title of professor. Through Goldschmidt, Riemann gained exposure to the field of cartography in which conformal mappings had long played a central role. He also presumably learned at this time about Gauss's prize-winning paper [Gauss 1825a], which was published in Volume 3 of *Astronomische Nachrichten*.

Oddly enough, though, he only got to read that paper in Berlin. As a matter of fact, the first two volumes of that journal, but not the third, are listed as belonging to the Gauss Bibliothek, today located in the old library. Riemann must have known this, as he later wrote home that he had not been able to get his hands on this third volume until he came to Berlin (see the letter to his father in Section 2.3). This illustrates only one of several instances showing how closely he followed in the tracks laid down earlier by Gauss, someone he imagined rather more as a daunting competitor as opposed to a helpful mentor. Klein's colorful imagery, portraying Gauss as the highest peak in an impressive chain of mountains [Klein 1926, 62], surely reflected the earlier contemporary view of him, at least from the vantage point of Göttingen.

When Riemann returned from Berlin, he began studying the volumes in the series *Results of Observations of the Magnetic Union* [Gauss/Weber 1837–43]. Two years later, in May 1851, just months after Goldschmidt's tragic death,



D. E. Rowe, *Bernhard Riemann: His Life and Wondrous Mathematical Legacy*,
https://doi.org/10.1007/978-3-032-25457-3_11

Riemann gave a lecture on Gauss's bifilar magnetometer drawing on the explanations in [Gauss 1840b]. By this time, he had begun working closely with Wilhelm Weber, who occasionally told Riemann about Gauss's research interests. Paging through the long list of works Riemann borrowed from the University Library, his early interest in terrestrial magnetism appears particularly striking. Even in later years, he borrowed the famous *Atlas des Erdmagnetismus* [Gauss/Weber 1840] on several occasions, a clear indication that this fascination had not subsided, despite his many other pursuits.

Riemann's borrowings confirm that pure mathematics, particularly number theory and algebra, held relatively little attraction for him, whereas mathematical physics and natural philosophy in general dominated his attention during many phases of his life. For some time, he immersed himself in the works of Euler and Cauchy as well as Newton and Leibniz. In the summer semester of 1856, Riemann taught a 4-hour course on the elasticity theory of solid bodies. Appropriately enough, he went to the library to fetch a copy of Gabriel Lamé's *Leçons sur la théorie mathématique de l'élasticité des corps solides* [Lamé 1852]. For years afterward, he must have spent countless hours studying Lamé's book, which he borrowed over and again. The fact that he did so points once again to the importance of Riemann's lecture courses. Indeed, browsing through his *Werke*, one would never encounter Lamé's name.

As noted earlier, Gauss never acted as a mentor to Riemann in the sense that term is used today. But this does not mean, of course, that Gauss had little or no influence on him; in fact, the very opposite was the case. For although he cited other authors sparingly, if at all, Riemann was never hesitant when it came to Gauss, whose works he named more often than those of any other mathematician.[1]

## 10.1 Precision

The Introduction to this book sketched the early career of Gauss, highlighting the period when he collaborated with Wilhelm Weber. Few sources survive that document how Riemann's career later crossed those of Gauss and Weber, but the evidence available strongly indicates that physics and philosophy then dominated his attention. It should be emphasized as well that Riemann's scientific outlook was strongly influenced by a general trend toward methods based on precise measurements and systematic mathematical methods for handling data. Gauss's method of least squares was only the most striking example of the latter (for which, see [Olesko 1997]). The problem of coordinating celestial and terrestrial data had an even older tradition in Göttingen going back to the statistical methods developed by Tobias Mayer for his study of lunar libration [Stigler 1986].

---

[1] For an older, but still insightful assessment of the intellectual ties linking Gauss and Riemann, see [Scholz 1980, 26–30, 53–55].

During the 1820s, Gauss turned the art of surveying into the science of higher geodesy [Scholz 1992]. Drawing on the work of the historian Hans-Ulrich Wehler, Scholz took note of the fact that the massive surveying and cartographic projects in the German states that took place over the course of several decades were only partially motivated by the need for better military maps. The government of Hanover, in particular, had an even greater interest in utilizing this data to assess the value of real estate formerly in the hands of the landed aristocracy. This land reform movement reflected the shift in ownership from a feudal society to a modern capitalist system, a trend that accelerated over the course of the first half of the nineteenth century. The state thus needed to survey these lands to establish ownership claims, on the one hand, but also to assess real estate taxes, on the other.

For these purposes, the refined methods developed by Gauss were, of course, largely superfluous. His aims were scientific; indeed, his higher geodesy was intimately connected with his work on the differential geometry of surfaces in [Gauss 1828b]. Moreover, one of his major goals was to investigate the shape of the geoid modeled on a rotational ellipsoid, part of the larger project aimed at studying the physical properties of the earth.[2] Using data from the Hanoverian surveys, Gauss was able to show that this model gave a surprisingly accurate fit for the geoid [Scholz 1992, 634]. Riemann surely knew that Gauss's study of conformal mappings [Gauss 1825a] ended by presenting a conformal map from a rotational ellipsoid to a sphere. What he may not have known, or perhaps only learned later, was that Gauss had used such a mapping for geodetic purposes years before [Scholz 1992, 634–635].

In his impressive survey of *Electrodynamics from Ampère to Einstein* (2000), Olivier Darrigol entitled his second chapter "German Precision" and his Section 2.3 carries the heading "The Gaussian spirit" [Darrigol 2000, 49–54]. What he described therein constituted key elements and formative influences in Riemann's educational experience. With respect to terrestrial magnetism (see Section 9.2), Gauss made his case with these words in 1833 :

> Magnetic experiments are becoming capable of a precision which far surpasses everything that went before, and its fundamental laws can have a truly mathematical precision, so that the separation between actual so-called physics and applied mathematics here too (as in the theory of motion and optics long ago) begins to disappear, and the thorougher treatment begins to fall to the mathematician. (quoted from [Jungnickel/McCormmach 1986, 1: 70])

One should not, on the other hand, identify this Gaussian spirit with Göttingen alone, as the scientific career of Alexander von Humboldt clearly shows. Indeed, this same emphasis on mathematical precision went hand-in-hand with the training students received in the famous seminar founded in Königsberg in 1834 by Franz Neumann and Carl Jacobi in alliance with the astronomer Friedrich Wilhelm Bessel [Olesko 1991]. Gauss had been instrumental in promoting Bessel's early career, which led in 1838 to the first reliable determination

[2] The geodetic project promoted by Johann Jacob Baeyer, which Riemann studied in the early 1860s (see 4.4), had a similar scientific purpose.

of the distance to a star when Bessel published the value of stellar parallax for 61 Cygni. Riemann was an early participant in the seminar founded by Stern and Weber in 1850. Its members did not experience the same kind of stringent regimen as those who attended the Königsberg seminar, and yet there was a striking parallelism between these two institutions. Neumann and Weber were highly successful physics teachers, both of whom utilized higher mathematics in their work. This placed both in an advantageous position with their colleagues in astronomy, who were theorists with great mathematical prowess. Neither Bessel nor Gauss, however, taught in these two seminars, which trained many young leaders of the next generation.

The Königsberg mathematics/physics seminar, in particular, significantly improved the standards of education for those who would teach those subjects at secondary schools. For the first time anywhere in Germany, students enrolled in the seminar followed a a well-structured plan of studies with strong doses of mathematics and physics. Neumann's student Gustav Kirchhoff already undertook original research on electrical circuitry; he later joined Hermann von Helmholtz and Robert Bunsen on the faculty in Heidelberg. Alfred Clebsch was another important graduate of the Königsberg seminar, which played a significant role in the long-term emergence of theoretical and mathematical physics in Germany [Jungnickel/McCormmach 1986].

Precise measurement clearly required instruments that could meet the demands of researchers. In Göttingen the firm owned by J. Philipp Rumpf specialized in manufacturing state-of-the-art scientific instruments. After Rumpf's death in 1833, his company was purchased by Moritz Meyerstein, who later also acquired the title of University Mechanic [Hentschel 2005]. Riemann almost certainly knew him, since he frequently visited the observatory to install or care for the devices he sold to the University. He also produced equipment for Weber's laboratory experiments. Beyond that, he was a good friend of Riemann's two teachers, Goldschmidt and Stern, all three of whom came from Jewish families. Meyerstein's workshop typically employed a half dozen technicians, who specialized in constructing geodetic and astronomical instruments, including leveling instruments, theodolites, meridian circles, but also magnetometers, galvanometers, heliostats, and spectrometers. Since Riemann's principal interest concerned geomagnetism, this section focuses on the early development of magnetometers.

During his long stay in France, Alexander von Humboldt used an instrument for measuring magnetic strength that was developed by Henri Gambey. Gauss's praise for Humboldt's early investigations of the varying intensity of the earth's magnetic field was quoted in Section 1.3. When he wrote those words, however, he followed them with an account of a new instrument that met a far higher standard of precision. Humboldt used a compass needle suspended from a thread around which it rotated freely and would align with the local geomagnetic field. If displaced from this equilibrium position, the needle oscillated about it and the oscillation period was then used to find the strength of the magnetic field acting on it by taking the torsional moment of inertia into account. This method gave reasonably accurate results for a specific locality over a short range of time, but it assumed that the strength of the magnetized needle remained constant. This raised an obvious problem for comparative measurements, since the magnetic strength was affected by changes in temperature.

Fig. 10.1: Placement of a magnetometer in the Göttingen Magnetic Observatory.

Gauss wanted to find a precise way to measure the earth's magnetic field in absolute units that did not depend on the measuring apparatus itself (see [Darrigol 2000, 51–53]). This meant finding a way to measure magnetic intensity in given absolute units for distance, mass, and acceleration. The magnetic mass in absolute units corresponds to fixing the constant in Coulomb's force law so that

$$F = \frac{m \cdot m'}{d^2}.$$

Gauss described how this can be done in *Intensitas vis magneticae terrestris ad mensuram absolutam revocata* [Gauss 1833/1995], a work Riemann already borrowed from the library on 23 July 1846. His method involved counteracting the effects of the first magnet by employing a second one; this involved taking several different measurements from different relative positions, from which one could then calculate the intensity $T$. The apparatus he used is often called a unifilar magnetometer to distinguish it from the bifilar case that Riemann was familiar with (Fig. 10.1).

This employed a twenty-pound bar magnet suspended horizontally by a single thread that allows it to rotate freely with negligible torsion effect. Part of the measurement then determines the period of oscillation $\tau$ for a standard displacement from equilibrium. This magnet is enclosed in housing to prevent air currents from affecting the measurements. A small mirror was attached to the magnet and this reflects a scale into a telescope located six meters away. This enables an observer to detect the angle of rotation very precisely using scale markings on the telescope lens. Once the oscillation period $\tau$ is found, one could easily calculate $TM$, where $M$ is the magnetic moment of the needle. The second magnet was mounted horizontally on the base of a tripod that can be rotated using a goniometric scale. By placing this second magnet in different positions relative, one can measure its effect on the first magnet. When the two are in equilibrium, one can then determine the value of $T/M$, and thus determine $T$, the horizontal intensity.

This was essentially the method Gauss and Weber employed up until 1837. On September 19th of that year, Gauss held a public lecture in the Royal Society in which he described the advantages of his new device, the bifilar magnetometer [Gauss/Weber 1837–43, 1838: 1–19]. As the name suggests, this device utilizes a bar magnet suspended by two vertical threads. Gauss could then show that the restoring torque depended directly on the weight of the magnet and was proportional to the magnetic torque itself. This enabled him to eliminate the problem of finding the oscillatory period $\tau$, which involved making observations over an interval of time. Moreover, with the bifilar magnetometer one could immediately find the equivalent of $TM$, making it possible to gain more accurate measurements of magnetic fluctuations. An unanticipated spinoff from this invention was Weber's electrodynanometer, which was inspired by an analogous principle [Darrigol 2000, 57–60].

Gauss's bifilar magnetometer did not displace its forerunner completely. The former was mainly advantageous for measuring horizontal magnetic intensity, whereas the unifilar magnetometer was used to find declination and its variations. Prior to Gauss's two inventions, there was possibility of finding a system of absolute measurement for magnetic field strength. In tandem with Gauss, in the *Resultate* for 1840 Weber introduced the first absolute electric current unit, which served as the basis for the absolute electromagnetic system he later developed.

For the year 1837, Gauss reported there were seven coordinated observation dates. Sixteen stations sent data in time to appear in the annual report, though some did not take readings on all seven dates and some only for a few [Gauss/Weber 1837–43, 1838: 130–138]. One also notes that a fair amount of this activity took place at localities that could be described as belonging to Gauss's personal scientific network: e.g., Encke in Berlin, Gerling in Marburg, and Möbius in Leipzig. Furthermore, Gauss listed the names of all those who took part in this enterprise to the extent that information was known to him. In Göttingen these were 19 such individuals under Weber's direction, several of them Riemann knew well: Goldschmidt, Listing, Meyerstein, Sartorius, Stern, and Ulrich. He, of course, also knew the leader of the group in Berlin, namely Franz Encke, with whom he had studied. For all of these localities, very precise information was given regarding the timing of the readings, yet Gauss remarked that he would refrain from summarizing the results, since he anticipated that the picture would change in the future as more data was processed [Gauss/Weber 1837–43, 1838: 136].

When Wilhelm Weber took up his interim professorship in Leipzig in 1843, his interest in terrestrial magnetism continued unabated. The Ministry in Saxony even provided funding for building an iron-free magnetic observatory, like the one he and Gauss had designed in Göttingen (Fig. 9.1). Moreover, he hoped to publish a seventh volume of the *Resultate* as a follow-up to the already existing six volumes [Gauss/Weber 1837–43]. However, this project never materialized, since the number of magnetic observatories had in the meantime increased so drastically that no single person could handle all the observational data. Moreover, Gauss's assiduous collaborator, Benjamin Goldschmidt was also now separated from Weber, whereas Gauss himself was by now losing interest in magnetism as he turned his attention to purely mathematical researches.[3] In the meantime, Johann von Lamont, director of the Astronomical Observatory in Munich-Bogenhausen, became the leading center in Germany for recording geomagnetic observations.

As described in Section 9.2 and elsewhere in this book, Riemann's early exposure to the Gauss-Weber tradition in Goldschmidt's course on geomagnetism laid the groundwork for much of his subsequent work. The scientific values he imbibed in Göttingen had much to do with the ethos of precision that were embedded in that culture.

## 10.2 Geometry of the Complex Numbers

Standard accounts that touch on the history of the complex numbers usually focus on a handful of names as co-inventors of the complex plane, often called the Gaussian number plane. As the term implies, Gauss often gets this prize, though some have argued that the Genevan amateur mathematician Jean-Robert Argand deserves the title of first inventor. Far more significant and certainly of much greater interest is the larger story concerning the origins and early development of complex analysis, a field that virtually came to dominate research during the second half of the nineteenth century. By longstanding consensus, mathematicians and historians have come to regard Cauchy, Riemann, and Weierstrass as its three principal creators, though only Riemann pursued geometric function theory in its full glory. Cauchy took significant steps in that direction, to be sure, whereas Weierstrass famously rejected appealing to geometric arguments in analysis [Bottazzini & Gray 2013].

These general remarks should not, however, leave the impression that the eighteenth century left no mark on modern analysis. Riemann, in fact, made a careful study of major works by Euler and Lagrange, the two most formidable

[3] See Gauss's letter to Weber from May 21, 1843, quoted in the Introduction.

representatives of algebraic analysis.[4] Its practitioners prized formal manipulations that treated infinite expressions as though these were akin to finite polynomials. Such reasoning often disregarded questions involving convergence, one of the main issues that distinguishes analysis in the tradition of Cauchy and Weierstrass from their predecessors.

Riemann undoubtedly knew Euler's classic *Introductio in analysin infinitorum* [Euler 1748] from an early age and drew on many of its key results, not least his remarkable findings connected with the zeta function (see Section 8.2). Euler introduced the letter $e$ for the base of natural logarithms and famously showed that

$$e^{i\phi} = \cos\phi + i\sin\phi,$$

only one among numerous other related results [Euler 1748, 104]. In the spirit of Euler, one can easily confirm this formula by means of power series:

$$e^{ix} = \sum_{n=0}^{\infty} \frac{(ix)^n}{n!} = \sum_{n=0}^{\infty} \frac{(-1)^n x^{2n}}{2n!} + +i \sum_{n=1}^{\infty} \frac{(-1)^{n-1} x^{2n-1}}{(2n-1)!} = \cos x + i \sin x.$$

Since these series converge absolutely, this formal argument is completely sound.

Euler's result clarified an old dispute between Leibniz and Johann Bernoulli, dating back to 1712: how to interpret the number $\log -1$? In fact, since $e^{(\pi+2n\pi)i} = -1 \cdot 1 = -1$, we have $\log -1 = (\pi + 2n\pi)i$.

Riemann would eventually distance himself from this Eulerian approach to analysis, in part by passing over to complex analysis, which in this case leads to a very simple proof going in the opposite direction:

$$\begin{aligned}
z &= \cos\phi + i\sin\phi \\
dz &= (-\sin\phi + i\cos\phi)\, d\phi \\
&= iz d\phi \\
\int \frac{dz}{z} &= \int i d\phi \\
\log z &= i\phi \\
z &= \cos\phi + i\sin\phi = e^{i\phi}.
\end{aligned}$$

In his standard introduction to the course he regularly taught on complex functions Riemann emphasized how his approach to analysis differed from the older tradition of Euler:

> Earlier, a function of a variable quantity was defined as an analytic expression containing this variable ... (Euler). In the application of analysis to the problems of physics, the need first arose to replace this concept of a function with a broader one: namely, to understand the function of a real quantity $x$ as any other quantity $y$ that, for each value of $x$, also has a definite value that varies continuously with that of $x$ – without presupposing an analytic expression of the dependence between the two. The same

[4] The influence of algebraic analysis on the young Riemann is discussed in [Laugwitz 1996/1999, 43–52].

need as in physics arises in analysis wherever one must consider functions that are not explicitly given. [Neuenschwander 1996, 23]

One sees from this that Riemann conceived of this development as part of a dialectical process, one in which problems stemming from mathematical physics had played a leading role. Undoubtedly, he had in mind here the methods of potential theory and partial differential equations, both of which were promoted by Gauss (see Section 9.1).

In his publications, Riemann often pointed to specific works by Gauss, but an even more telling reference to [Gauss 1831] can be found in the various extant manuscripts documenting his lectures on the foundations of complex function theory [Neuenschwander 1996, 22]. In all his courses on complex function theory, Riemann began with introductory lectures so that students could grasp the foundational principles. Neuenschwander's critical edition was based primarily on a manuscript prepared by Ernst Abbe (see Section 3.5). This *Nachschrift* documents Riemann's lectures on elliptic and Abelian functions from the summer semester 1861, the final time he taught that course. A modern reader, when confronting the opening lecture, will likely be struck by the contrast between the elementary level of the mathematics, on the one hand, and the philosophical language used to describe it, on the other. One need only reflect, however, that the geometric representation of complex numbers was still quite novel, enough so that Riemann cited [Gauss 1831] when he introduced it. To motivate the concept of complex numbers, he first considered how negative integers arise as an extension of the natural numbers.

The first step is to look for examples of such numbers; then one can ask the question: under what conditions is it generally possible to think of the opposite of a quantity? – and in this case, the answer is: if what is measured is not a substance per se, but a relation between two substances, $A$ and $B$, then from the relation $B$ to $A$ one can also conceive of the opposite. ...

The same applies to complex quantities. First, we look for examples of such quantities in which $\sqrt{-1}$ has a conceivable meaning. This is the case when we consider a complex quantity as the measure of a change of position in the plane, determined in terms of magnitude and direction. [Neuenschwander 1996, 21–22]

Riemann went on to define $\sqrt{-1}$ as the mean proportional of 1 and $-1$, and then introduced the four units in the Gaussian plane $1, i, -1, -i$. He then commented:

The geometric interpretation of complex quantities is of great importance, apart from providing an example of their reality, in that it provides the means for a spatial representation of their values and thus provides a guide for abstract investigation. (*ibid.*)

Riemann was surely aware that Gauss had long hesitated before making his views on imaginaries known. When he finally did, his lengthy remarks had the tone of a ceremonial unveiling:

One soon recognizes after this, that one can only break into this rich area of higher arithmetic by completely new paths. The author had already given an indication in the first work that to do so a remarkable extension of the whole field of higher arithmetic was essentially necessary, but without at the time explaining more closely what this consisted of: the present work has the intention of bringing this matter to light. For

> the true foundation of the theory of biquadratic residues, this is none other than the extension of the field of higher arithmetic, which has otherwise only been extended to the real whole numbers, to also include the imaginaries as well, for these must be granted exactly the same status as the others. As soon as this has once been observed, that theory appears in a completely new light and its results take on a highly surprising simplicity [Gauss 1831, 171]

Riemann also took to heart what Gauss wrote about earlier traces of imaginaries in his work:

> Regarding the reality of negative numbers the situation has been clear for some time. It is only the imaginaries – formerly and sometimes still called impossibles [*unmögliche*] – which standing opposite the real numbers still remain less accepted. Merely tolerated, they thus appear more like an in itself contentless sign-game to which one categorically denies any intelligible substrate, without however wanting to disdain the rich tribute that this sign-game affords among the wealth of relationships between the real numbers.
>
> For many years the author has studied this highly important part of mathematics from a different standpoint, whereby the imaginary numbers can be used in connection with an object just as well as the negatives. Until now, however, there has been no appropriate occasion for publicly stating this in a definite manner, although the attentive reader can easily find traces of it in the work of 1799 on equations and in the *Preisschrift* on transformations of surfaces. [Gauss 1831, 175]

Gauss went on to argue that much of the confusion surrounding imaginary numbers was merely a matter of terminology. He pointed out that had the quantities $+1, -1, \sqrt{-1}$ been named direct, inverse, and lateral units – an allusion to their geometric properties under multiplication – the metaphysical doubts regarding the status of imaginaries never would have arisen.

In Riemann's *Grundlagen* [Riemann 1851/1876], he mentioned two important works by Gauss ([Gauss 1825a] and Art. 13 of [Gauss 1828b]). He cited these mainly for their conceptual importance, as they supported his general approach to complex functions. The first paper dealt with conformal mappings or, in the language of the day, mappings that preserve infinitesimal similarity.[5] The second reference pointed to Gauss's sphere mapping in surface theory. It should be emphasized that in Gauss's various works related to complex analysis one finds nothing that could be regarded as an anticipation of the idea of Riemann surfaces.

## 10.3 Hyperelliptic Functions

Little has been said up until now about [Riemann 1857a], the extremely important study Riemann wrote about functions that arise as solutions to the hypergeometric differential equation.[6] At the outset, Riemann gave a clear indication of the motivation behind it:

[5] An account of the likely influence of Gauss's paper on Riemann's early thinking can be found in [Laugwitz 1996/1999, 71–73].

[6] See the discussion in connection with Riemann's 1856/57 lecture course in Section 3.5.

> In the following treatise I have treated certain transcendental [functions] using a new method, which essentially remains applicable to any function that satisfies a linear differential equation with algebraic coefficients. Employing this, the results which were previously found partly through rather laborious calculations can be derived almost directly from the definition. This has been done in the present part of this treatise, mainly with the intention of giving a convenient overview of the possible representations useful for the many applications of such functions in physical and astronomical investigations. [Riemann 1857a, 62]

This paper might have been discussed earlier in conjunction with the one on Abelian functions [Riemann 1857b], as these were Riemann's two outstanding contributions to complex analysis building on the ideas in his dissertation [Riemann 1851/1876]. For two reasons, however, I chose instead to consider it only briefly in this section. The first has to do with the fact that Jeremy Gray long ago dealt with the whole story surrounding the hypergeometric differential equation from Gauss to Riemann and from Fuchs to Poincaré in such a broad and informative way that I only wish to point the interested reader to his impressive study [Gray 2008].

The second reason stems from wanting to emphasize that Riemann himself saw this purely analytic work as holding great importance for mathematical physics. In his prepublication announcement, Riemann wrote the following:

> This treatise is dedicated to a class of functions, which are useful for solving certain problems in mathematical physics. The series arising from these perform the same service for difficult [physical] problems as do the so often used series involving sine and cosine functions in the case of simpler problems. These applications, in particular to problems in astronomy, ... seem to have led Gauss to investigate them and to publish part of his findings in 1812 .... The now customary name hypergeometric series was already proposed earlier by Johann Friedrich Pfaff for the more general series in which the quotient of one term into the following is a rational function of the place value; whereas Euler, following Wallis, understood it to mean a series in which this quotient is an entire first-degree function of the place value. ...
>
> The unpublished part of Gauss's investigations on this series, which was found among his papers, has meanwhile been supplemented in 1835 by the work of Kummer in volume 15 of Crelle's Journal. ...
>
> Kummer succeeded in developing Euler's method into a procedure by which all transformations could be found; however, its actual implementation required such detailed treatment that he refrained from carrying out the transformations of the third degree and contented himself with completely deriving the transformations of the first and second degree and those composed from them.
>
> In the treatise under review, a method is applied to these transcendentals based on a principle presented in the author's inaugural dissertation (Article 20), which allows all previously found results to be obtained almost without calculation. The author will soon be able to present some further results obtained by means of the same method to the Royal Society. [Riemann 1876/1892/1902, 1876: 79–80]

Unfortunately, Riemann never kept that final promise, though many years later the substance of his lectures on hypergeometric functions from 1858/59 came to light (see Section 12.5).[7]

---

[7] For an overview of those lectures and much more, see Appendix 2 in [Gray 2008].

Riemann's original innovation in this study was to show that one could go beyond the known special cases of functions satisfying the hypergeometric differential equation by considering complex-valued functions with any three branch points $a$, $b$, $c$ that fulfilled certain additional properties. In the classical hypergeometric case $a = 0$, $b = \infty$, $c = 1$, whereas Riemann generalized this by associating to each of the points $a$, $b$, $c$ two exponents, $\alpha, \alpha'; \beta, \beta'; \gamma, \gamma'$ in order to define a $P$-function with this corresponding data:

$$P \begin{Bmatrix} a & b & c & \\ \alpha & \beta & \gamma & x \\ \alpha' & \beta' & \gamma' & \end{Bmatrix}.$$

Such a function must then satisfy three conditions: 1) $P(x)$ is single-valued and finite except when $x = a,\ b,\ c$; 2) if $P', P'', P'''$ are any three branches, then these satisfy a linear homogeneous equation $c'P' + c''P'' + c'''P''' = 0$; and 3) at each of the branch points $x = a,\ b,\ c$, $P$ can be expressed in the form $c_\alpha P^{(\alpha)} + c_{\alpha'} P^{(\alpha')}$, where $P^{(\alpha)}$ is branched like $(x-a)^\alpha$ and $P^{(\alpha')}$ like $(x-a)^{\alpha'}$. The same conditions also apply for the exponents $\beta$, $\beta'$ and $\gamma$, $\gamma'$. Riemann further demanded that the three differences $\alpha - \alpha'; \beta - \beta'; \gamma - \gamma'$ be integers and that the sum $\alpha + \alpha' + \beta + \beta' + \gamma + \gamma' = 1$.

Much of what then follows was devoted to the transformation properties satisfied by such $P$-functions. Riemann noted to begin with that all $P$-functions with the same exponents $\alpha, \alpha'; \beta, \beta'; \gamma, \gamma'$ can be transformed to the hypergeometric case:

$$P \begin{Bmatrix} 0 & \infty & 1 & \\ \alpha & \beta & \gamma & x \\ \alpha' & \beta' & \gamma' & \end{Bmatrix}.$$

In the third section, Riemann discussed the global properties of a $P$-function using methods that look very much like monodromy groups. At the end of the section following, he showed that for two branches of a $P$-function whose quotient is not constant their linear combinations determine all other $P$-functions with the same exponents. Starting with the fifth section, Riemann began examining three different classes of transformations that act on $P$-functions. Along the way he derived the 24 solutions of the hypergeometric differential equation obtained by Kummer, and by the end of this tour de force he summed up all the distinct types of integral representations for a given $P$-function without any explicit formalism. Nor did he derive the general linear differential equation satisfied by these functions, which was first published by Erwin Papperitz in 1889.

An important study that served to link Riemann's paper with subsequent work of Klein and others was [Schwarz 1872].

Methodologically, it is interesting to compare Riemann's opening remarks, cited above, with the motivation he later gave for his paper on shock waves [Riemann 1860]:

> This study does not claim to provide useful results for experimental research; the author wishes that it only be regarded as a contribution to the theory of nonlinear partial differential equations. Just as the most fruitful methods for the integration of linear partial differential equations were not found by developments aimed at the general problem, but rather emerged from the treatment of special physical problems, so the theory of nonlinear partial differential equations also seems to be best promoted by a detailed treatment of special physical problems that take all secondary conditions into account. In fact, the solution of the specific problem which forms the subject of this paper required new methods and approaches that led to results that will probably also play a role in more general tasks. [Riemann 1876/1892/1902, 1876: 79–80]

## 10.4 Non-Euclidean Geometry and Higher Manifolds

For centuries prior to the discovery of non-Euclidean geometry, countless numbers of writers attempted to prove Euclid's famous parallel postulate. Some tried to prove it outright; others to replace it with a more self-evident principle from which it could be deduced. Practically no one imagined it might not be true. Gauss refused to write about the subject, but privately he had long expressed his views to some of his colleagues, who were mainly fellow astronomers. The issue at stake was not so much whether physical space was Euclidean – as had always been presumed – but rather whether one could somehow prove this. In the meantime, two other mathematicians – the Hungarian János Bolyai and the Russian Nikolai Lobachevsky – had pursued the same ideas and had written about a new theory of geometry in which Euclid's parallel postulate was false. Today, this is known as hyperbolic geometry, a name first introduced by Felix Klein.

Gauss called this anti-Euclidean geometry, and he saw that it was logically defensible. He was also pleased to learn that Friedrich Wilhelm Bessel took the same view:

> The ease with which you delved into my views on geometry gives me real joy, given that so few have an open mind for such. My innermost conviction is that the study of space is a priori completely different than the study of quantities; our knowledge of the former is missing that complete conviction of necessity (thus of absolute truth) that is characteristic of the latter; we must in humility admit that if number is merely a product of our minds, space has a reality outside our minds whose laws we cannot a priori state .... (Gauss to Bessel, 9 April 1830, in https://gauss.adw-goe.de)

Two years later, Gauss received Janos Bolyai's appendix to his father's book, *Tentamen*, published in 1832. The author, Farkas Bolyai, was an old friend of Gauss from their student days, and so he wrote to ask Gauss for his opinion of his son's new theory. Before answering, Gauss wrote the following to Christian Gerling in Marburg:

> In addition I note that in recent days I received a small work from Hungary on non-Euclidean geometry in which I find all of my ideas and results developed with great elegance, although in a concentrated form that is difficult for one to follow who is not familiar with the subject. The author is a very young Austrian officer, the son of a friend of my youth with whom I had often discussed the subject in 1798, although my ideas at that time were much less developed and mature than those obtained by this young man through his own reflections. I consider this young geometer, v. Bolyai, to be a genius of the first class. (Gauss to Gerling, 14 February 1832, in https://gauss.adw-goe.de)

Unfortunately, the tenor of Gauss's response to Farkas Bolyai was quite different, as he wrote: "To praise it would amount to praising myself. For the entire content of the work ... coincides almost exactly with my own meditations which have occupied my mind for the past thirty or thirty-five years" (Gauss to Farkas Bolai, 6 March 1832, in https://gauss.adw-goe.de). This letter sparked a great deal of acrimony, though this was surely not Gauss's intention.

In the meantime, Nicolai Lobachevsky had been publishing his results on non-Euclidean geometry in Russian journals. In 1837, a summary of these appeared in Crelle's Journal, making these ideas accessible to the German-speaking mathematical world. According to Sartorius von Waltershausen, Gauss made use of his measurements of a huge terrestrial triangle to test whether the angle sum might deviate slightly from two right angles [Sartorius 1856/2012, 81].[8] The result was negative, but this empirical test only lent support for the hypothesis that the geometry of space was *locally* Euclidean. Riemann's lecture, on the other hand, underscored that the global question was completely open.

Since Riemann's posthumously published lecture played a major role in the reception of non-Euclidean geometry [Volkert 2013], one naturally has to ask: what did Riemann know about earlier investigations dealing with alternatives to Euclidean geometry? Could he have spoken about these things with Gauss, whose geometrical work ? If so, not a trace of evidence survives that suggests they ever discussed these matters. Nor is this terribly surprising, considering that Gauss had become increasingly remote with age. Moreover, there is no mention of works by Bolyai and Lobachevsky anywhere in Riemann's various writings.

Regarding the fertility of investigations of the Euclidean axioms, Erhard Scholz cited this passage from one of Riemann's unpublished texts:

> If the possibility of this treatment of geometry is also of interest, carrying it out would be extremely unfruitful. For by doing so we would not find any new theorems, whereas the simple and most comprehensible features of space would by this means become complex and difficult. [Scholz 1982b, 218]

Taking his unpublished reflections on analysis situs and foundations of geometry into account, one can hardly imagine Riemann writing such words had he been aware of the works of Bolyai and Lobachevsky. The fact that he mentioned

[8] On the importance of Gauss's work in geodesy for his understanding of geometry, see [Scholz 1992].

only spaces of constant *positive* curvature and neglected negative curvature (the hyperbolic case) supports this same conclusion. Moreover, the hypotheses Riemann has in mind reflect different types of mathematical and physical principles, but the central concern in the end concerns the properties of physical space.

In a study of Riemann's early reflections on the concept of a mathematical manifold, Scholz shed new light on the nature of his interest in foundational issues in geometry [Scholz 1982b]. A fragment on analysis situs, which Riemann wrote before composing his qualifying lecture, begins by asserting:

> The concept of a manifold of higher dimension exists independent of our perceptions (*Anschauungen*) of space. The [notions of] space, plane, and line are only the most visible (*anschaulichste*) examples of a manifold of three, two, or one dimension. We could develop all of geometry without even the slightest appeal to spatial perception. [Scholz 1982b, 228]

Here the context Riemann had in mind was akin to an affine manifold as opposed to a differentiable manifold. Yet in both cases his point of departure was an analytic structure (a *Zahlenmannigfaltigkeit*), a standpoint that did not require the axioms employed in synthetic geometry.

Regarding Riemann's interest in higher-dimensional manifolds, Paul Stäckel pointed out that he might easily have learned about a specific instance of this through his friend August Ritter [Stäckel 1922, 55–56]. In 1850/51 Ritter attended Gauss's standard lecture course on the method of least squares, in which Gauss treated the problem of determining the least value of a sum of $n$ squares subject to $m$ linear inequalities [Gauss 1863–1933, 10(1): 473–482]. In all likelihood, Stäckel contacted Ritter in the late 1890s when he was still teaching in Aachen as Professor of mechanics. Presumably he was curious to know whether Ritter could tell him anything about Gauss's influence on Riemann. Ritter had taken his doctorate under Gauss in 1853 with a dissertation on the Gaussian principle of least constraint. Stäckel selected three passages from it as well as the relevant part of Ritter's text from Gauss's lecture for [Gauss 1863–1933, 10(1): 469–482], identifying these as contributions to "manifolds of $n$ dimensions." Assuming these Gaussian ideas were, indeed, related to Riemann's famous lecture, Stäckel surely must have asked whether Ritter had spoken to Riemann about them. He and Riemann had even lived together in the same boarding house on Wenderstrasse during the winter semester of 1852/53, and they usually took their main meals there, too. But that was more than forty years in the past, making it highly unlikely that Ritter remembered anything about what they had spoken about.

## 10.5 Topology

In his history of knot theory, Moritz Epple gave an account of the several ways in which Gauss took up topological ideas in his physical researches [Epple 1999]. Much of this work was never published, of course, but Epple highlighted the fact that two of Gauss's earlier students, August Ferdinand Möbius and Benedict Listing, had both made pioneering contributions to topology. The former became a well-known astronomer in Leipzig, whereas the latter assumed Wilhelm Weber's position in 1837 and afterward taught physics alongside him when Weber returned in 1849. Epple also noted that Paul Stäckel had earlier taken up the problem of assessing Gauss's influence on the emergence of *geometria situs*, the earlier term for topology [Epple 1999, 80].

In the cases of Möbius and Listing, Stäckel presented compelling evidence of Gauss's influence, whereas he readily admitted that the situation with Riemann was far less clear, since he had few meaningful contacts with Gauss. As shown in Section 2.5, Gauss knew nothing about Riemann's dissertation before the time it was submitted. Stäckel was also of the opinion that Gauss never contemplated an approach to complex functions based on an anologue of Riemann surfaces [Stäckel 1922, 56]. His suggestion that Riemann may have heard about Gaussian topological ideas through the grapevine – he named August Ritter and Wilhelm Weber as two possible conduits – was only a guess. One might more likely have thought of Listing, the author of *Vorstudien zur Topologie* (1848), since during the early 1850s Riemann had ample opportunities to speak with him in Weber's Institute. Yet his name seems to have never come up, nor die Riemann ever take any of his courses so far as is known. Clearly, Gauss and Riemann shared a strong interest in the connectivity of surfaces, but as noted in Section 11.1 the idea that Riemann learned about the properties of cross-cuts from Gauss would seem highly implausible.

Gauss's doctoral dissertation [Gauss 1799], which was discussed in Section 2.4, gave what many long considered the first rigorous proof of the fundamental theorem of algebra. As Gauss explained fifty years later in [Gauss 1863–1933, 3: 73–102], he originally formulated this as asserting that every polynomial equation can be factored into terms of the first and second degree, since he wanted to avoid making reference to imaginary quantities. This, however, was no longer of any concern:

> At present, when everyone is familiar with the concept of complex quantities, it seems more appropriate to leave that form alone and to express the theorem in such a way that that function can be decomposed into n simple factors, whereby the constant parts of these factors do not need to be real quantities, but also any complex values must be permissible for them. With this arrangement, the theorem itself becomes more general, because the restriction to real values does not need to be assumed for the coefficients either. [Gauss 1863–1933, 3: 74]

This being the case, in 1849 Gauss formulated the theorem to assert that every polynomial equation over the complex numbers has a solution in complex numbers.

Riemann surely agreed that the proof was essentially complete, though Gauss himself freely admitted that it contained a gap.[9] From a modern point of view, there were several gaps, for in those days one lacked appropriate tools for carrying out sound topological arguments. Riemann surely noted what Gauss wrote

[9] Gauss also claimed this gap could be filled without difficulty, though he never showed how.

about the nature of his topological proof, which brings to mind the remarks Riemann made at the outset of his Habilitation lecture (see 6.2):

> I will present the proof in a form taken from the geometry of position because this gives it greatest clarity and simplicity. In principal, though, the actual content of the argument belongs to the higher realm of general abstract quantities, independent of space. This studies quantitative objects from the standpoint of continuity, a field which is currently not very well developed and which requires a language borrowed from spatial images. [Gauss 1863–1933, 3: 79]

The gist of Gauss's argument was this: Let $f(z) = 0$ be a polynomial of degree $n$ and consider the two real algebraic curves given by the equations $Re(f(z)) = 0$ and $Im(f(z)) = 0$. Gauss considered the intersection properties of their branches in order to prove the existence of $z_o \in \mathbb{C}$ with $Re(f(z_o)) = Im(f(z_o)) = 0$. For this purpose, one chooses a circle $|z| = r$ with sufficiently large radius $r$ that it meets the two curves in $2n$ points. Gauss noted that the respective intersection points will alternate on the boundary of the circle. He then claimed that if one began at any of the $2n$ intersection points and followed the corresponding branch into the interior of the disk, the curve would eventually leave the circle at a different intersection point. Assuming this to be the case, Gauss could then show that the alternating placement of the intersection points on the boundary implied that $Re(f(z)) = 0$ and $Im(f(z)) = 0$ had to intersect somewhere in the interior of the circle.

This was the idea behind the first of Gauss's four proofs of the fundamental theorem of algebra. It was, of course, a pure existence theorem since nothing is asserted as to how one could calculate a solution to an equation based on the data given by its coefficients. If, however, one reads further in [Gauss 1863–1933, 3: 73–102], his commentary from 1849, it quickly becomes apparent that Gauss has a deep interest in the probably of calculating the roots of a polynomial equation. Riemann no doubt appreciated this as well, as he himself had a deep interest in concrete methods for handling analytic expressions. Gauss's existence proof hinged on the Jordan curve theorem, which asserts that a simple closed curve $C \subset E^2$ separates the plane into two regions both of which have $C$ as their common boundary. In this setting, a connected curve cannot pass from the interior region to the exterior without meeting the boundary curve, in this case Gauss's circle $|z| = r$. Leibniz observed that Euclid implicitly invoked this property in his proof of Prop. 1 in Book I of the *Elements*: to construct an equilateral triangle on a given line segment.

For Riemann, Gauss's argument was surely compelling, just as the result was highly significant for integration theory. Consider, for example, an integral of the form $\int \frac{dx}{P_n(x)}$, where $P_n(x)$ is an $n$th-degree polynomial with real coefficients. Then, by the fundamental theorem, $P_n(x)$ breaks down into linear and quadratic factors. One can then use the method of partial fractions to rewrite $1/P_n(x)$ as a sum of rational expressions in which the denominators are either linear or quadratic. Integration leads to so-called elementary functions (polynomials, logarithms, and arctangents), illustrating that the integral of a rational function never requires more than this class of functions. To use the language that became common after Riemann's work, the rational functions constitute the elementary part of the genus $p = 0$ case. More generally, if $f(x, y) = 0$ is a second-degree curve, then $\int R(x, y)dx$, i.e., the integral of a rational function whose variables are constrained to satisfy a quadratic condition, can be expressed as a sum of elementary functions because the corresponding Riemann surface has genus $p = 0$. This can be interpreted as the simplest case of Abel's addition theorem (see Section 7.1).

Proving a mathematical object must exist became increasingly important over the course of the nineteenth century as part of the shift away from concrete formulas to a broader understanding based on conceptual principals. Detlef Laugwitz made this one of the central themes in his biography of Riemann (see in particular [Laugwitz 1996/1999, 293–340]).

## 10.6 Beyond Gauss's Shadow

Many mathematicians, starting with Ernst Schering, have commented on the nature of Gauss's influence on Riemann, though without reaching any clear consensus. Felix Klein's influential views have often been cited, even though his heroic picture of Gauss and Riemann was all too obviously self-serving. Laugwitz went out of his way to polemicize against Klein's claims regarding Riemann, rejecting his treatment of Gauss's influence on him as exaggerated necromancy [Laugwitz 1996/1999, 214].[10] Yet, while dismissing Klein's mystification of their relationship, Laugwitz also wrote:

> The shy young student was certainly unable to establish personal relations with Gauss; after all, Gauss taught very reluctantly, took little interest in most of his students, and was quite unapproachable. We must nevertheless call Riemann a student of Gauss, in fact, the only true student of Gauss who grasped his inner [!] ideas, as we are now gradually getting to know them in outline from the *Nachlass*. (*ibid.*)

The above remarks underscore the mysterious quality of what was plainly a strong intellectual influence, but which lacked the kind of immediate personal contacts that might have given the Gauss-Riemann relationship a more intimate character. It seems most unlikely that Riemann ever expressed disappointment

[10] Laugwitz wrote, furthermore: "Klein had access in Göttingen to all the sources we have relied on; he made incomplete use of them" (*ibid.*). Laugwitz, too, overlooked important source material, including the essays and sources in [Riemann 1990]. It should be noted here that, biased as Klein's opinions certainly were, he himself took an active part in supplementing Riemann's Nachlass with additional texts of his lectures, those Max Noether and Wilhelm Wirtinger later utilized for the 1902 edition of [Riemann 1876/1892/1902]. Moreover, Klein's wartime lectures were intended merely as an overview of nineteenth-century mathematics. The oft-cited, posthumously published version [Klein 1926] had to be edited after his death; it remains unclear whether Klein ever authorized such a publication. In any event, he never considered his sketchy lectures as a serious historical study (rather, they were a bit of a "pious fraud" [Klein 1926, 2]).

over this, as Gauss's distant behavior fell within the norms of his day.[11] In 1850, Riemann tried to engage Gauss's interest in an older study by Johann Friedrich Pfaff that Gauss had summarized in [Gauss 1815].[12] The long-deceased Pfaff, who taught in Helmstedt and had formally approved Gauss's dissertation, was during his lifetime a leading authority on partial differential equations. Riemann may very well have received a reply from Gauss, but unfortunately no evidence for this has survived.

Nevertheless, many years later Heinrich Weber found a draft of Riemann's letter and sent it to Adolf Mayer. He received this reply on 29 October 1894:

> I enjoyed reading Riemann's letter, which I am sending back to you enclosed. Riemann raises exactly the same objections against Gauss's representation of Pfaff's method, which I have justified in detail, without specifically naming Gauss, in [Mayer 1880].[13] Although for this reason publication of this letter could only be pleasant to me, I still think that it would be better not to do so. The letter refers to a previous oral or written communication and is therefore difficult to understand without lengthy comment: I myself am unable to state exactly which passage in the introduction Riemann refers to in his objections. He probably sent the letter as a feeler to see whether Gauss would be hurt by his comments and only hinted at what he meant.[14]

Another interesting piece of anecdotal information comes from the winter semester of 1861/62, when Riemann taught a course on complex functions. He began with a standard proof of the law of inertia for quadratic forms, which he had first learned back in 1846/47 when he took Gauss's course on the method of least squares. Commenting on this some fifteen years later, Riemann remarked that the proof was due to Gauss, who placed this result at the very heart of that course. But to this he added "[Gauss] never included this in his treatise because he always followed the maxim: one should always tear down the scaffolding and leave the building alone standing" [Riemann 1876/1892/1902, 1902: 59]. Riemann probably never heard Gauss say this; he was merely repeating what he had read in the biography of Gauss written by Wolfgang Sartorius von Waltershausen [Sartorius 1856/2012, 82].[15] Ironically enough, this saying could just as well be applied to Riemann's own works.

---

[11] Twenty years later, Weierstrass broke with such social conventions when he took on the Russian student Sofia Kovalevskaya, who became one of his most intimate friends.

[12] Riemann to Gauss, 12 June 1850, Cod. Ms. Riemann 2: 92r–93v, SUB Göttingen.

[13] Mayer, in fact, did note Gauss's resumé, but he only commented that it provided an excellent account of Pfaff's method without mentioning the conditions under which it applied [Mayer 1880, 527].

[14] Cod. Ms. Riemann 2: 79r–81v, SUB Göttingen.

[15] Man dürfe einem guten Bauwerke, pflegte er zu sagen, nach seiner Vollendung nicht mehr das Gerüste ansehen. (He used to say that one should no longer see the scaffolding of a good building after its completion.)

# Part III
# Riemann's Legacy

# Part III

# Chapter 11
# Riemann's Pupils and Followers

## 11.1 Enrico Betti and Felice Casorati

Histories of modern mathematics in Italy invariably take the year 1858 as a pivotal turning point that dovetails nicely with the culminating epoch of the *Risorgimento*. For it was in 1858 that Enrico Betti, Francesco Brioschi, and Felice Casorati traveled to Germany and France, hoping to establish lasting scientific relationships with leading mathematicians in those two countries [Tazzioli 2012, 467–470]. They met with a friendly reception in Berlin, where Weierstrass had recently begun teaching, and also in Göttingen. Brioschi was an algebraist with limited interest in Riemann's work on complex analysis, but both Betti and Casorati were deeply influenced by his novel ideas.

After returning to Pisa, Betti published an Italian translation of Riemann's dissertation in *Annali di matematica pura e applicata* [Riemann/Betti 1859].[1] The next year Betti delivered his public inaugural lecture, in which he spoke about the historical development of elliptic functions. On that occasion, he commented as follows on Riemann's general point of view:

> Soon after the publication of Weierstrass's method Mr. Riemann published the principles of his much more general approach. He undertakes to study a function having an algebraic derivative, thus an Abelian integral, but he defines it by means of characteristic properties, i.e., by data necessary and sufficient for its determination. These are its discontinuities and the quantities by which its value changes when the complex variable traverses closed lines around given points. [...] This method has the advantage over all others of its immense generality and of fully satisfying the main requirements of modern analysis, since the apparatus of calculation hardly enters into it at all; it is almost entirely a magnificent work of pure thought. Yet the strength of this eminent geometer's intellectual conception is matched by his concise and obscure manner of presentation. Thus, up until now, it is almost as if his works did not exist in the scientific world. [Bottazzini 1977, 28]

Betti (Fig. 11.1)made serious efforts to change that situation. The following year, he published [Betti 1862], a study that utilized more conventional algebraic

[1] Judging from the letters he sent to Placido Tardy in October 1863, Betti evidently had not fully grasped Riemann's topological methods until the latter explained these to him in Pisa.



D. E. Rowe, *Bernhard Riemann: His Life and Wondrous Mathematical Legacy*,
https://doi.org/10.1007/978-3-032-25457-3_12

Fig. 11.1: Enrico Betti (1823–1892).

methods rather than the obscurer Riemannian approach. After sending him a copy, he received this reply from Riemann, dated 9 June 1863:

> I wish we had it in German, as it would contribute significantly to making this theory more accessible. My representation of the bifurcation of functions by surfaces in the form in which I presented it in Borchardt made it difficult even for German readers. In my lectures I have succeeded in making this mode of representation clear and familiar to my listeners by means of simple, appropriately chosen examples. But in any case, your derivation of the algebraic results without this auxiliary means and without the help of integrals in a purely algebraic way is very desirable. [Bottazzini 1977, 29]

When Riemann returned to Pisa later that year, he brought with him some of Gauss's fragmentary studies related to elliptic functions. He showed these to Betti during the course of his visit, as we know from a letter Betti wrote to Tardy on 4 January, 1869. Betti recalled seeing these fragments, which had since come out in Volume 3 of the Gauss edition: "I saw these works on elliptic functions when I was with Riemann; there were many detached sheets on which Gauss had written down the day and year on almost every page" [Bottazzini & Gray 2013, 284]. Ernst Schering, as the editor of that volume, noted that Riemann wanted to handle these documents as part of his responsibility for preparing the Gauss edition [Gauss 1863–1933, 3: 492]. His comments left the impression that Riemann never found the time to work on them. What actually happened, however, was later exposed by Karl Hattendorff in [Hattendorff 1869] (see Section 12.1).

During Riemann's first stay in Pisa in 1863, he had the chance to explain to Enrico Betti the key topological methods underlying his approach to complex analysis (see [Epple 1999, 80]).[2] These ideas were new to Betti, despite the fact he had already produced an Italian translation of Riemann's doctoral dissertation for *Annali di matematica pura e applicata* [Riemann/Betti 1859]. In two important letters sent to his friend Placido Tardy, Betti explained Riemann's method of cross cuts to establish the connectivity properties of surfaces and higher-dimensional manifolds. Writing from Florence on 6 October 1863, Betti began by saying:

> I have newly talked with Riemann about the connectivity of spaces, and have formed an accurate idea of the matter. A space is called SC [simply connected] when every closed surface contained in it makes up the whole boundary of part of the space, and every closed line contained in it is the whole boundary of a surface also contained in it.
> ...
> The generalization to higher dimensions is easy; the importance of this theory, for the theory of multiple integrals, is obvious. What gave Riemann the idea of the cuts was that Gauss defined them to him, talking about other matters, in a private conversation. In his writings one finds that analysis situs, that is, this consideration of quantities independently from their measure, is "wichtig"; in the last years of his life he has been much concerned with a problem in analysis situs, namely: given a winding thread and knowing, at every one of its self-intersections, which part is above and which below, to find whether it can be unwound without making knots; this problem he did not succeed in solving except in special cases ... [Weil 1979, 95]

One must imagine that the conversation with Gauss mentioned here took place when Riemann submitted his dissertation, thus twelve years before he spoke to Betti about this. Riemann already had introduced cross-cuts in his dissertation, of course.

By the mid-1860s, interest in Riemann's theory of complex functions had begun to grow, aided by the publication of several monographs that were far easier to read than Riemann's original works.[3] The first of these, [Durège 1864], written by the German analyst Heinrich Durège, was an authoritative account that eventually went through four editions. Luigi Cremona, one of several Italian mathematicians who read it, first heard about Durège's book from Placido Tardy. Before studying it carefully, he sent these remarks to Wilhelm Fiedler:

> But what you write to me, namely that [Durège] has managed to give a representation of the theory of functions with complex variables according to Riemann, interests me to the highest degree. I am very curious to see this presentation because Riemann's ideas have always seemed rather difficult and obscure to me. The ninth section of this textbook [Durège 1864] contains the first account of Riemann's theory of the connectivity of surfaces. [Confalonieri 2019, 192]

One year later, Carl Neumann published [Neumann 1865/1884], a book Gustav Roch criticized for evading the use of Dirichlet's principle, which the author treated separately in [Neumann 1865]. Much more substantive was volume one of Felice Casorati's *Teorica delle funzioni di variabili complesse* [Casorati 1868], even though he never published the intended sequel. Fittingly enough, what

---

[2] For an account of Betti's work on higher dimensional manifolds, see [Scholz 1980, 243–112].

[3] For an overview of the textbook literature on complex analysis from the 1860s and 70s, see [Bottazzini & Gray 2013, 695–710].

held him back was his inability to come up with a convincing proof of Dirichlet's principle [Bottazzini 1977, 33]. Casorati made some significant contributions to the theory, in particular the Casorati-Weierstrass theorem. In Casorati's formulation, this states that at a point of (essential) discontinuity, a holomorphic function assumes all possible values. Despite this rather ambiguous statement, Casorati's proof makes clear what he intended to demonstrate, namely that at such a point the function comes arbitrarily close to any complex number.

In 1864, Casorati visited Berlin, where Weierstrass told him that Riemann's disciples were wrong to attribute everything to their master. Many of his results, he said, were due to Cauchy, whereas Riemann only dressed them up in his own way and for his own convenience. Casorati apparently heard far more from Kronecker:

> ...the mathematicians are somewhat arrogant in their use of the concept of function. Even Riemann, generally very exact, is not beyond reproach in this respect. If a function increases and then decreases, or vice versa, Riemann claims there must be a minimum or a maximum (see the demonstration of the so called Dirichlet'sche Princip); whereas one should restrict that conclusion to the sphere of so-called reasonable functions.
>
> ...Betti has not comprehended the spirit of Riemann's things. He finds the number of the equations and without fail derives the number of arbitrary constants without showing that the determinant of the said linear equations is nonzero. Moreover, it is not in the determination of the number of constants that the spirit of Riemann's researches lies. As for his $\Theta$-series, Kronecker said that Riemann let these rain down from the sky. [Bottazzini 1977, 34–35]

At a later meeting in September 1869 on Lake Como together with Cremona, Casorati recorded Kronecker's skepticism with regard to Riemann surfaces. As Bottazzini pointed out, this was mainstream opinion at that time. When Charles Briot and Claude Bouquet brought out the second edition of their textbook (first published as [Briot/Bouquet 1859]), they mentioned "the fine results achieved by Riemann" using his geometrical methods, but they found these presented many difficulties; moreover "they did not seem to us to offer any advantage for the object we had in mind."

In 1869, Eugenio Beltrami offered these remarks about Casorati's book, but also with regard to the rivalry between French and German mathematicians shortly before the outbreak of the Franco-Prussian War:

> The research to which Casorati devoted this work, in particular that of Riemann, has brought great and unexpected appreciation to the heritage of science [...]. One of the principal merits of the new book is that it takes equal account of the contributions both schools [of Cauchy and Riemann] made to this new direction. In Germany, there is now widespread research in the field initiated by the man whose premature end his followers deplore, whereas the conviction that Cauchy's work offered extremely fertile and well-prepared ground for that research is perhaps less deeply rooted. Indeed, it is unfortunate to note that when Germans criticize these works, they sometimes exceed the bounds of reasonableness and appropriateness [...]. In France [...] the echo of these Göttingen dictates seems to have reawakened fervor [...] for the works of that very industrious analyst, without however inducing persuasion that the rich material he accumulated could now be constrained within a more orderly and simpler framework. (translated from [Bottazzini 1977, 33–34])

Casorati probably never met Riemann in Italy, but he eagerly sought out contact with the latter's disciples. One of these was Friedrich Prym (see Section 11.3), to whom he wrote on 10 November 1864:

> During my stay in Berlin, in the second half of October, I expressed to Mr. Kronecker the desire to know you personally. Last year I had obtained your inaugural dissertation; now I wished to obtain your lithographed notebook of Mr. Riemann's lessons and everything that you would subsequently have the kindness to make known to me, relating to the admirable works of the Professor of Göttingen. [Neuenschwander 1978, 23]

Casorati had hoped to meet Prym in Vienna, where the latter was working at a bank, but he had to cut his trip short and return to Pavia. One can see from Casorati's letter how interest in Riemann's ideas had begun to spread during the early 1860s. Evidently, he had heard that Prym owned copies of the lectures Riemann had delivered during the summer of 1861. In fact, when Prym arrived in Göttingen for the semester that followed, he borrowed this transcription from Karl Hattendorff.

## 11.2 Karl Hattendorff

Given these and other similar circumstances, it would seem puzzling that the single most active transcriber of Riemann's lectures, the above-named Karl Hattendorff, has virtually disappeared from history. Hattendorff was born in 1834 in Hanover, where his father worked as a bookbinder. He later studied in Göttingen, where he followed Riemann's lectures closely while taking his doctorate under M.A. Stern in 1862. After habilitating two years later, he began offering lectures, beginning with a 5-hour course on elliptic and Abelian functions in the 1864/65 semester. To judge from the catalog, Hattendorff's course offering may have raised the ire of Ernst Schering, who announced a 4-hour course on the same topic convening at 3 PM, exactly one hour before Hattendorff's lectures were scheduled to begin! By this time, probably no one in Göttingen anticipated that Riemann would be healthy enough to teach again. Beginning in 1869, Hattendorff began teaching at a school in his native Hanover, though only for one year.[4] In the meantime, he gained an appointment at the newly opened Polytechnic in Aachen, where he taught up until his premature death in 1882.

If few of his contemporaries seem to have appreciated Karl Hattendorff, it seems Riemann was one who did. In most situations, it was a high honor for a student to prepare an authorized version of a professor's lectures. Hattendorff's transcriptions turned out to be particularly important because they eventually appeared in print, starting with Riemann's lectures on partial differential equations from 1860/61. These served as the basis for [Riemann/Hattendorff 1869],

[4] Considering what he wrote about Schering in [Hattendorff 1869], one might conjecture that this conflict may have induced Hattendorff to leave Göttingen; see Section 12.1.

which came to be widely known as Riemann-Hattendorff.[5] As noted in its preface, Hattendorff also made use of the manuscript Riemann prepared when he taught this course in 1854/55 (see Nr. 30 in Cod. Ms. B. Riemann). He used this for the introductory section, which presents a nearly verbatim transcription of the words Riemann spoke when he taught this course for the very first time (see Section 3.3). Hattendorff only modified a single passage, which began:

> The subject of these lectures is the treatment of partial differential equations and their application to physical problems. ... Therefore, it will be appropriate to preface this with some introductory remarks on the relationship between the theory of partial differential equations and physics. [Riemann/Hattendorff 1869, 1]

Replacing the ellipses, what Riemann actually said was: "Since this subject has up until now not been taught here in Göttingen, it will be appropriate ...."

Hattendorff's original edition of Riemann's lectures on partial differential equations went through two nearly unchanged editions during his lifetime (1876, 1882). During those years, this book had few serious competitors. It had long been out of print when Heinrich Weber decided to renew this project by paying homage to Riemann with his highly successful textbook [H.M. Weber 1900], popularly known as Riemann-Weber. Although Vieweg marketed this as the fourth edition of Riemann-Hattendorff, Weber's highly technical account drew little inspiration from its forerunner.[6] Seen in this larger context, though, Riemann's lectures lived on as part of a conscious Göttingen tradition in mathematical physics. Richard Courant later famously promoted this cause with his *Methoden der mathematischen Physik* (Courant-Hilbert), first published in [Courant/Hilbert 1924] but afterward expanded and reworked in several editions.

Riemann also asked Hattendorff to write up his older unfinished paper on minimal surfaces, which came out in two different versions [Riemann 1868/1876]. Wilhelm Weber knew about the surrounding circumstances, and he gave his approval for the paper on minimal surfaces to appear in Volume 13 of the *Abhandlungen*, alongside the two others from Riemann's Habilitation that Dedekind had edited. For this original version, Hattendorff wrote an historical preface describing earlier works on minimal surfaces, but Heinrich Weber chose to drop this part from the second version, which appeared in the *Werke*. Weber also realized that the original version needed certain revisions, a matter he tried to negotiate with sensitivity [Scheel 2015, 352–357].

In the course of their correspondence, Hattendorff eventually sent Weber some details about how the manuscript came to be written. The circumstances he related show the difficulties Riemann faced in trying to complete this task. Only toward the end of his life did he reveal to Hattendorff that he had long ago

---

[5] Hattendorff's edition of Riemann's lectures on gravity, electricity, and magnetism [Riemann/Hattendorff 1876] proved far less influential.

[6] Reinhard Siegmund-Schultze followed the subsequent history of these Vieweg volumes, which took a far more applied direction with the seventh and eighth editions, published in 1927 and 1943 by Philipp Frank and Richard von Mises; see [Siegmund-Schultze 2007, 32–36].

worked on minimal surfaces, in particular on what came to be called Plateau's problem. Writing to Weber on 27 October 1875, Hattendorff recalled that:

> Riemann spoke to me twice about this problem. The first time was probably during the summer holidays of 1861 .... He said at the time: "The theory of functions of complex variables allows for an interesting application to the problem of minimal surfaces." The second statement dates from April 1866. When Riemann handed me the manuscript, which was already as yellowed then as it is now, he said: "I would like this work to be completed. It is, of course, already five or six years old." From that statement, I believed I could date the origin of the manuscript to the years 1860 or 1861. [Scheel 2015, 355]

Riemann had thus postponed his decision regarding the fate of this work until it was nearly too late to salvage anything. He and Elise had arrived in Göttingen in October 1865 from their second sojourn in Italy. Not until the following April, however, did Riemann inform Wilhelm Weber that he wished to enlist Hattendorff's support in a final effort to complete the manuscript. The main ideas had been clear to him for a long time, but he was far too physically weak to put these down on paper. During this last phase of his life, he also relied on his wife's assistance when he needed to dictate a letter. Hattendorff's account of what transpired afterward provides a vivid picture of Riemann's state at this time:

> The minimal surfaces were to be the starting point. He believed we could finish this in about eight days, and then more important things would follow. However, it soon became clear that, unfortunately, he had misjudged the time as well as his strength. He was often completely inaccessible for several days at a time. He was unable to speak at all. His voice was completely toneless, and when he tried to whisper something to me, he was interrupted by the most violent coughing. His communication therefore took place in writing. With a stylus on a blackboard, he briefly sketched what the manuscript was supposed to explain; he wrote each time until I nodded to him that I understood. He also wrote with a pencil on paper a few times, but always only aphoristically. In the intervals between these conferences, which rarely lasted more than half an hour, I wrote down what we had discussed in a coherent manner. Thus, Riemann saw and approved most of the treatise. In June, he was seized by an irresistible longing for Italy. On June 16th, he departed. This is the story of the manuscript's creation. However, it is not easy to present it in a form suitable for the public! For even the slightest hint of the difficult final months before Riemann's death would awaken painful memories in his surviving family. [Scheel 2015, 355–356]

Hattendorff's final remark was perhaps only an impression, but even so, it accords well with what we know about Elise Riemann's concerns over her deceased husband's public image. It was around this very time that she conveyed to Dedekind the pain she felt when people referred to Riemann as an "incurable hypochondriac" [Scheel 2015, 310].

On 9 November, Hattendorff wrote Weber again. The latter had in the meantime received two long letters from Hermann Amandus Schwarz filled with detailed criticisms of Hattendorff's new version of Riemann's paper, which would soon appear in the *Werke*.[7] Weber successfully thwarted off Schwarz's urge to interfere with the planned publication, but at the same time he felt compelled

[7] These letters can be found in [Scheel 2015, 369–375].

to inform Hattendorff that Schwarz was none too happy about the situation. Based on Hattendorff's response, he likely misinterpreted this to mean that Schwarz's unhappiness stemmed from the paper's lack of acknowledgment of Karl Weierstrass's concurrent work on minimal surfaces [Scheel 2015, 356–357].

In any event, Heinrich Weber decided to omit the earlier seven-page historical introduction, adding only Hattendorff's brief explanatory note:

> This treatise is based on a manuscript by Riemann, which, according to the author's own statement, was written in 1860 and 1861. This manuscript, written in a very concise form with only formulas and no text, was entrusted to me by Riemann in April 1866 for further work. The treatise that resulted from this work was submitted by me to the Royal Society of Sciences in Göttingen on January 6, 1867, and was published in Volume 13 of the Society's proceedings. This treatise is now being published here for the second time in a carefully revised version. [Riemann 1868/1876, 283]

## 11.3 Friedrich Prym

Friedrich Prym was mentioned above in connection with Felice Casorati. He was born in 1841 in Düren, the hometown of Gustav Dirichlet. Originally, he studied in Berlin, where he came under the influence of Elwin Bruno Christoffel, who later advised him to attend Riemann's lectures in Göttingen. Arriving for the winter semester of 1861/62, he had the opportunity to hear his last lectures on Abelian functions while sitting alongside Riemann's faithful amanuensis, Karl Hattendorff. He also heard Riemann's lectures on partial differential equations, the course that led to the monograph [Riemann/Hattendorff 1869]. This encounter with Riemann's ideas turned out to be the pivotal experience in Prym's young career; thenceforth, he made a name for himself as a leading authority on Riemannian function theory.

This began when he returned to Berlin, where he completed his dissertation and took his doctorate in February 1863. One should note that this took place at the very time when Karl Weierstrass began offering lectures on complex analysis based on algebraic methods. Later, Weierstrass would consciously eschew Riemann's global approach and instead introduce analytic functions via local power series and his principle of analytic continuation. In view of these later developments, the opinion of Prym's dissertation expressed by his thesis adviser, E.E. Kummer, makes for a striking contrast:

> Prym has adopted a method for treating algebraic functions and their integrals that is entirely Riemann's own, based on geometrical considerations. Riemann's method is still not widely used. Studying it from Riemann's treatises is very difficult because it is not discussed in the desired detail. Therefore, the detailed application of it to a specific and limited problem, which the candidate has posed, is a useful work that can be employed by many mathematicians. [Biermann 1988, 94]

During the early spring of 1865, Prym was invited to stay for a number of weeks with the Riemanns in Pisa. According to his devoted pupil, Adolf Krazer, Prym's visit gave him the chance to learn firsthand about Riemann's theory of theta characteristics [Krazer 1917, 2]. This topic became a central focus for Prym's work in Würzburg, where he began teaching in 1869. Communicating with Riemann was already very difficult by this time. This becomes evident from the following anecdote, which only surfaced many years later. During Prym's stay with the Riemann family in Pisa, he one day sent him the following query:

> Dear Professor!
>
> A friendly morning greeting and my best wishes for your well-being today. I hereby take advantage of the permission you granted me to write you, and should this not trouble you in any way, I ask that when we dine you give me an explanation for the above.[8]

Prym's query had to do with a special type of hyperelliptic integral with two first-order branch points that have coalesced into a single branch point of order two. Prym wondered what such a Riemann surface would look like, or if this was even possible, since usually three leaves would meet at such a point. That same day, Riemann wrote up his reply. These exchanges only came to light much later, however, when Max Noether and Wilhelm Wirtinger were preparing the Nachträge for the third edition of Riemann's *Werke*. Wirtinger found a draft of Riemann's answer to Prym, who afterward sent him the final version of Riemann's letter.[9]

More than fifteen years later, Prym published a booklet on the main topic he discussed with Riemann in Pisa, Riemann's theta formula (*Untersuchungen ueber die Riemann'sche Thetaformel und die Riemann'sche Charakteristikentheorie* [Prym 1882]). In the preface he recalled that time:

> In the spring of 1865 I was lucky enough to be able to spend a few weeks in Pisa with my highly honored teacher Riemann, where he was staying for health reasons. On this occasion, Riemann gave me a formula ... which he regarded as fundamental for the theory of theta functions, and at his suggestion I derived a proof for this formula, which met with the approval of my teacher. However, I was unable to make use of this formula at the time, first because a deterioration in Riemann's condition made further discussions impossible, and, second, because more detailed ... investigations, which seemed necessary to me to carry out beforehand, took up all my time. [Prym 1882, v]

The following year he published another proof of Riemann's theta formula in *Acta Mathematica* [Prym 1883]. These highly technical works came out just as Klein and Henri Poincaré were promoting a new approach to Riemann surfaces generated by discrete groups that act on a fundamental domain [Scholz 1980, 198–222]. The latter ideas proved to be not only fertile but also far more accessible than the theory of theta functions. Prym was by no means the most active of Riemann's students, and yet he outlived most of the others, including the far more prolific Gustav Roch, who died in the same year as Riemann. Roch's importance for Riemann's legacy will only become fully apparent when we turn to the prehistory of *Riemanns Werke* [Riemann 1876/1892/1902] in Section 12.2.

[8] Prym to Riemann, Pisa, 27 March 1865, Nachlass Riemann 2: 66r, SUB Göttingen.

[9] See [Riemann 1876/1892/1902, 1902: 103–105].

## 11.4 Gustav Roch

The name Roch is familiar to many mathematicians today, though only because of a single result: the Riemann-Roch theorem. He deserves to be far better known. The Riemann-Roch theorem was so named by Max Noether and Alexander Brill in a jointly written paper from 1874. There they reinterpreted this key result, which played an important role in their new theory of algebraic curves [Gray 1998]. Roch's contribution in [Roch 1865c] amounted to a refinement of the Riemann inequality. As discussed in Section 7.3, this gave a lower bound $h = m - p + 1$ on the dimension of the space of meromorphic functions with simple poles at $m$ points on a Riemann surface of genus $p$. Roch showed that one can turn this into an equality $h = m-p+1+q$, where $q$ is the dimension for those integrals of the first kind that vanish at the $m$ given points. To illustrate this, Roch gave an example based on a 5th-degree nonsingular curve $F = 0$, thus with $p = 6$. In this case, the linear space of integrals of the first kind is identical with polynomials of the second degree. Roch took a rational function $s'$ of two linear expressions, where the denominator meets $F = 0$ in five points. Then the three arbitrary constants in the numerator serve as parameters for the linear space of integrals of the first kind that vanish at all five points. This yields $h = 5 - 6 + 1 + 3 = 3$ as the dimension of the corresponding space of meromorphic functions [Roch 1865c, 375].

Gustav Roch grew up in Dresden, where he came under the influence of Oskar Schlömilch, editor of *Zeitschrift für Mathematik und Physik.* In 1859, he entered the University of Leipzig, where he took a wide range of courses. During his two years there, Roch became particularly interested in mathematical studies related to electricity and magnetism, a topic he dealt with in four papers published in Schlömilch's journal. Based on his impressive progress as a student in Leipzig, he obtained a scholarship to continue his studies in Göttingen and Berlin.

Starting in April 1861, Roch spent three semesters in Göttingen studying under Wilhelm Weber, but also attending lectures given by Riemann before going on to Berlin. In the Prussian capital, he came into contact with Kummer, Kronecker, Weierstrass, and Borchardt. He then returned to Leipzig, where he was awarded a doctorate in 1863 for his dissertation on a topic in potential theory. His thesis advisers were Moritz Drobisch and the physicist Wilhelm Hankel, father of Hermann Hankel, who also studied briefly under Riemann. From Leipzig, Roch went on to habilitate in Halle with a thesis (*De theoremate quodam circa functiones Abelianas* [Roch 1863b]) that took up Riemann's ideas. One year later, he submitted his most famous paper [Roch 1865c], which turned the Riemann inequality into an equality: the Riemann-Roch theorem. This, however, was but one of several papers Roch wrote on Riemannian topics, several of which appeared in Schlömilch's *Zeitschrift für Mathematik und Physik.*[10]

By 1866, however, Roch was in very poor health. He took off teaching during the winter semester of 1866/67 and left for Venice, where he hoped the warmer

[10] His two articles [Roch 1863a] and [Roch 1865a] were based directly on Riemann's lectures; see also [Roch 1863a], [Roch 1863b], [Roch 1866a], [Roch 1866b].

weather would help him recover. Instead, he died there of consumption at just 26 years of age. Roch's name would live on through the fundamental Riemann-Roch theorem, but had he lived one can easily imagine that the reception of Riemann's ideas would have gone far more smoothly. He was one of the few in Germany who possessed the talent and energy required to promote the Riemannian legacy. Those who eventually did – first and foremost Alfred Clebsch and Felix Klein – not only never knew Riemann, they also approached his work from rather different points of view.

## 11.5 Alfred Clebsch

As a student of Otto Hesse in Königsberg, Alfred Clebsch (Fig. 11.2) developed a strong affinity for Hesse's analytic methods in his studies of algebraic curves and surfaces. Many of their properties had been uncovered earlier by Julius Plücker [Wiescher 2026] and Jakob Steiner, but Hesse's approach was far more powerful and elegant. By the time Clebsch arrived in Giessen in 1863, he had immersed himself in the symbolic methods Siegfried Aronhold had begun to introduce in invariant theory [Parshall 1989]. This phase in his career, from 1863 to 1868, coincided with Clebsch's fruitful collaboration with Paul Gordan, which led to their influential monograph *Theorie der Abelschen Funktionen* [Clebsch/Gordan 1866].

During the early 1860s, Clebsch had come to realize the importance of Riemann's monumental study of Abelian functions and Abel's theorem as tools for deriving deep results in algebraic geometry. Igor Shafarevich called Clebsch's lengthy study [Clebsch 1864] the "birth cry" of modern algebraic geometry [Shafarevich 1983, 136]. Shafarevich remarked further that Clebsch's long-term reputation may have suffered not only due to his early death – he succumbed to an attack of diphtheria in November 1872 at the age of 39 – but also because his influence spread through the work of his many students. Still, Clebsch was widely credited (alongside Riemann) with having inaugurated a new subdiscipline: birational geometry.

In Riemann's theory, one studied functions and their integrals related to an algebraic equation $f(s, z) = 0$, which determines the associated Riemann surface $S$. Riemann connected the topology of $S$ with the invariant $p$ associated with Abel's theorem; more precisely, $p$ was the number of independent Abelian integrals of the first kind that can exist on a given Riemann surface $S$. By this means, Riemann could discuss complex integration for multi-valued algebraic functions. Such a function can be represented by a surface, say with $m$ leaves and $w$ branch points, in which case

$$p = w/2 - m + 1,$$

[Riemann 1857b, 104]. Clebsch wanted to avoid using complex analysis, and he later found a way to define $p$ without it. He called $p$ the genus (*Geschlecht*) of the

Fig. 11.2: Alfred Clebsch (1833–1872).

curve $f(s, z) = 0$, and in 1866 he and Paul Gordan proved this was a birational invariant [Clebsch/Gordan 1866, 15]. If a curve has only simple singularities ($d$ double points and $r$ cusps), then

$$p = \frac{1}{2}(n-1)(n-2) - d - r.$$

Clebsch approached this terrain from the standpoint of projective algebraic geometry and the classical invariant theory of the projective group. Within that setting, these ideas had a familiar ring. In 1865, Arthur Cayley referred to $p$ as the *deficiency* of an algebraic curve $C_n$, an allusion to the maximum possible number of double points [Cayley 1865, 2]. Thus, for a nonsingular quartic, for which $p = 3$, allowing $d = 3$ double points leads to vanishing deficiency. If a quartic $C_4$ thus has $d = 4$, it can no longer absorb them and the curve reduces to two conics: $C_4 = C_2 \cup C_2'$.

Clebsch took on the challenge of applying Riemann's theory of Abelian integrals and their functions, though his introductory remarks in [Clebsch 1864] clearly reveal that he remained a partisan of Jacobi's analytical methods. To adapt Riemann's theory to projective algebraic geometry, Clebsch needed to homogenize the equations in the Riemannian formalism. As a first step he replaced the algebraic relation $F(s, z) = 0$ of degree $n$ in $s$ and $m$ in $z$ by a homogeneous polynomial $f(x_1, x_2, x_3) = 0$ of degree $n$.[11] Clebsch viewed $f = 0$ as defining a ground curve $C_n \subset P^2(\mathbb{C})$ with double points and cusps as its only singularities. In Riemann's theory, $F(s, z) = 0$ is associated with $p$ linearly independent integrals of the first kind

$$\int \frac{\phi(s, z)dz}{\frac{\partial F}{\partial s}},$$

with $\phi$ a polynomial of degree $n - 2$ in $s$ and $m - 2$ in $z$ that vanishes for points at which $\frac{\partial F}{\partial s} = \frac{\partial F}{\partial z} = 0$. If $r$ is the number of such points, then Riemann showed that $p$ is invariant under a rational mapping between $F(s, z) = 0$ and $F_1(s, z) = 0$, where here $p = (n - 1)(m - 1) - r$ [Riemann 1876/1892/1902, 118–119].

In Clebsch's formulation these $p$ Abelian integrals can be written:

$$\int \phi_k \, d\tilde{w}, \quad k = i, \ldots, p,$$

where the $\phi_k$ are homogeneous polynomials of degree $n - 3$ that vanish at the singular points of $C_n$ and

$$d\tilde{w} = \frac{x_k \, dx_l - x_l \, dx_k}{\frac{\partial f}{\partial x_m}},$$

is an expression independent of the permutations of $k, l, m$. The corresponding definite integrals can be written

$$(u_k)_y^x = \int_y^x \phi_k \, d\tilde{w}, \quad k = i, \ldots, p.$$

This leads to Clebsch's version of Abel's theorem: Let $C_m$ and $C_m'$ cut out $mn$ pairs of points on $C_n$, then

$$\sum_{i=1}^{mn} (u_1)_{y_i}^{x_i} = \sum_{i=1}^{mn} (u_2)_{y_i}^{x_i} = \cdots = \sum_{i=1}^{mn} (u_p)_{y_i}^{x_i} = 0.$$

[11] For details, see [Lê 2020, 80–83].

More importantly, Clebsch proved the converse, namely that the vanishing of these $p$ sums implies that the $mn$ pairs of points lie on two curves of order $m$ [Clebsch 1864, 197–198].[12]

As emphasized by Bottazzini and Gray, Riemann's novel approach to complex analysis – including his appeal to Riemann surfaces, which cannot be embedded in $\mathbb{R}^3$ – posed immense challenges for those who sought to understand him [Bottazzini & Gray 2013, 331–334]. They noted further that Alfred Clebsch was among the first to bring out the fertility of Riemann's ideas for algebraic geometry. His seminal paper, "On the Application of Abelian Functions in Geometry" [Clebsch 1864], paved the way for several papers by Felix Klein. To be sure, Klein drew on many different sources in his early work, which reflects a strong visual orientation almost entirely absent in the publications of Clebsch and Riemann [Rowe 2024].

## 11.6 Felix Klein

Felix Klein has often been remembered as perhaps the leading promoter of Riemannian ideas in complex analysis. By the late 1870s, he had begun to link Riemann surfaces with new methods in the theory of transformation groups. Today, this topic is often associated with his "Erlangen Program" [Klein 1872/1893], which he wrote toward the end of his brief collaboration with Sophus Lie (see [Rowe 2025]). During his years in Erlangen, however, Klein's research aimed mainly at developing a broader theory of Riemann surfaces adapted to his visual approach to algebraic geometry, which he later called *anschauliche Geometrie*. In this context, Riemann's ideas were far less relevant than those of Klein's mentors, Julius Plücker and especially Alfred Clebsch, both of whom did pioneering work in the field of projective algebraic geometry. Klein's papers from the years 1873 to 1876 followed in the footsteps of his two mentors, whereas these works owe almost nothing to the ideas in his "Erlangen Program." Nor do they reflect the strong direct influence of Riemann; it was only in the 1880s that Klein became an outspoken advocate of Riemann's mathematics. Not coincidentally, this was the period when he and Henri Poincaré competed to develop the new theory of automorphic functions [Gray 2008].

Klein's *anschauliche Geometrie* was mainly concerned with methods for visualizing the real and imaginary parts of low-degree algebraic curves given by equations with real coefficients [Rowe 2024]. For this purpose, he invented a new type of Riemann surface that enabled him to interpret the genus of the curve in a visually convincing manner. This approach was essentially an elaboration of methods introduced earlier by Clebsch in projective algebraic geometry; as such, Klein made clear that his new surfaces were independent of Riemann's original motivation. To illustrate this theory, he first took up the familiar case

[12] The values of the integrals are given only up to multiples of their periodicity module, which we ignore here.

of real cubic curves [Klein 1874]. These had already been carefully studied by Isaac Newton, who undertook an affine classification in which he derived all cases from five basic types by means of projections. Two of these five types were nonsingular, whereas the other three had a singular point: either an ordinary double point, a cusp, or an isolated double point. These standard results represent the curve as a locus of points, whereas Plücker explained the relationship between point curves and their duals, which treat curves as envelopes of their tangent lines. Plücker was also the first to observe that a nonsingular point curve always has tangential singularities. In the case of a cubic, these are its nine inflection tangents, three of which are real with points of tangency lying on a line. In fact, all nine inflection points lie in threes on twelve lines; this is Otto Hesse's famous inflection point configuration.

In classical algebraic geometry, one referred to the order and class of a curve. Thus, a curve $C$ has order $n$ and class $k$ if a generic line meets it in $n$ points and a generic point lies on $k$ tangents. For the nonsingular case $k = n(n-1)$, a cubic thus has order 3 and class 6, while its dual has order 6 and class 3. If a point curve has singularities, then these will lower the class $k$. For a curve with $d$ double points and $r$ cusps, Plücker showed that $k = n(n-1) - 2d - 3r$, a formula that dualizes when taking tangential singularities into account. When doing so, double points will correspond to bitangents and cusps to inflection tangents, leading to the Plücker formula $n = k(k-1) - 2t - 3w$, where $t$ and $w$ are the number of bitangents, resp. inflection tangents. Clebsch, as noted above, showed that for curves with these simple types of singularities, the genus $p$ is given by $p = 1/2(n-1)(n-2) - d - r$.

When one dualizes a cubic, the inflectional tangents pass over into cusps – which are *not* tangential singularities. The curve $C$ with two components shown in Fig. 11.3 is thus a nonsingular curve of class three. This means that from each point $P$ in the plane $\mathbb{R}^2$, where $P \notin C$, one can draw three tangents to the curve. Since the coefficients of the curve's equation are real, the coordinates of imaginary tangents appear as conjugate imaginary numbers. One then easily sees that for points outside and inside $C$ all three tangents are real, whereas for those in the annular region only one *real* tangent can be drawn to the inner component of the curve. The two remaining imaginary tangents thus have conjugate complex coordinates. At each such point $P$, Klein constructed two equidistant points on the normal line at $P$, one above and one below the plane containing $C$. He then imagined moving $P$ throughout the annulus, thereby sweeping out two leaves that fall together on the boundaries as in Fig. 11.3. By this means the nonsingular cubic appears as a torus embedded in projective 3-space. Note that the tangents at the three real cusps meet in a point, dualizing the fact that the three real inflection points lie on a line. Clebsch had shown how to parameterize the curve by means of an elliptic function with periods $\omega, \omega'$. Using this, Klein was able to identify the placement of the remaining six imaginary inflection points, three of which are shown.

The duals of Newton's two types of nonsingular cubics are shown in Figure 11.4. Since Klein's construction takes place in projective space the two leaves outside the triangular figure on the right join at "infinity" to form a topological

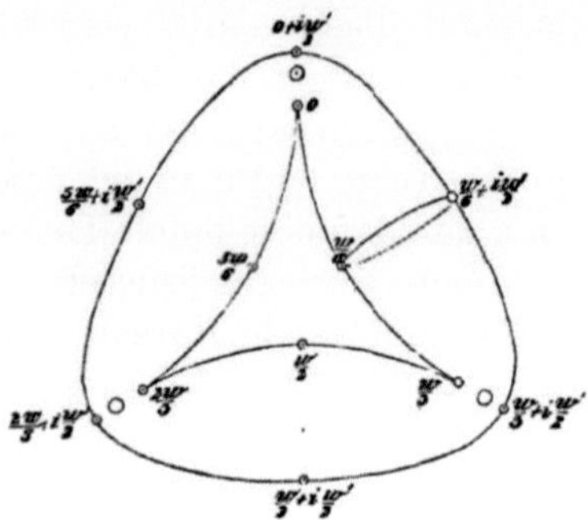

Fig. 11.3: Klein's picture of a nonsingular cubic class curve.

Fig. 11.4: Two nonsingular curves of class 3.

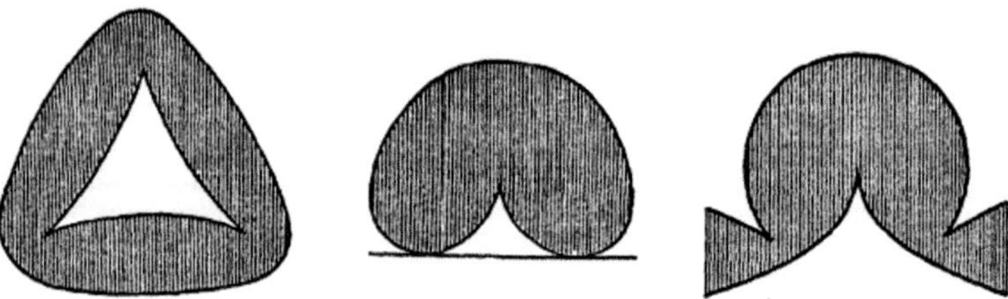

Fig. 11.5: Transition between the two types of nonsingular cubic class curves via a double tangent singularity.

torus, just as for the curve on the left. But what happens when one of these curves takes on a singularity, e.g. a double tangent? Klein dealt with this again by way of a "proof" by pictures (see Figure 11.5). He imagined a transition between the two cases with $p = 1$, which he carried out by crushing the bottom region of the left curve until the two cusps disappeared. This deformation leads to a surface of genus $p = 0$, where the real cusps are replaced by a double tangent.

Now imagine thickening up the region around its two points of tangency so that the double tangent disappears. The two missing cusps then reemerge, but now the real curve has a single real component instead of two. These deformations thus provide a way to visualize the real and imaginary parts on all possible types of cubic curves, which form a single connected manifold. Klein expended even more effort applying these ideas to quartic curves, which are far more complicated.

Over the course of Klein's early career, he steadily moved away from the influential ideas of Clebsch as he came to regard his own work as rooted in the fertile soil of Riemann's theory of complex functions, in particular the theory of Riemann surfaces. In looking back on his early work, Klein thought it owed too much to Clebsch's influence, which had deflected him from what he called the "true Riemann" [Klein 1921–23, 2: 5]. Here he pointed especially to the interpretation of Riemann surfaces he sketched in his monograph *Über Riemanns Theorie der algebraischen Functionen und ihrer Integrale* [Klein 1882/1923]. Aside from a section on symmetric Riemann surfaces, Klein's presentation in this booklet departed entirely from the approach in his earlier papers. Even when he dealt with symmetric Riemann surfaces, he *derives* the portion of the Riemann surface $S$ corresponding to the points on a real curve from a special property, namely these constitute the fixed points under a conformal mapping of $S$ of order two. The fact that this leads to an equation with real coefficients is thus merely a consequence of a geometric property [Klein 1921–23, 3: 566–567]. In the third and final section of his booklet, Klein drew far-reaching consequences by considering the $\rho$ parameters for the conformal self-mappings of a surface $S$ of genus $p$.

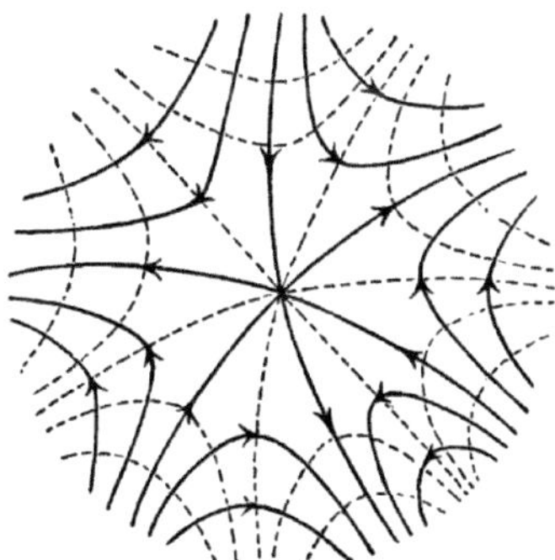

Fig. 11.6: A crossing point of multiplicity $\alpha = 2$.

Klein was himself fully aware that Riemann's whole approach relied on showing the existence of analytic functions in a bounded region for which the function takes on fixed continuous values on the boundary. Since Riemann's argument based on Dirichlet's principle had fallen into disrepute, Klein wanted to skirt this problem by appealing to physical principles that could account for the main results.[13] To do so, he introduced complex-valued functions of position on closed surfaces $S$ of known genus $p$. Citing the work of Camille Jordan, he introduced these as canonical objects obtained by attaching $p$ handles to a sphere.

Before proceeding further, Klein first imagined two families of curves in the plane corresponding to two conjugate harmonic functions. These represent a uniform complex function $w(z)$ in certain regions where this function is everywhere regular. Such regions are formed around so-called crossing points, where $w'(z) = 0$. Such a point is shown in Fig. 11.6, where $w'(z) = w''(z) = 0$, thus a crossing point $z_o$ of multiplicity $\alpha = 2$. The darker curves correspond to a stationary flow, whereas the dotted curves are equipotential lines in this interpretation. One can easily see that crossing points correspond to the branch points of $w$ since the three flow lines $C_i$ passing through $z_o$ divide the region around it into sectors. A curve $C_{z_o}$ encircling $z_o$ thus takes on $w$ values in each sector, which correspond to the leaves of a Riemann surface. When $C_{z_o}$ crosses over one of the $C_i$ it passes to a new leaf, so to speak. Starting with this rather elementary observation, Klein went on to derive key results from Riemann's paper on Abelian functions [Riemann 1857b].

He next imagined covering an arbitrary closed surface $S$ with a conducting material. His idea was to introduce a stationary current flow on the surface by placing two poles of a galvanic battery at two arbitrary points on $S$, producing a flow with two sources of equal but opposite strength. This corresponds to the case of two logarithmic´ singularities where the sum of the residues vanishes. Depending on the type of singularity to be represented, Klein joined these two points by one or more adjacent but non-intersecting curves, which have constant electromotive force. This gave rise to a stationary flow for which the two points are vortices for the flow; for its conjugate they are sources and sinks of equal and opposite intensity.[14] To obtain more complicated cases, Klein appealed to the principle of superposition.

These flow curves correspond to harmonic functions, which arise from solutions to a generalized form of the usual Cauchy-Riemann equations in the complex plane, thereby yielding functions of position on closed surfaces of a given genus $p$. In Figs. 11.7 and 11.8, Klein indicated how one can introduce two logarithmic singularities ($m$ and $n$) on a genus one surface by appropriately deforming the current lines on a surface of genus $p = 2$ until its homology is reduced to a surface with $p = 1$.

Klein noted, furthermore, that one can also merge the two logarithmic singularities into a simple pole. The upshot of these deformation arguments was to show how the number of poles can be directly related to the genus $p$ of the surface, since the number of branch points (here 2) remains constant through-

[13] Klein alluded to the new methods developed by Neumann and Schwarz, which circumvented Riemann's flawed argument based on Dirichlet's principle, but he chose to ignore these techniques, which were neither simple nor transparent.

[14] Helmholtz noted the analogy between electric and hydrodynamic flows in [Helmholtz 1858, 27], where he cited Riemann's definition of connectivity in [Riemann 1857b, 108].

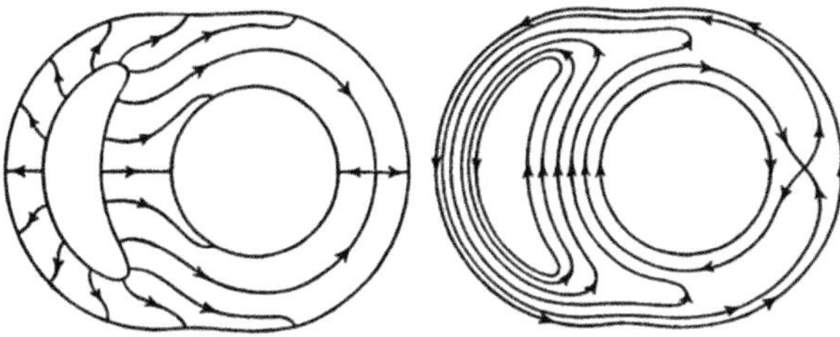

Fig. 11.7: Flow curves on top portion of a Riemann surface of genus $p = 2$ with two crossing points (one visible).

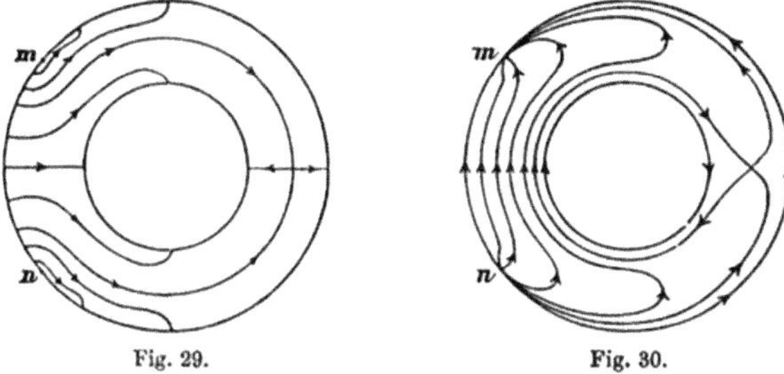


Fig. 11.8: After deformation a Riemann surface arises with $p = 1$ and two logarithmic singularities.

out. This visualization process thus also served to confirm Riemann's finding for the number of branch points, namely, $w = 2m + 2p - 2$ [Riemann 1857b, 113], a result Riemann explicitly proclaimed to be alone due to the topology of the surface (see [Scholz 1980, 67–68]).

Klein's approach to Riemann surfaces originally met with a mixed reception. Carl Neumann, as the author of [Neumann 1865/1884], an early monograph on Riemann surfaces, apparently found Klein's booklet superficial and unimpressive [Tobies 2021, 254]. In 1894, Alexander Brill and Max Noether published their detailed report on the history of developments that led to the modern theory of algebraic functions. One of their main aims was to explicate various attempts to rigorize this theory after 1871. In this connection they claimed that all serious efforts to pursue this goal departed from Riemann's original methods. In their view, his approach suffered not only from the problematic character of Dirichlet's principle but also because of the inherent vagueness surrounding the branch cuts one employed in defining the genus of a Riemann surface [Brill/Noether 1894, 285].

These criticisms help to explain why Brill and Noether barely mentioned Klein's booklet [Klein 1882/1923] or his published lectures [Klein 1892/1986], which strongly defended the primacy of intuition over rigor. On the other hand, it was truly ironic that Klein's booklet appeared in the very year that Richard Dedekind and Heinrich Weber found an entirely new way to ground Riemann's theory in the conceptual setting of Dedekind's theory of ideals [Dedekind/Weber 1882/2012]. Brill and Noether instead placed great stock in Weierstrass' approach, which was based on the representation of analytic functions by local power series. They by all means recognized the legitimacy of other approaches – the algebraic line pursued by Kronecker and Hensel as well as the arithmetical theory developed by Dedekind and Weber – even though they neglected to describe these alternatives in their report. Even if their own geometric-algebraic theory, which they had pursued based on earlier work by Clebsch, received pride of place, their dismissal of Riemannian methods reflected prevailing opinion during the 1890s. By the same token, Klein's earlier works on the visualization of real curves by means of projective Riemann surfaces received only the slightest mention in [Brill/Noether 1894]. Two developments, however, completely overturned the longstanding doubts about the soundness of Riemann's methods.

First, in 1901 David Hilbert introduced a subtle new argument that enabled him to salvage Dirichlet's principle. This breakthrough led to a great number of subsequent investigations by other analysts, including Richard Courant and Hermann Weyl. The latter gave a new and far simpler proof of Dirichlet's principle in his lecture course from 1911/12. Second, these lectures served as the basis for the first truly modern textbook on Riemann surfaces [Weyl 1913]. As a private lecturer in Göttingen, Weyl prepared this text under the watchful eye of Klein, to whom the book was then dedicated.

Although he drew on a great number of recent developments in topology and analysis, Weyl shared Klein's deep reverence for Riemann's geometrical methods. He also admired Klein's own novel approach to Riemann surfaces as set forth in [Klein 1882/1923], a work others appreciated long afterward as well [Springer 2001, 1–41]. Weyl's preface thus paid due homage to Klein's general views:

> Here and there one encounters the view that the Riemann surface is nothing more than an "image," a (one admits: very useful, very suggestive) means of visualizing and illustrating many-valued functions. This view is altogether mistaken. The Riemann surface is an indispensable substantive component of the theory and its very foundation. It ... must certainly be regarded as the prius, as the mother soil out of which the functions can grow and flourish in the first place. It must be admitted, however, that Riemann himself had somewhat obscured this true relationship of the functions to the Riemann surface through the form of his presentation ... as he did not make use of the more general ideas (only later developed with transparent clarity by Klein) .... And yet there can be no doubt that it is only in Klein's conception that Riemann's basic ideas come to full fruition in their natural simplicity and in their living, penetrating power. The present work is based on this conviction.

Weyl's text soon became a classic. For several years afterward, it was reprinted unchanged before Weyl finally decided to prepare a new edition in 1955, the year of his death. Klein thus enjoyed the privilege of seeing his faith in Riemannian principles vindicated, a turn of events that he and Hilbert celebrated together in Göttingen.

# Chapter 12
# Editing Riemann's Collected Works

The preceding chapter touched on the work of several key figures who played various roles in the reception of Riemannian ideas. None of them, however, was in a position to grasp the full breadth of his accomplishments, which awaited the editing of Riemann's posthumous papers. Although these began to appear in 1868, several of Riemann's texts and all of his philosophical writings only became widely accessible after the publication of his Collected Works in 1876. How that project came about constitutes the main topic of the present chapter. To explore this properly, however, requires picking up some threads related to Riemann's lecture course on complex functions from 1861/62, parts of which were included in the first edition of Riemann's *Werke*. A larger, but related question concerns the importance of Riemann's ideas for developments in algebraic geometry, as evidenced by several papers written by Alfred Clebsch and Felix Klein (Sections 11.5 and 11.6). Here one should not overlook that algebraic geometry would only become a well-grounded discipline well after Riemann's death. A giant step toward creating that field was taken by Richard Dedekind and Heinrich Weber when they published "Theory of Algebraic Functions of a Single Variable."[Dedekind/Weber 1882/2012] How that highly influential paper emerged from their collaboration in editing Riemann's Collected Works is described in the closing paragraphs of Section 12.3.

## 12.1 Elise Riemann's Message for Dedekind

After her husband's death, Elise Riemann met with Richard Dedekind and conveyed to him Riemann's request that he go through the papers in his estate to assess what might be salvaged for publication (see the end of Chapter 4, 4.7). Dedekind kindly agreed to do so, despite knowing full well that Riemann's manic work habits were such that these papers were bound to be in a



D. E. Rowe, *Bernhard Riemann: His Life and Wondrous Mathematical Legacy*,
https://doi.org/10.1007/978-3-032-25457-3_13

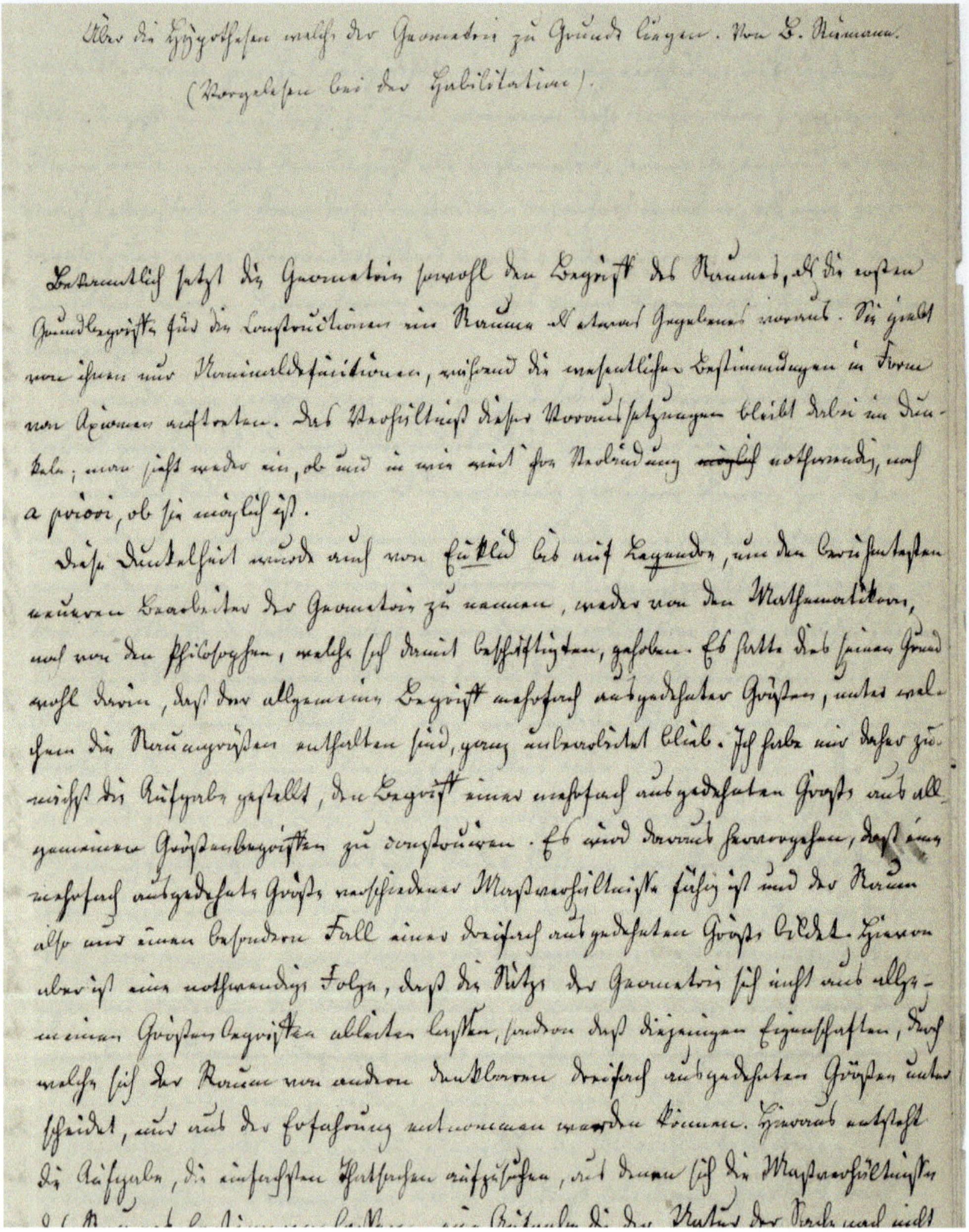

Ueber die Hypothesen welche der Geometrie zu Grunde liegen. Von B. Riemann.
(Vorgelesen bei der Habilitation).

Bekanntlich setzt die Geometrie sowohl den Begriff des Raumes, als die ersten Grundbegriffe für die Constructionen im Raume als etwas Gegebenes voraus. Sie giebt von ihnen nur Nominaldefinitionen, während die wesentlichen Bestimmungen in Form von Axiomen auftreten. Das Verhältniss dieser Voraussetzungen bleibt dabei im Dunkeln; man sieht weder ein, ob und in wie weit ihre Verbindung ~~möglich~~ nothwendig, noch a priori, ob sie möglich ist.

Diese Dunkelheit wurde auch von Euklid bis auf Legendre, um den berühmtesten neueren Bearbeiter der Geometrie zu nennen, weder von den Mathematikern, noch von den Philosophen, welche sich damit beschäftigten, gehoben. Es hatte dies seinen Grund wohl darin, dass der allgemeine Begriff mehrfach ausgedehnter Grössen, unter welchen die Raumgrössen enthalten sind, ganz unbearbeitet blieb. Ich habe mir daher zunächst die Aufgabe gestellt, den Begriff einer mehrfach ausgedehnten Grösse aus allgemeinen Grössenbegriffen zu construiren. Es wird daraus hervorgehen, dass eine mehrfach ausgedehnte Grösse verschiedener Massverhältnisse fähig ist und der Raum also nur einen besondern Fall einer dreifach ausgedehnten Grösse bildet. Hiervon aber ist eine nothwendige Folge, dass die Sätze der Geometrie sich nicht aus allgemeinen Grössenbegriffen ableiten lassen, sondern dass diejenigen Eigenschaften, durch welche sich der Raum von andern denkbaren dreifach ausgedehnten Grössen unterscheidet, nur aus der Erfahrung entnommen werden können. Hieraus entsteht die Aufgabe, die einfachsten Thatsachen aufzusuchen, aus denen sich die Massverhältnisse des Raumes bestimmen lassen – eine Aufgabe, die der Natur der Sache nach nicht

Fig. 12.1: Dedekind's transcription of Riemann's famous lecture: On the Hypotheses underlying the Foundations of Geometry.

truly chaotic state.[1] Dedekind was thoroughly familiar with Riemann's working manner, which was nothing like his own; he was extremely fastidious himself and wrote in a beautiful hand (see Fig. 12.1). He later informed Heinrich Weber that he had spent a great deal of time trying to salvage and restore some of the fragments [Scheel 2015, 44].[2] Dedekind found only three manuscripts in suitable condition for publication: the texts from Riemann's Habilitation ([Riemann 1868/1876] and [Riemann 1868/1876]) and a short note on electrodynamics [Riemann 1867a/1876]. Riemann had actually submitted that third paper, but he then withdrew it from publication for unclear reasons.

When Dedekind received these and other papers from Riemann's widow, he may well have been the first person to read "On the Hypotheses underlying the Foundations of Geometry" [Riemann 1868/1876] since the time it was written thirteen years earlier. He immediately realized that Riemann had left out nearly all the mathematical machinery related to his new theory of curvature in higher dimensional differentiable manifolds. Recognizing that Riemann's text would be unintelligible without appropriate technical details, Dedekind set about trying to produce a convincing commentary. Before he could complete it, though, he felt urged to prepare clean copies of Riemann's postdoctoral dissertation as well as his Habilitation lecture. Thus, on July 17, 1867, he sent Wilhelm Weber these two texts dating from the time of Riemann's Habilitation. The very next day he informed his good friend Jakob Henle, Professor of Anatomy in Göttingen:

> ... Riemann's papers have constantly occupied me; finally, yesterday I sent two essays to Hofrat Weber, with the request that he place them before the Royal Society. These are the two papers that Riemann wrote in 1854 for his Habilitation. My immediate work on them consisted only in copying the manuscripts, which I was reluctant to entrust to the printer. The second was only read aloud by him at the time and therefore contains only results, but no proofs. My original intention was to elaborate on the latter by following the hints he gave and then adding these as notes to the text. For various reasons, however, I have decided to write up a separate essay based on these, which I hope to complete before the end of the semester. I have often despaired over my slow progress, and therefore I wanted to get at least these two essays published. Now that they have been sent, I feel somewhat relieved. I constantly think Frau Riemann might be angry with me and regret having entrusted me with [her husband's] papers; and yet I can assure you that my days and nights are filled with thoughts of this work alone. The coming holidays, during which I have been ordered to travel, will bring an interruption; my doctor ... recommends a trip to Switzerland; I would be even more inclined to persuade my parents and sister to travel as far as the Italian lakes. ... I hope to be able to visit Göttingen after the end of our semester, if possible with my addition to Riemann's treatise. ... [I] received your mailing of Riemann's essay on acoustics,[3] for which I express my sincere thanks. I was already familiar with the content, but I am very pleased that such an orderly whole has emerged from the manuscripts; I envy you for this and wish only for a similar success. I am still discouraged, but I hope gradually to fulfill the task that has been given to me. [Dugac 1976, 171–172]

[1] He later gave a vivid account of the condition of Riemann's *Nachlass* in a report on the evolution of the first edition of *Riemann's Werke* in [Dedekind 1876b].

[2] This can easily be confirmed from various transcriptions, many of which are available online by searching for "Riemann" at https://gdz.sub.uni-goettingen.de/.

[3] This refers to Riemann's incomplete paper on the mechanics of the ear [Riemann 1867b/1876]; see the end of Section 4.6.

Dedekind attached a footnote to the version of [Riemann 1868/1876] which appeared in Volume 13 of the *Göttinger Abhandlungen*, stating that he planned to publish mathematical commentary on the text. As he described in his letter to Henle, such commentary was needed because Riemann had suppressed all technicalities in keeping with the general audience to whom he spoke. This publication never appeared, however, for two reasons that Dedekind later explained in a letter to Heinrich Weber. First, he was deeply immersed in preparing the second edition of Dirichlet's lectures on number theory [Dirichlet 1871], which contains an early version of Dedekind's theory of ideals. The second reason had to do with the appearance of important new papers that came out shortly after [Riemann 1868/1876] and which dealt directly with some of the technical issues Riemann had suppressed. These were written by E.B. Christoffel ([Christoffel 1869]) and Rudolf Lipschitz ([Lipschitz 1869], [Lipschitz 1870]), both of whom only knew that Dedekind had announced a forthcoming commentary on the text.

In the meantime, Elise Riemann and her sister-in-law Ida had been sorting through Riemann's correspondence and other personal property. Ernst Schering, their neighbor at the astronomical observatory, was already preparing the public lecture he would deliver at the Royal Society on the 1st of December 1866, in honor of Riemann, no easy task. Elise offered to help by providing Schering with a detailed account of their three trips to Italy (see Sections 4.5, 4.6, and 4.7). Beyond that, she loaned him a large packet of letters Riemann had received, several of them written by distinguished fellow scientists. Schering not only made no use of these letters in his address, he later failed to return these documents to their rightful owner, despite her repeated requests. As for the lecture itself, one can hardly escape the conclusion that Schering's main intention was to suggest that Riemann was a talented epigone of Gauss, the great genius who went to his grave without imparting some of his greatest discoveries:

> Unfortunately, Gauss did not present his research findings to the public; these, like so many of his other discoveries, had to be rediscovered. This circumstance, however, enabled Riemann, even in his youth, to prove himself to be an independent and self-reliant thinker.
>
> But here, too, the same misfortune befell his science; for he himself was not granted the opportunity to present his collected results to posterity in a complete form.[4] While his discoveries had to remain largely a repetition of Gauss's works, Riemann's other activities constituted a very successful extension of the boundaries of the field of pure analysis, which Gauss had already opened. [Schering 1909/1990, 368]

To the above, Schering added a similarly ungenerous remark about Dirichlet's work in number theory, suggesting that he, too, never got far beyond Gauss's shadow.

Schering's words that day were probably soon forgotten, though he managed to publish a shorter obituary article the following year [Schering 1867]. His orig-

[4] The parallelism Schering constructed here is wholly misleading. Gauss, who enjoyed economic security and high esteem throughout his long life, voluntarily withheld important mathematical ideas that he could have easily communicated. Riemann had to cope with financial and health problems that hampered his scientific career.

inal intention was to publish lengthier memorial articles devoted to both Gauss and Riemann in the *Abhandlungen*, but the public lecture he delivered on Gauss the following year proved unsuitable for this; at any rate, he soon gave up this plan.[5] Dedekind and Wilhelm Weber both found Schering to be a difficult, self-centered personality. Whatever promises Schering may have made to the family in 1866, he afterward never found time to write about Riemann's work again. Moreover, as *de facto* editor of the Gauss edition, he went out of his way to ensure that Riemann's planned contribution to Volume 3 never appeared in print, an episode recounted below.

Over time, Schering managed to build a scientific fiefdom at the observatory, where he dominated the Royal Society's ongoing Gauss edition. Wilhelm Weber referred to this situation in a letter to Dedekind, written on 28 December 1871, when he sent him page proofs for parts of Volume 2. Weber recalled that his colleague Moritz Abraham Stern had originally been part of the editorial team,

> ...but Stern himself very quickly relinquished this right, and since then Schering has operated (*schaltet und waltet*) without any control. Such oversight could certainly be taken by Clebsch, but reestablishing such a system would create circumstances one would prefer to avoid. [Dugac 1976, 174]

Only a few months later, Wilhelm Weber must have decided in favor of avoidance, as he arranged with Elise Riemann that Clebsch be appointed editor of Riemann's Collected Works, including suitable materials from his *Nachlass*, which was then in Dedekind's hands.

Relations between Dedekind and Schering had soured long before this, beginning with Dedekind's appointment as a corresponding member of the Göttingen Royal Society in 1862 (Schering had been named an Assessor in 1860 and a full member in 1862). In a letter to Rudolf Lipschitz from 5 April 1878, Dedekind recalled the situation with the Gauss edition fifteen years earlier, when

> ...I participated with Riemann, Stern, and Schering, and became embroiled in the most unpleasant and upsetting personal conflicts with the latter, through no fault of my own (as I remain convinced to this day). As a result, I soon withdrew from the entire undertaking. [Scharlau 1986, 90]

Schering, for his part, certainly made no secret of his antipathy for Dedekind, which later in 1885 created a problem in Göttingen for his colleague, Hermann Amandus Schwarz. Stern's retirement created a vacancy, and both Schwarz and Schering were determined that Felix Klein would not become his successor, a fight they eventually lost.[6] Schwarz wrote to Weierstrass on 5 July that he would be delighted to have Dedekind as a colleague, not least because of his character: "infinitely modest, kind, and communicative." Knowing Schering's attitude

[5] Schering's Gauss lecture was published posthumously in [Schering 1909, 385–392], immediately after the lecture honoring Riemann.

[6] Their plans backfired when Weierstrass's former pupil, Georg Hettner, turned down this position, probably to avoid being used as a pawn in a scheme hatched in Berlin. For a detailed account of the maneuvering behind the scenes that led to Klein's appointment, see [Tobies 2021, 316–320].

toward Dedekind, though, Schwarz did not want to risk alienating his ally, and Weierstrass concurred that it would be better not to nominate Dedekind for the vacant professorship [Dugac 1976, 143–144].

Ernst Schering's treatment of Dedekind fit a pattern which, viewed retrospectively, looks obvious: he wanted to make himself the spiritual successor of Gauss in Göttingen. In fact, to a considerable extent he achieved that goal during his lifetime, if one considers the praise heaped upon him by distinguished mathematicians throughout Europe, cited in the laudatory obituary [Schering, K. 1909]. In 1868, Ernst Schering, following his promotion to full professor, moved into the west wing of the observatory, where he thereafter occupied Gauss's former living quarters. That same year he was appointed director of theoretical research, which essentially meant continuing the original Gaussian program devoted to geodesy, magnetism, and precision measurements.

In Schering's mind, he neither needed nor wanted others to assist him with the Gauss edition, the Royal Society's first and most important long-term project [Gierl 2004, 152–153]. Since the Society appointed a Gauss Commission to oversee this work, Schering could not simply take control directly, but he managed to achieve his objective by taking advantage of two factors: first, his work and living quarters gave him immediate access to Gauss's estate; second, he had an uncanny ability to engender disgust among those with whom he was supposed to cooperate on the project. Dedekind and Stern were simply *rausgeekelt*, the colorful German expression for behavior of this kind.[7]

In the case of Riemann, who was of course deathly ill, Schering took a different tack. As noted in Section 11.1, while staying in Pisa, Riemann showed Enrico Betti unpublished source material from Gauss's estate relating to special types of elliptic functions (see Section 7.1). Betti recalled seeing these pages years later when he opened a copy of Volume 3 of the *Gauss Werke*. In the commentary, Schering mentioned Riemann's interest in these materials:

> Riemann requested the editing of Gauss's unpublished manuscripts on elliptic functions, which the Royal Society handed over to him for that purpose. Unfortunately, he left behind neither written nor oral results from his studies. It was only after Riemann's unfortunate and untimely death that I took up the Gaussian manuscripts for my editorial work on the present third volume, excepting those parts already printed. [Gauss 1863–1933, 3: 492]

This statement enraged Riemann's loyal supporter, Karl Hattendorff (Section 11.2), whose pamphlet [Hattendorff 1869] aimed to clarify the record. In it, he described how Riemann had begun work on the Gauss materials, but due to his failing health soon came to realize he would not be able to complete this task on his own. After returning to Göttingen in June 1863, Riemann received permission from Wilhelm Weber to enlist Hattendorff's support for this work,

---

[7] Up until his death in 1897, Schering watchfully guarded his position at the observatory and within the Royal Society, where he and other senior members warded off Klein's efforts to reform the latter institution [Gierl 2004, 180–182]. Much to his dismay, however, in 1873 the Gauss Commission decided to publish Volume 4 without delay, thereby excluding materials on non-Euclidean geometry that Schering had been preparing [Schering, K. 1909, 462].

an arrangement Schering surely learned about that same summer. Hattendorff then transcribed the two longest texts and gave these manuscripts to Sartorius von Waltershausen, who in 1864 visited the Riemanns in Pisa.

After this, Hattendorff began working on some shorter Gauss texts. Then, toward the end of the summer, Schering approached him to ask, for whatever reason, whether he could borrow some of the Gauss material that Riemann was responsible for editing. Months after he loaned out parts of Gauss's papers, Hattendoff began pestering Schering to return this source material, but without success. Instead of doing so, Schering asked Hattendorff to let him have *all* the Gauss papers in his possession. By the summer of 1865, Hattendorff realized he would never again see these original sources.This meant that Riemann's contribution to Volume 3 would necessarily remain incomplete. After reporting what had transpired to Weber, Hattendorff felt impotent to press his case further. He portrayed this conflict as a matter that could only be resolved by full members of the Royal Society (he himself became an Assessor in 1867).

Writing three years after the publication of Volume 3, Hattendorff underscored how Schering's statement, cited above, seemed to imply that he had *obtained* the Gaussian manuscripts after Riemann's death, though it carefully avoided actually asserting this. In telling his side of the story, Riemann's former student made clear that those sources had long been in Schering's possession, and he spelled out further how that had occurred. When Riemann returned to Göttingen for the last time, he demanded to know what had happened. Hattendorff wrote nothing about how Riemann took the sad news, but he did write that if Schering wished to know about Riemann's reaction, he need only inquire. After citing the above passage from Volume 3 for the second time, Hattendorff suggested the following as a more accurate version:

> Riemann reserved the editing of the Gaussian works on elliptic functions especially for himself. But I already found a way to get all the manuscripts into my possession a year and a half before Riemann's death. So it happened that Riemann was *unfortunately* unable to leave behind a complete work. Although the publication of the third volume was considerably delayed as a result, my conscientiousness forced me to wait until Riemann's unfortunate, all-too-early death before starting my own official work. [Hattendorff 1869, 15]

Given the seriousness of these charges, which Hattendorff supported by citing numerous deviations he found between Schering's printed text and the version he had prepared on behalf of Riemann, one is left to wonder: did Riemann tell his wife about this? Hattendorff likely told no one other than Weber, who presumably had no interest in exposing what Schering had done. Only after leaving Göttingen did Hattendorff decide to air what had happened behind the scenes. Felix Klein, who would later clash with Schering in the Royal Society,[8] knew full well about this. Presumably, he took pleasure in recalling Hattendorff's attack as well as Schering's mistakes in his comments on the work of Gauss and Riemann in [Klein 1921–23, 3: 578].

[8] This concerned Klein's proposal to fund the acquisition of texts documenting Riemann's lecture courses, about which see [Tobies 2021, 366–367].

In keeping with his promise to Elise Riemann, Dedekind invested a great deal of time and energy editing various fragments from Riemann's papers. Before he could submit them, however, Alfred Clebsch informed him in early 1872 about a plan to publish Riemann's Collected Works, including manuscripts from his estate. Dedekind agreed to back this project, leaving Clebsch, who had assumed Riemann's former chair in 1868, to do the lion's share of the work. Supported by Dedekind as well as Elise Riemann, Clebsch arranged to have the Teubner publishing house in Leipzig issue a contract for the volume, after which he pursued this undertaking with great fervor.

By the fall of 1872, Clebsch delivered to Teubner a paginated manuscript containing 23 works, to which he planned to add three short supplementary papers. Then, in November of that year, all these plans came to a sudden halt following Clebsch's sudden death from diphtheria. Like Riemann, this came just short of when he would have celebrated his fortieth birthday. Plans to produce a collection of Riemann's works based on materials in his estate thereafter came to a standstill. To better appreciate Heinrich Weber's role in restarting this project (Section 12.3), it will be useful to begin with a brief discussion of a special topic from Riemann's unpublished lectures.

## 12.2 On Algebraic Functions and Theta Characteristics

Even in the early 1890s, there was no established consensus for the foundations of algebraic functions, as can readily be seen from the report on the subject compiled by Alexander Brill and Max Noether [Brill/Noether 1894]. This vast, but only partial report omitted the developments related to algebraic function fields, thus works stemming from [Dedekind/Weber 1882/2012] as well as Leopold Kronecker's programmatic study [Kronecker 1882].[9] No one would have denied the importance of Riemannian ideas for the modern theory of algebraic functions, but his work clearly had far more relevance for complex analysis than for algebraic geometry.[10]

For Heinrich Weber, the principal editor of Riemann's works, a key issue concerned Riemann's investigations on the bitangents to quartic curves, one version of which appeared posthumously in the 1876 edition of the *Werke*.[11] This text and the one that preceded it (Nrs. XXIX and XXX) were *not* written by Riemann, but rather came from a manuscript compiled by Gustav Roch in 1861/62, when he attended Riemann's course on Abelian functions. Others, of course, heard those lectures, but reconstructing this material would pose many difficulties. These cannot be addressed here, but since such matters are of real

[9] Kurt Hensel had originally agreed to report on these developments, but he was unable to fulfill this plan due to other responsibilities that arose after Leopold Kronecker's unexpected death in December 1891.

[10] See the list in [Bottazzini & Gray 2013, 311–312].

[11] For a brief summary of Riemann's presentation, see [Gray 2008, 149–151].

relevance in assessing Riemann's legacy, they should at least be noted (see the discussion of the later editions in Section 12.5).

Although we cannot exclude the possibility that Clebsch had known about Riemann's work on the bitangents to quartics, there appears to be no extant evidence suggesting that he did. Assuming that Clebsch's knowledge of Riemann's theory of Abelian functions was limited to [Riemann 1857b], one can hardly escape the conclusion that his study [Clebsch 1864] was an impressive and *largely independent* creative achievement. Around the time that Wilhelm Weber convinced Clebsch to undertake an edition of Riemann's papers, the latter sought to promote research on the general theory of genus three Abelian functions. Clebsch did so by suggesting that the Philosophical Faculty announce this topic for its Benecke Prize, which was later won by Heinrich Weber. These circumstances show that both Clebsch and Weber were keenly interested in this particular aspect of Riemann's theory. Before turning to the main subject of the present chapter, we first briefly review the state of this knowledge prior to November 1874, the month in which Weber agreed to become editor of the Riemann edition. In particular, it will be useful to sketch what was known about theta-characteristics during the period immediately preceding Riemann's death.

As noted in Section 7.2, Riemann had carefully studied Georg Rosenhain's prize-winning memoir for the Paris Academy [Rosenhain 1851/1895]. This text was published in German translation in 1895 along with editorial notes by Heinrich Weber, who had long before established his expertise in this area, in part through his editorship of *Riemanns Werke*. Weber's final note for [Rosenhain 1851/1895] contains a table of notations for theta-characteristics, a theory generally attributed to Riemann, but which already surfaced in Rosenhain's study, at least implicitly.[12]

A key figure for understanding Riemann's interests in this direction was Gustav Roch. In 1866, Roch published a detailed study of the bitangents to a quartic curve based on theorems and formulas derived by Riemann. In that paper, Roch alluded to the latter's lecture course, noting that Riemann had showed how one can forgo the use of $\theta$-functions in obtaining the same results [Roch 1866d, 97]. What he meant by that remark, however, remains unclear. Roch's own paper from 1866 appears to be the first place in which Riemann's notion of *characteristics* of $\theta$-functions appeared in print [Roch 1866d, 101].[13] Moreover, his presentation drew a sharp line between Riemann's methods and those employed by Clebsch, making clear the larger agenda Roch had in mind. Clebsch also dealt with the characteristics of double tangents, though he did not employ that terminology as such. Moreover, he handled this topic in connection with Hesse's results on the 63 systems of conics tangent to a quartic curve at four points [Clebsch 1864, 209–210].

---

[12] Rosenhain's study utilized 16 $\theta$-functions to solve Jacobi inversion for $p = 2$; see further [Hudson 1905/1990, 141–147].

[13] Riemann had employed these in 1862 [Riemann 1876b] and also even earlier in his lecture course from 1861/62 (see page 7 from the supplement in the 1902 edition of Riemann's *Werke*).

Roch filled his paper with precise references to Riemann's results on Abelian functions in [Riemann 1857b], but without making even a single direct reference to Clebsch's publications. Instead, he began by alluding to the many related results on algebraic curves published by Hesse and Jakob Steiner. Only later did he mention Clebsch's work in passing [Roch 1866d, 108], stating in effect that Riemann's theory of $\theta$-functions leads directly to the principal tool utilized by Clebsch in his work on algebraic curves, namely the addition theorem for Abelian integrals. Roch's apparent animus against Clebsch may well have come in response to remarks the latter penned at the outset of his groundbreaking study [Clebsch 1864, 189–190], which placed his work firmly in the soil of Jacobi's methods as opposed to those of Riemann.

Roch was already long dead when Clebsch proposed tackling the $p = 3$ case for Abelian functions as the Benecke Prize. Unfortunately, Clebsch himself was no longer alive to read Heinrich Weber's award-winning manuscript, which was eventually published by the Berlin Academy in [H.M. Weber 1876].[14] This study offered only a partial solution to the problem posed, but it nevertheless gave a nearly exhaustive analysis of the various cases pertaining to the bitangents of a quartic curve. Weber evidently drew on Riemann's initial work, but he cited it only sparingly; more surprising still, he made no reference whatsoever to Roch's 1866 study. At one point he referred to derivations Riemann presented in unpublished lectures, but with no concrete indication of the source for this information [H.M. Weber 1876, 101].[15]

Adding to these mysteries, it remains very unclear what Clebsch may have known about Riemann's unpublished investigations on this topic. It would seem hardly accidental that he proposed the $p = 3$ case for the Benecke Prize at the very time when he began editing Riemann's posthumous papers, but unfortunately the extant correspondence appears to contain nothing relating to these matters. We are far better informed about other issues that arose in the correspondence between Richard Dedekind and Heinrich Weber, as presented in [Scheel 2015]. The following section draws heavily on Katrin Scheel's book, which brought to light some of the behind-the-scenes activity leading up to the publication of Riemann's *Werke*.

## 12.3 Dedekind Wins Over Heinrich Weber

After the death of Clebsch, Wilhelm Weber hoped to avert the collapse of the Riemann project by turning to Carl Neumann, Clebsch's old friend from their student days in Königsberg. Neumann had taken a position in Leipzig around the

[14] When Dedekind read that Weber had won the Benecke Prize, he sent him his congratulations on 26 March 1875 [Scheel 2015, 67].

[15] Weber's monograph was later studied by Klein, who recommended it to at least one of his students. The copy available online at the Internet Archive was originally acquired by Klein's American doctoral student Mellen W. Haskell in 1888 when he was studying in Göttingen (see https://archive.org/details/theoriederabelvom00weberich/page/n19/mode/2up).

same time that Clebsch gained Riemann's chair. Together they then launched the new journal *Mathematische Annalen*, which later became a crown jewel for the publishing house of B.G. Teubner in Leipzig. Clebsch had thus found it easy to arrange a contract with Teubner for the publication of *Riemanns Werke*. When he suddenly died in November 1872, all these plans, including the future of the *Annalen*, were suddenly in doubt. As co-editor of the journal, Neumann already looked to other members of the Clebsch school for assistance, in particular to young Felix Klein. He apparently also gave serious consideration to assuming the editorship of Riemann's works, but personal considerations caused him to postpone and then eventually decline accepting that responsibility.

Dedekind probably sensed Carl Neumann's reluctance to step in for Clebsch, but he patiently awaited word from Wilhelm Weber about Neumann's decision. During the fall 1873 semester break, Dedekind visited his old haunts in Zurich, where he once taught from 1858 to 1862. This gave him the opportunity to meet the multitalented mathematician Heinrich Weber (Fig. 12.2), who joined the faculty at the ETH in 1869. A native of Heidelberg, Weber largely identified with the Königsberg mathematical tradition. His teachers in Heidelberg – Otto Hesse, Gustav Kirchhoff and Hermann Helmholtz – all came from Königsberg, and they encouraged him to habilitate there. In fact, Weber would soon leave Zurich in 1875 to join the faculty in Königsberg, where he remained until 1883. Two of his students from that later period were David Hilbert and Hermann Minkowski.

While getting to know one another for the first time, Dedekind brought up the fact that plans to publish an edition of Riemann's works were in a state of limbo. Dedekind made no secret of his own personal interest in promoting this project, and he was surely pleased to learn that Weber, too, had a keen interest in helping out. A year went by, however, before Dedekind finally got word that Neumann had indeed declined, at which point he turned to Heinrich Weber. In a letter from 1 November, he recalled their conversation from a year earlier and expressed his hope that his younger colleague would now agree to take on this task. Dedekind also explained some of the circumstances that had led to this project:

> Soon after Riemann's death, [his widow] handed over to me all the mathematical papers found in his estate, with the task of identifying those suitable for publication (only the papers relating to the problem of minimal surfaces – if any existed at all – were entrusted to Mr. Hattendorf; a short essay on the mechanics of the ear has also been published by Messrs. Henle and Schering). I found only three essays (on trigonometric series, on the hypotheses underlying geometry, and one on electrodynamics) in such a finished form that I was able to have them printed soon afterward. Everything else was in a chaotic state, and only with a great expenditure of time have I been able to compile a few fragments that I consider worthy of publication.

As Dedekind recalled, these events had taken place between late 1866 and early 1872, at which point he learned from Wilhelm Weber about a more ambitious plan. As it happened, Weber had convinced Clebsch that he should prepare a complete edition of Riemann's works. Soon afterward, Alfred Clebsch visited Dedekind in Braunschweig, where they quickly reached agreement that all the material Dedekind had prepared should be sent to Göttingen. During the months that followed, Clebsch scrutinized much of this material carefully and prepared it for publication. Dedekind remarked further that

> ... it was only after much resistance that I was persuaded by [Clebsch] and W. Weber to take further part as co-editor. Shortly afterward, ... I was elected director of our Polytechnicum for three years, and although I had no idea of the heavy burden this would entail, I immediately explained that my involvement in the publication could only involve occasional revisions of the proofs. [Scheel 2015, 43–44]

This meant, of course, that Clebsch would have to do the lion's share of the editorial work. As Dedekind reported to Heinrich Weber, this arrangement worked very well, since Clebsch pursued the task at hand with great fervor. At the time of his death, he had already delivered a paginated text to Teubner containing 23 works, to which he planned to send three more short papers.[16]

Having summarized this prehistory, Dedekind asked Heinrich Weber whether he would agree to take over the editorship at this stage. On 5 November, Weber replied from Zurich, promising to do whatever he could to bring this worthy project to completion. Dedekind soon realized that these were no empty words, as only a few weeks later Weber reported that he had already oriented himself with regard to the unpublished papers in Riemann's estate. These exchanges marked the beginning of a remarkable collaboration between these two distinguished German mathematicians, who formed a deep friendship rarely seen at this time. As Weber himself remarked, they were surrounded by colleagues who constantly promoted academic infighting and even public outbursts of animosity.[17]

In the meantime, Elise Riemann conveyed her thanks to Dedekind for making these arrangements. She had received the various packages Clebsch had prepared and promised to send these immediately to Heinrich Weber. Her list of these contained thirteen items clearly identifiable from the edition and another four whose contents remain obscure. Frau Riemann also mentioned that Neumann had received certain documents, whereas Hattendorf held those related to Riemann's paper on minimal surfaces. She herself had certain materials written by Dirichlet, which she planned to return to Wilhelm Weber. Most interesting of all, she indicated that she would keep Riemann's personal course notes, some of which were written in Berlin, as well as manuscripts he prepared for his own lectures [Scheel 2015, 307–308]. It remains unclear whether Elise Riemann's descendents received some of these documents after her death.

[16] Up to this time, eleven articles had been published during Riemann's lifetime and another seven had appeared posthumously [Scheel 2015, 385].

[17] By the end of their work on the Riemann edition, they were on a "Du" basis and began their letters with the salutation "Lieber Freund!" Not long after arriving in Königsberg, Weber wrote on 6 November 1875: "The conditions at the university are, as now unfortunately almost everywhere, very unpleasant. Disputes and factions disrupt all impartial communication for no good reason. These disputes, which are now rampant at almost all German universities, represent a deeply concerning cultural phenomenon that deserves serious attention" [Scheel 2015, 77].

Fig. 12.2: Heinrich Martin Weber (1842–1913).

In view of Weber's own recent research for [H.M. Weber 1876], we can easily understand why he was eager to recover Riemann's work on Abelian functions of genus $p = 3$ and the related theory of $\theta$-characteristics. Unfortunately, what Weber found in Riemann's estate were only unordered pages of diverse calculations in a hopelessly chaotic state. Luckily, due to his intimate familiarity with this topic, plus the fact that he possessed corresponding lecture notes prepared by Gustav Roch, he "already recognized the essential core (*das Wesentliche*) of his very beautiful and important investigation for the Abelian functions" [Scheel 2015, 52]. Surely hoping for Dedekind's approval, he added that no one would probably take issue with Weber's planned use of Roch's *Vorlesungsheft*.

As a matter of editorial policy, Weber regarded this unpublished material from Riemann's lectures as far too important to be neglected. Clearly, he had no intention of making a systematic search of other lecture notes, knowing full well that any such endeavor would have delayed publication of the edition by many months, if not years. The question of whether to include material from Roch's lecture notes came up more than once in their correspondence, but it seems Weber was never able to elicit a reaction from Dedekind. Weber surely had a major stake in preparing this topic for the edition, but as months went on he developed serious doubts about the feasibility of this plan. In March 1875, he wrote Dedekind:

> ... regarding the investigations on double tangents, I have become doubtful again about whether it is appropriate to take these up. Among [Riemann's] papers, one finds only formulas, and I must take a great deal from [Roch's] *Heft* in order to deliver something comprehensible, which requires going into some lengthier elaborations [Scheel 2015, 65].

Finally, in a letter from 8 July 1875 [Scheel 2015, 70–72], Weber asked for Dedekind's opinion as to whether two topics in Roch's lecture notes could be published in Riemann's *Werke*. If Dedekind ever replied, no evidence of this survives, though ultimately these were published as numbers XXIX and XXX in the original edition from 1876. In the *Vorrede*, Weber mentioned that these texts were based on Roch's *Vorlesungsheft*, which presumably no longer exists. Weber surely knew that Roch had been entrusted at some point by Riemann to publish parts of an earlier lecture course, as these had appeared in [Roch 1863a] and [Roch 1865a]. He therefore had good reason to believe that Roch had conscientiously written down what Riemann presented, which included a definition of even and odd characteristics for general theta-functions [Riemann 1876/1892/1902, 1876: 456–458]. This established Riemann's priority, in some sense, but without clarifying when and how the theory of characteristics came into circulation.

In view of the fact that Gustav Roch passed away already in 1866, just four months after Riemann's own death, one naturally wonders how Weber ever acquired access to this notebook. The answer can be found in one of his letters to Dedekind, where Weber mentioned that he had obtained Roch's *Heft* from Hermann Amandus Schwarz, his then colleague in Zurich [Scheel 2015, 71]. One can also easily surmise the circumstances that enabled Schwarz to acquire Roch's lecture notes. Probably this occurred not long after Roch's death in Venice, where he succumbed to tuberculosis on 21 November 1866. Shortly before he left for Italy, Roch had been promoted to an associate professorship in Halle, and following his death, the Prussian ministry appointed Schwarz as his successor. Roch surely left behind many of his personal belongings when he set off for Venice, so it seems likely that his notes from Riemann's lectures were among these.

Schwarz may have inquired about Roch's estate or perhaps colleagues in Halle engaged him to investigate whether his scientific papers contained documents of potential value. In any event, it appears likely Schwarz acquired Roch's notes not long after the young man's death. In fact, Schwarz remained only a short time in Halle; in 1870 he was appointed to a professorship at the ETH in Zurich, where he was later joined by Heinrich Weber [Frei/Stammbach 1994, 41]. Both remained there until 1875, when Weber accepted the professorship in Köngisberg formerly held by Jacobi's star pupil, Friedrich Richelot. Once there, he put the final touches on the first edition of Riemann's *Werke*. As mentioned above, this volume excluded material from Riemann's lectures – with the exception of Roch's notes – as well as documents from his scientific correspondence other than excerpts from Riemann's letters that had previously found their way into print. Thus, Weber felt driven to include Roch's lecture notes only because he realized the hopelessness of piecing together a publishable account based on the scattered documents in Riemann's estate.

In the meantime, Schwarz had left Zurich for Göttingen, where he took the chair formerly occupied by Riemann and Clebsch. Schwarz continued to take an avid interest in the Riemann editorial project, and he soon found himself in an ideal position to monitor its progress: his neighbor on Weender Chausee in Göttingen was none other than Elise Riemann (see his letters to Weber from 1875–76 in [Scheel 2015, 360–378]). He took charge in editing the fragment on the representation of quotients of hypergeometric series by means of continued fractions, which was written in Italian [Riemann 1876/1892/1902, 1876: 400–406]. As an expert on minimal surfaces, Schwarz also weighed in heavily regarding Karl Hattendorff's revision of the paper on this topic that he had originally published as [Riemann 1868/1876]. Weber had asked Hattendorff to make certain revisions for the version in the *Werke* [Riemann 1876/1892/1902, 1876: 283–315], but Schwarz found these wholly inadequate and pestered Weber with complaints about this work. Weber, who knew about Schwarz's pedantic tendencies, wrote him that he couldn't bother with such quibbles (*Spitzfindigkeiten*) [Scheel 2015, 370–374].

On 10 July 1876, Weber wrote to Dedekind, who had recently received a request from Schwarz to write a review of the forthcoming volume. Weber explained that he had not heard from Schwarz in quite a while and suspected he was angry about the manner in which Weber had rebuffed him. Weber's wife had also written to Frau Schwarz, but never received a reply, another sign that the family was offended by these recent events. As Weber explained to Dedekind:

> When I had the proof sheets . . . sent to him at his request, he sent me back a long list of comments on it, partly accompanied by rather bitter words directed at Hattendorff. Some of these complaints were, of course, well founded, but others were petty. I wrote to him openly about this, while at the same time requesting that he not damage the entire undertaking by voicing an unfavorable judgment. This may have made him unhappy and therefore unwilling to write me. In any case, I won't inquire about this, and if you answer him you can greet him from me. But I know his type (*ich kenne meine Pappenheimer*). [Scheel 2015, 150]

Fifteen years later, however, when Weber prepared the second edition of the *Werke*, circumstances had changed. Hattendorff was no longer alive, whereas Schwarz had solidified his reputation as the leading German authority on minimal surfaces. In view of this, Weber decided to add some commentary on Hattendorff's text, following suggestions he had received from Schwarz [Riemann 1876/1892/1902, 1892: 334–337].[18]

Reading Weber's letters to Dedekind, one is struck by the fact that the earlier work undertaken by Clebsch was rarely mentioned. Evidently, Heinrich Weber thought he needed to start from scratch, as it were, and Richard Dedekind was very pleased that he took this attitude. Weber realized that the disparate materials he wanted to include in Riemann's *Werke* required careful organization and, in some cases, technical commentary. He divided its thirty papers into three sections, presenting those in the first two sections strictly chronologically. Section one contains the eleven papers Riemann himself published, four of which were merely announcements or relatively minor works. The second section presents seven more papers that only appeared in print after his death. The final group of twelve papers from Riemann's estate were altogether new and posed many challenges for Weber as the principal editor.[19] Some of these documents were given with dates, but in several cases it was impossible to discern when they were written. To these twelve new works, Weber added an appendix devoted to Riemann's philosophical and speculative writings, followed by Dedekind's biographical essay. Dedekind offered a few brief remarks about the importance of Riemann's philosophical speculations, but these have rarely received serious attention from subsequent commentators.[20]

Several of the new texts that appeared in the third section were based on fragments that required extensive elaboration by Weber. This being the case, several fall in the genre of works by Riemann-Weber. In fact, in some cases it is impossible to discern the degree to which they reflect Weber's own contributions. For example, as noted in Section 5.1, Weber's rendition of the fragment "Equilibrium of Electricity on Cylinders with Circular Cross-Section and Parallel Axes" in [Riemann 1876/1892/1902, 1876: 413–416] raises many questions about its origins and significance within Riemann's oeuvre.

Among the papers that required additional commentary, one of the most challenging proved to be Riemann's "Commentatio." No one apparently knew about the existence of this text until Dedekind discovered a draft of it in Riemann's estate (see Section 8.3). Later, Heinrich Weber tried to understand the chain of arguments in its second part, which contained evident gaps. He wrote about this to Dedekind on 28 February 1875:

> What's still bothering me now is the Paris paper, which I've been exclusively occupied with for some time now, and which still presents some tough nuts to crack. I had hoped that printing could begin at the beginning of summer, since I'm pretty much finished with the rest. But now I don't know how long it will keep me. I have, however, sorted out some of the results. [Scheel 2015, 53–54]

[18] For a summary of work on minimal surfaces by Riemann, Weierstrass, and Schwarz, see [Bottazzini & Gray 2013, 540–546].

[19] He gave a summary of these papers in his review article [H.M. Weber 1877].

[20] See, however, the discussions in [Scholz 1982a], [Laugwitz 1996/1999, 277–292], and [Ferreirós 2006].

Weber pointed to the report in the *Comptes Rendus* (1861, p. 1165) of the Academy, explaining why Riemann's paper was not awarded the prize: "One of the competitors limited himself to outlining a method whose consequences were not yet fully developed."[21] At first, he imagined the draft in Riemann's estate was incomplete, so he hoped to recover the actual text Riemann had submitted to the Paris Academy in 1861.

The very next day, Weber wrote again to confirm that the matter was being pursued by the Göttingen Royal Society. In the meantime, Wilhelm Weber had contacted his colleague Friedrich Wöhler, Secretary of the Royal Society, who would soon obtain the original manuscript. As it turned out, this version differed little from the draft Weber had been working with, so he still had "some hard nuts to crack" [Scheel 2015, 323]. Wilhelm Weber also mentioned that Riemann had spoken of *two* different submissions to the Royal Society, which confused matters further. A few days later, on 4 March, Elise Riemann was able to clarify what had happened in a letter to Heinrich Weber:

> Riemann did not send a second paper to Paris, but he did consider following the first with further elaborations, since the Academy considered the problem solved but withheld awarding the prize because the method by which the result was found was not clarified; they therefore granted an extension of the deadline. In the winter of '62, which we spent in Messina for Riemann's health, he still wanted to work on it and took the relevant papers with him. However, it became so cold in December that we were unable to maintain a warm room, and so, unfortunately, work at home had to be stopped before the deadline expired on January 1, 1863. Riemann did not work on this problem again afterward. [Scheel 2015, 322]

Dedekind sent Weber a very long letter on 14 March 1875, in an effort to clarify a number of difficult problems. One of these concerned the various philosophical fragments that Weber ultimately published as an appendix. Some of this material was written during the early 1850s before Riemann had habilitated (see Section 2.7), but there were no apparent threads connecting the various pieces.[22] Clebsch had originally considered the possibility of presenting these fragments using a smaller font. He also consulted with his younger colleague, the philosopher Carl Stumpf, who made suggestions for ordering the material.[23]

Dedekind concurred with the opinion of Clebsch, who thought these materials could not be omitted from the edition. At the same time, Dedekind underscored the importance of Riemann's fundamental physical views:

> As far as I am concerned, I am extremely passionate about the continuous material plenum of space and the explanation of the phenomena of gravity and light. For I know from years ago that in 1856 and 1857, when I had a lot of contact with Riemann, that he attached great importance to these ideas. During his stay in Harzburg, he showed me Newton's letter, ... in which he spoke about the philosophical impossibility of direct

[21] "L'un des concurrents c'est borné, à esquisser une méthode dont le temps ne lui a pas permis de développer les conséquences ..." [Scheel 2015, 69].

[22] Weber would later change the ordering for the second edition to better reflect the chronology; see [Riemann 1876/1892/1902, 1892: 508].

[23] In response, Weber indicated that he himself asked the advice of Wilhelm Wundt, whose suggestions agreed essentially with those made by Stumpf [Scheel 2015, 64].

> action at a distance, despite the great success of his [Newton's gravitational] hypothesis. Riemann took up these thoughts very early on, not just in his later years, and if I'm not mistaken, the remark "Found on March 1, 1853" appears on a nasty-looking folio sheet.[24]

Dedekind then added this remark, which captures the essence of Riemann's program for the foundations of physics:

> His aim was undoubtedly not to overturn mechanics but rather to establish the most general principles for a new and more natural conception of physical processes. These should be understood as based fundamentally on the principle of conservation, by means of which changes in states depend on immediately neighboring states in time and space, as expressed through partial differential equations. [Scheel 2015, 59–60]

In closing, Dedekind heaped praise on Heinrich Weber, whose diligence truly impressed him:

> Believe me, I recognize in the highest degree your laudable zeal for the cause of the deceased, because I have come to know so intimately the great difficulty of the undertaking and appreciate the sacrifice and self-denial required for it. If only you maintain your courage! [Scheel 2015, 62]

Weber responded in an equally long letter, written on 22 March. He related what Elise Riemann had conveyed about the Paris prize paper, but also his fears that the original manuscript would not contain much more than the draft (which happened to be true). Weber thus foresaw that he would still face considerable difficulties in trying to prepare a readable commentary for the "Commentatio." Up until this time, Dedekind had written very little about his own earlier efforts to decipher this text, but he now must have felt the need to offer whatever support he could. Thus, on 26 March he wrote back with the following background information:

> The connection between this work (Pars II) and the lecture on the hypotheses of geometry prompts me to add the following here. When publishing this latter work, I expressed the intention of supplying the analytical investigations later, and I occupied myself with this subject for a long time in the following years (mainly in 1867, I believe), but later abandoned publication altogether, partly because others (Christoffel, Lipschitz, Beltrami) had taken up this subject, and partly because in 1869 I was forced by the indispensable preparatory work for the second edition of Dirichlet's number theory to devote myself to a completely different field, namely the creation of a general and complete theory of ideal numbers. I yesterday searched through my then very extensive papers and found among them three drafts, some very precisely executed, for such a supplementary commentary. I will send them to you (along with other papers to be mentioned shortly) in the next few days, although I suspect that they will bring you little new information. [Scheel 2015, 66]

Christoffel, of course, had no knowledge of Riemann's "Commentatio" when he wrote [Christoffel 1869]. A number of writers have assumed that since he cited

[24] In the original edition, this appears on [Riemann 1876/1892/1902, 1876: 502], but a colleague later pointed out that the text Riemann referred to predated the far more famous speculations on gravity and light. For this reason, Weber changed the ordering of this material, as he explained in [Riemann 1876/1892/1902, 1892: 508].

Riemann's recently published "On the Hypotheses underlying the Foundations of Geometry" [Riemann 1868/1876], he must have studied it carefully and that doing so had inspired his paper. A careful reading of [Christoffel 1869] suggests, however, that this interpretation is very improbable. Christoffel's main result, his reduction theorem, transformed the classification of differential invariants for quadratic forms to a related problem for algebraic invariants. Even if he had taken inspiration from Riemann's Habilitation lecture, one can hardly imagine that he hoped to fill in its technical gaps as Dedekind had hoped to do.

To summarize, it was not until Heinrich Weber published the "Commentatio" with mathematical commentary in [Riemann 1876b] that the mathematical world learned about its contents (see Section 8.3). Secondly, he revised the original commentary when he prepared the second edition of the *Werke* [Riemann 1876/1892/1902, 1892: vii]. Thirdly, following the lead of Dedekind, Weber assumed that interest in the text centered almost exclusively on its relevance for understanding the ideas Riemann only alluded to in [Riemann 1868/1876].

By August, when Heinrich Weber was preparing to depart for Zurich, he sent Dedekind part of the manuscript for his approval. Later that month they spent a few days together vacationing in Bönigen on Lake Brienz near Interlaken, Switzerland. Weber was already missing Zurich when he wrote to Dedekind on 18 January 1876, with an update on the third section of the *Werke*. He had recently sent Teubner the first eight of its documents, so only four remained. One of these consisted of a fragment in two parts that Dedekind had been struggling with for some time. Weber urged him to send his commentary soon, as he hoped this would prompt Teubner to continue with the printing. Thus, he wrote: "I would also like to ask you to send me your comments on the above-mentioned [two-part] fragment ... as soon as possible, as I will soon be in a position to have the remainder ready for printing" [Scheel 2015, 95–96]. For the next several weeks, Dedekind received messages repeating this request, to which he eventually responded in February.[25]

Although these fragments cast little new light on Riemann's unpublished work, their mysterious content haunted Dedekind for several months. Eventually he found an ingenious way to test the validity of Riemann's formulas and wrote this up for Weber. The first fragment was dated 1852 and written in Latin as a kind of supplement to § 40 of Jacobi's *Fundamenta nova* [Jacobi 1829]. It began with seven identities expressing various integrals as formal infinite series. Riemann used these to consider their convergence properties for certain rational arguments. The second fragment was undated and consisted in a long series of identities that led to eight different cases. Dedekind transcribed most of this from a single very pale page written in pencil (reproduced from Cod. Ms. Riemann 14, p. 18v in [Haffner 2021, 34]). Neither Dedekind nor Weber offered substantive clues to account for the motivation behind these thorny calculations. They only speculated, partly based on the assumption that the document was written in

[25] Dedekind's commentary can be found in Cod. Ms. B. Riemann 4, pp. 25-29; this document can be found at https://eman-archives.org/DWR/files/show/69.

1852, that Riemann undertook these calculations in order to find examples of infinitely oscillating functions, a theme in his postdoctoral thesis (see Section 6.1). This second fragment still occupied Dedekind's full attention even up to the time when the manuscript for *Riemanns Werke* was sent off to Teubner.

As Emmylou Haffner pointed out, Dedekind was drawn to study Riemann's text because of its connection with a very different topic. As he wrote:

> The method used here by Riemann to determine the behavior of the modular functions occurring in the theory of elliptic functions in the case that the complex period ratio
>
> $$\omega = \frac{K'\,i}{K} = \frac{\log q}{\pi i}$$
>
> approaches a rational value, permits at the same time a very interesting application to the so-called theory of the infinitely many forms of $\theta$ functions, namely the determination of the constants occurring in the first-order transformation, which, as is well known, Jacobi and Hermite reduced to Gaussian sums, thus to the theory of quadratic residues. Since I only made this observation during the last days before printing, not enough time remained to check whether Riemann's formulas were exactly correct for the real parts. Yet because all of them must result from the investigation that follows, hopefully their publication without this confirmation will prove to be justified [Riemann 1876/1892/1902, 1876: 438]

Dedekind's closing remarks in this passage were, in fact, somewhat misleading, as can be seen from the many letters he and Weber exchanged in an effort to confirm the correctness of Riemann's calculations. In one of those letters, Dedekind offered these interesting comments:

> ...this pale manuscript is really strange. Judging by the number of things crossed out and by all the side calculations, one would think that Riemann worked these out for the first time on this piece of paper. Yet, on the other hand, in the progression from one equal sign to the next there are gaps that are difficult to fill in just by using one's head. For that reason, I suspect he did the actual calculations on a slate tablet (which he sometimes used) and only wrote the more important results of those calculations on the paper. At times it has been difficult for me to fill in these gaps .... [Scheel 2015, 121–122]

When in mid-February Dedekind sent Weber his commentary, he emphasized that he had ignored the first fragment, about which he had nothing to add that went beyond Weber's own guess, namely that both fragments were connected with Riemann's work on his postdoctoral thesis. Dedekind's commentary was thus only concerned with the *mathematical content* of the second fragment, and he also realized his explanations were inspired by ideas that had no direct bearing on Riemann's text:

> At first I wrote down my commentary with the intention of recommending it to you for printing. But since it has become far too long, and especially since I was unable to examine Riemann's formulas due to lack of time, I consider it expedient to print Riemann's formulas ... adding a note that expresses roughly what appears in the first paragraph of my commentary ... i.e., mainly the reservation due to possible inaccuracy of the formulas. If the calculation were not so meticulous and not so susceptible to errors, I would have asked you to give me a little more time; but I despair doing this within a few days, and to be honest, this matter has become almost unbearable for me. [Scheel 2015, 104]

Dedekind then appended his own independent derivation of formulas that could, in principle, be used to confirm whether or not Riemann's results were correct.

Based on the above and the rather sheepish tone of his letter, one must imagine that Dedekind expected Weber would be disappointed by what he received. Yet not long afterward, he received this surprising response:

> First of all, thank you very much for your text and for all the effort you put into the whole thing. My intention now is to have your commentary printed in full ... since without it the second fragment would be completely incomprehensible, and furthermore your treatment of the subject is so interesting and beautiful that it is entirely in place. But I have also included the formulas that you sent me in your letter, which I have recalculated and found to be completely correct, and have accordingly formulated the conclusion a little differently. These formulas contain the preferred explanation of Riemann's fragment.[26] [Scheel 2015, 107]

Neither one of them could have suspected that Dedekind's ideas would eventually set off an avalanche of work in the field of modular functions.

Weber hereafter threw himself headlong into the unresolved problems raised by Dedekind's commentary. His enthusiasm was such that before long they were exchanging long letters not only about the mathematics underlying this fragment but also the new theory of modular functions Dedekind had foreseen. As Weber quickly realized, the latter's commentary developed an entirely new independent theory based on $\eta(\omega)$, which later became known as Dedekind's eta-function. Already in April, Dedekind was writing up parts of his "valence theory" [Dedekind 1877], as he called it, and he was astonished to see how quickly Weber grasped the main ideas and reacted to them. It was Weber who suggested that he send this famous theory to Borchardt [Scheel 2015, 119–120], who had only recently published portions of letters from Lazarus Fuchs to Charles Hermite, which also occupy a central place in this story.[27]

Some months later, Dedekind read and reacted to the extract in [Fuchs 1877]. He then sent Borchardt a portion of the theory already sketched in his commentary on item XXVII. For his part, Borchardt requested that Dedekind submit a full account of Fuchs' letter explaining how the ideas Hermite had been pursuing related to Dedekind's theory. This led to [Dedekind 1877], a classic work submitted on 12 June 1877. At one point in this paper, Dedekind noted how the properties of $\eta(\omega)$ had enabled him to grasp the mathematics behind Riemann's fragment. Placing this in the context of the limit values of modular functions, he noted that the results disproved a contention in [Fuchs 1877] concerning the natural circular boundary of the functions he investigated. Dedekind noted that one needed to distinguish between functions that approached a dense set of points on the boundary and those with the full circle as their limit set.

A striking feature of [Dedekind 1877] is the interplay between complex analysis and number theory, on the one hand, and the appeal to Riemannian ideas, on the other, even though Dedekind's paper contains not even a single geometric figure. He introduced his valence function $v(\omega)$ as an existence theorem in

[26] These formulas appear on [Riemann 1876/1892/1902, 1876: 446–447].

[27] For an overview of these developments, see [Gray 2008, 101–114].

the spirit of Riemann's dissertation [Dedekind 1877, 274], noting that it was uniquely determined by taking

$$v(\rho) = 0, \quad v(i) = 1, \quad v(\infty) = \infty,$$

where $\rho = \frac{-1+i\sqrt{3}}{2}$. This modular figure soon became very familiar beginning with Felix Klein's paper [Klein 1878]. In modern language, Dedekind described the fundamental domain of the modular group under which $v(\omega)$ is an absolute invariant. Klein would give a slightly different formulation in terms of the absolute invariant $J$, which he based on cross ratios, which are projective invariants. He took care to note that his own presentation was closely related to the approach in [Dedekind 1877].

For present purposes, the main point of interest stems from the fact that Dedekind's paper grew out of his collaboration with Heinrich Weber in preparing the extensive commentary to item XXVII in the original edition of *Riemanns Werke*.[28] As editors, both men were far more interested to excavate and then elaborate on Riemann's unpublished documents than to merely put them before the mathematical public. Their approach to editorial work clearly clashes with modern-day standards, but the larger question of motivation nevertheless remains. As creative mathematicians of the first rank, they took much inspiration from their mutual interest in uncovering whatever they could from Riemann's papers. Just as Dedekind took the opportunity to create his theory of ideals while editing Dirichlet's lectures on number theory [Dirichlet 1871], so did Weber take inspiration from Riemann and Dedekind to conceive of a firmly grounded conceptual approach to algebraic functions based on compact Riemann surfaces [Dedekind/Weber 1882/2012].[29]

Among the fruits of their collaboration, in effect, was Dedekind's mature theory of ideals, as presented in Supplement XI of his third edition of Dirichlet's lectures. Dedekind commented about these interconnections in the preface:

> Particularly worthy of mention is the broader presentation of the same ideal theory contained in the final Supplement [Nr. XI], which I first published in the second edition, but in such a condensed form that many have expressed the desire to have a more detailed treatment. I was all the more willing to comply with this request because an investigation carried out by my friend H. Weber in Königsberg in collaboration with me, and which will be published shortly, has shown that the same principles can be successfully transferred to the theory of algebraic functions. [Dirichlet 1879, vii–viii]

Indeed, all the while Weber worked on cracking those "hard nuts" for Riemann's *Werke*, he was receiving Dedekind's letters describing his theory of ideals. This theory first arose as a new treatment of Kummer's ideal numbers in algebraic number fields, but by the later 1870s it had evolved into a general algebraic theory. To Dedekind's great amazement, Weber not only absorbed this

[28] As noted in the following section, Dedekind revised his commentary for the second edition, while pointing to the extensive literature on modular functions published since 1876.

[29] For an assessment of its significance, see John Stillwell's introduction to the English edition of [Dedekind/Weber 1882/2012] and [Geyer 1981].

approach very quickly, he even saw how it could be applied to Riemann's theory of algebraic functions. Beginning with his letter to Dedekind, written on 21 February 1879, up until Dedekind's final response from 30 October 1880, their surviving correspondence documents beautifully the genesis of their famous paper [Dedekind/Weber 1882/2012], which paid the highest mathematical tribute to Riemann's original insights by showing they were, indeed, sound.

Weber broached this new idea when he first wrote:

> ...But now to another point, namely the ideals. First of all, I must ask for your forgiveness for interfering with your work, but the matter interests me so much that it leaves me no peace day or night. And so I came to the following, which seems to me to be of interest .... [Scheel 2015, 228]

Unfortunately, only a few fragments from Dedekind's side of the correspondence survive, but Weber's replies often provide clues as to what those lost letters must have contained. Only about two weeks later, on 5 March 1879, Weber gave this assessment of what he hoped might be achieved:

> We'll have to wait and see whether this whole thing will lead to something new. What we have so far is basically nothing original, but after all this is a very elegant and pretty way of expressing well-known theorems and in this respect it satisfies an aesthetic need. What I initially hope for from this is a strict or at least more general justification of Riemann's theory. [Scheel 2015, 230]

By the middle of March, Weber had already sketched out his provisional ideas for this new theory, having discovered along the way that a number of open problems still needed to be addressed. He therefore planned to visit Braunschweig during his forthcoming vacation break in April. On 22 March 1879, Weber sent Dedekind an extremely long letter in order to give his friend a clear idea of how far he had gotten [Scheel 2015, 231–237]. In it, he detailed several key points and addressed nine different issues he wanted to discuss with Dedekind when they would meet on 7 April. As always, their meeting went very smoothly, as they thoroughly enjoyed each other's company. Still, a number of problems remained to be resolved. On 22 June, Weber wrote to apologize for having sent a faulty proof of the Riemann-Roch theorem, one of the major foundation stones for the theory. Weber therefore attached a new proof that he hoped contained no holes, while trusting that if there were any, Dedekind would likely find them. As it turned out, Dedekind probably never scrutinized Weber's proof all too closely. Instead, he sent his own proof of Riemann-Roch, and on July 5 Weber acknowledged that it was superior to his own [Scheel 2015, 245–250].

This exchange, like several others, clearly reveals the underlying dynamic in their collaboration, which may have reminded Dedekind of his earlier, though far less fruitful mathematical relationship with Riemann. It was Weber who acted as the initiator, here by suggesting there was a deep analogy between algebraic number fields – a theory closely related to Galois's approach to algebraic equations – and Riemann's theory of algebraic functions based on the still informal notion of Riemann surfaces. The more skeptically minded Dedekind had to be persuaded, though he clearly took up Weber's ideas with gratitude, especially since this new theory appealed very strongly to his own intellectual sensibilities.

Direct mention of Riemann's theory and his sources came up in a letter Weber wrote on 28 July:

> The fundamental concerns you noted regarding Riemann surfaces, points, developability, etc. have been on my mind for a long time, and my suggestion of a joint publication involved the hope that we could still resolve these. I hoped and still hope that the treatise by Lagrange, "Nouvelle méthode pour résoudre les équations littérales pour le moyen des séries," ... might help. This is the treatise that I had in mind when we brought back a different one in Braunschweig. It is obviously the same one Riemann noted in [Riemann 1876/1892/1902, 1876: 105]. [Scheel 2015, 251–252]

Weber's view of Dedekind's role in their collaboration was clearly expressed in this same letter, which continued:

> However, I am sticking to the plan of our joint publication on the ideals, provided of course that the points you raised are dealt with satisfactorily. Your contribution to the matter is much greater than you represent in your modesty, and I would not have been able to do anything without your suggestions and constant help. [Scheel 2015, 253]

There was, in fact, much still to do before [Dedekind/Weber 1882/2012] was finally completed the following year, but Weber's persistence never ceased. When they were done, he wrote the foreword to their long paper, dated 22 October 1880, and sent this on to Dedekind. Eight days later, Dedekind wrote this moving reply:

> ... I take this opportunity to express once again my deepest gratitude to you for the entire, almost two-year-long work in which you have put in so much effort; to have taken part in this has brought me the greatest joy and a significant enrichment of knowledge. It is a particularly beautiful feeling to meet each other in this way while exploring the truth, which Pascal expressed so well in his first letter to Fermat: "For I would now like to open my heart to you, if I may, because I am so happy to see our agreement. I see clearly that the truth is the same in Toulouse as in Paris."[30] I have often thought about this passage in connection with the progress of our work, which, after many oscillations, is still increasing and has taken on the character of inner necessity. I should also be very happy if the matter receives some recognition, although I'm not counting on that too much for the time being because the boring modules will certainly discourage most people. [Scheel 2015, 271]

As John Stillwell noted in his introduction to [Dedekind/Weber 1882/2012], their paper gave the first rigorous proof of the classical Riemann-Roch theorem. In doing so they introduced the concept of "polygon" instead of the modern term "divisor," which became the standard term in the twentieth century [Dedekind/Weber 1882/2012, 2012: 27]. This shift in terminology is highly ironic in view of the fact that *Divisorentheorie* was created by Leopold Kronecker as an alternative to Dedekind's ideal theory, which employed infinite sets. Kronecker probably never set his eyes on proving the Riemann-Roch theorem, but his followers did and somehow the original notion of a polygon was replaced by the today ubiquitous concept of a divisor.

[30] Car je voudrais dśormais vous ouvrir mon coeur, s'il se pouvait, tant j'ai de joie de voir notre rencontre. Je vois bien que la vérité est la même á Toulouse et á Paris.

## 12.4 Dedekind's Biography of Riemann

As Dedekind and Weber were putting the last touches on the Riemann edition, they continued to discuss what was to be its crowning contribution, namely Dedekind's biographical essay, written with the help of Riemann's sister Ida and especially his widow Elise. The present section describes how it came to be written. In fact, much like the editorial project itself, Dedekind's biographical essay came about through happenstance. Moreover, in both cases, Heinrich Weber acted as the main moving force. Already in a letter from 24 November 1874, he raised this possibility:

> I would think it a a nice enrichment of the work if it were preceded by a short biography of Riemann. One could simply reprint Schering's memorial speech [Schering 1867], or find someone to write something new on an ad hoc basis. I have already thought about whether Frau Riemann herself might be inclined to take on the matter, at least so far as the biographical part was concerned; I would be happy to hear your opinion about this, too. [Scheel 2015, 53]

Dedekind only got around to answering Weber's rather lengthy letter on 14 March 1875, at which time he sent back an even longer reply. In response to the above, he wrote:

> A biography of Riemann would be a wonderful addition; but it will be difficult to achieve this. I would completely reject Schering's account. I would prefer a brief outline of [Riemann's] personal fortunes, since the scientific side is represented by the work itself. Frau Riemann would certainly be happy to supply the required material, but you would have to reserve the right to select what is appropriate and design it accordingly. Maybe I can be of help with that. [Scheel 2015, 61]

Weber probably had no way of knowing that Schering was *persona non grata* in the eyes of Dedekind, who also rightly recognized that Elise Riemann's role would need to be a restricted one. Dedekind's general reluctance to get involved seems quite apparent, but he also was grateful to Weber for his support and hoped not to dampen his enthusiasm for the project. In any case, only a week later Dedekind received the following request from Heinrich Weber:

> As far as the sketch of Riemann's life is concerned, you would do me a great favor if you agreed to take over preparing that. I would be happy to ask Frau Riemann to put together material for it, but I wouldn't dare do anything about this myself, given my complete lack of acquaintance not only with Riemann but also with the situation in Göttingen, in general. [Scheel 2015, 65]

Dedekind could not easily evade this request, but he also did not wish to make any firm commitment until such time as the source material had been gathered. He thus replied on 26 March:

> I cannot yet confirm with certainty that I will take over preparing a sketch of Riemann's life. Let us try to collect material everywhere we can immediately, but with the assurance that the real use of it will depend on the overall nature of what flows in from various sides. For example, I want to write Professor [August] Ritter in Aachen, who was a friend of Riemann's for years, even going back to their school days, I think. The main material should be expected from Frau Riemann; if you yourself would like to ask her to

compile this and then send it to me, I would prefer that; otherwise, I can take that over, too. But only afterward, when all the material is in front of me, can I decide whether I will be able to make something reasonable out of it. Couldn't you also add a portrait of Riemann? There are some quite good photographs of him. [Scheel 2015, 67]

Fig. 12.3: Elise Riemann née Koch with her daughter Ida, ca. 1870.

Weber responded accordingly and wrote to Elise Riemann on 10 April 1875:

I have corresponded with Prof. Dedekind about this, and we both think it would be a nice addition to Riemann's works if it were possible to include a small biographical sketch. You, dear lady, would probably be able to provide the main material for such an undertaking. I would therefore like to ask you if it were possible for you to put together what you know about the external circumstances so that this could be used to create a short description of Riemann's life, which I hope Prof. Dedekind will take over. I would therefore ask you, whether you can fulfill my request by sending the relevant material directly to Prof. Dedekind. [Scheel 2015, 323]

In all likelihood, Frau Riemann had no inkling beforehand of such a plan, but she wasted no time in pooling together whatever resources she could get her hands on. In her cover letter to Richard Dedekind, written only three weeks later on 1 May 1875, she described several of the sources he would soon receive. There she mentioned a black book containing notes Riemann made during his trip to Paris in 1860, one of several items that later disappeared without a trace. Attached to her cover letter was a lengthy note describing the major events in Riemann's life, beginning with the years up until his return to Göttingen in 1849 after two years of studies in Berlin. This part was likely written or dictated by Ida Riemann, who was particularly familiar with this period in her brother's life [Scheel 2015, 312–313].

Elise then listed practically year by year the noteworthy events and related activities that followed from 1849 up until his death [Scheel 2015, 313–316]. Curiously, she recalled sending such information to Clebsch in Giessen in 1867, as he wanted to send relevant facts to Arthur Cayley in Cambridge. Yet she neglected to say that one year earlier she had written up quite lengthy notes on Riemann's life for Ernst Schering, who had requested such information in planning his memorial address, delivered on December 1st 1866. Those notes only resurfaced many years after Schering's death, when the editors of his Collected Works published them in [Schering 1909, 441–447]. Elise Riemann's recollections of Riemann's last years served as a principal source for the narrative in Sections 4.5, 4.6, and 4.7.

What Elise certainly had not forgotten were the several letters written to Riemann, which she had loaned to Schering many years earlier. Despite several requests that he return these, however, he never complied. Feeling the time had come to awaken Schering to the urgency of this situation, Frau Riemann adopted a different strategy, as she related in her letter to Dedekind:

> What will interest you more, and what I am now sending you, dear Professor, are the letters in the larger package, which I finally received back from Schering yesterday after a long wait. Since nothing else had worked, I sent my little daughter (Fig. 12.3) to him with the message, "Please, Herr Professor, give my mother back the letters written to my dear father." He then brought them to me himself yesterday. They are letters from important men of Riemann's acquaintance and relate to his work and various aspects of his life. [Scheel 2015, 311]

Unfortunately, the contents of all those letters to Riemann remain unknown to this day; if they still exist, their whereabouts is another mystery waiting to be solved. Neuenschwander eventually recovered a list of those who sent one or more letters to Riemann; this includes many names that appear elsewhere in the present book.[31]

Among those several correspondents, the algebraist Leopold Kronecker felt a special admiration for Riemann. Shortly before the latter's death, Kronecker

---

[31] Among the correspondents listed were E. Betti, A. Clebsch, G. Lejeune Dirichlet, J.F. Encke, C.F. Gauss, Ch. L. Gerling, J. Henle, W. Klinkerfues, R. Kohlrausch, L. Kronecker, E.E. Kummer, J. Ch. Poggendorff, F. Prym, G. v. Quintus-Icilius, A. Ritter, G.F. Stisser, W. Sartorius v. Waltershausen, W. Weber, K. Weierstrass, J.G. Westphal, and F. Wöhler. [Neuenschwander 1988, 104].

wrote the recommendation that led to Riemann becoming an external member of the Berlin Academy:

> Mr. Riemann was elected a corresponding member of our Academy in August 1859, a short time after his epoch-making work on Abelian functions was published. ... Since then, Riemann's ideas and methods have found acceptance everywhere; his teachings – both spoken and written – have penetrated ever-widening circles of the scientific world, stimulating and fostering new research. Indeed, the need to disseminate knowledge of his work even further has given rise to a number of commentaries and treatises designed to surmount the difficulties posed by Riemann's style of exposition – a style that, while consistently exemplary in its precision, is, by virtue of its extreme conciseness alone, not readily accessible to general understanding. Thus, the immense significance of Mr. Riemann's achievements for the advancement of science is already universally recognized, and his profound influence on the trajectory of modern analysis secures him a place among the foremost mathematicians of all time. It is this conviction of ours – regarding the high and enduring value of Mr. Riemann's contributions – that has given rise to the present proposal. [Biermann 2022, 29–30]

As mentioned in Section 4.6, Kronecker's younger brother Hugo had met the Riemanns during their second stay in Italy. In the fall of 1875, Leopold Kronecker visited the country himself and took this opportunity to visit Riemann's grave in Selasca. He afterward wrote Elise Riemann about this, and on 2 October 1875, she wrote him back to say how deeply moved she felt on reading Kronecker's letter:

> It was with tears of heartfelt emotion and a feeling of heartfelt gratitude that I received your dear writing from Lago Maggiore. Many of Riemann's acquaintances traveled to Italy and the silent question always hovered on my lips, "Didn't you go to see his lonely grave?" You, dear Professor, have fulfilled my wish. I know the dear Güllnos faithfully look after the grave for me, which encloses the most precious thing I owned, but it is different when a friend's eye looks at such an abandoned grave. Riemann considered you to be his loyal friend, dear Professor; he always spoke of you with the greatest respect and fondly remembered your kindness and friendliness toward him. How often do I think of the small, quiet place of worship, the church of Biganzolo near Selasca! I deposited a small sum with Güllnos and I know that as long as the good Güllnos live, the grave will be faithfully looked after. His earthly shell is dust, but praise God his works survive him and I have the happy hope that they will appear together in the near future. Professor Dedekind has been working through Riemann's papers separately and compiling them for years with the most touching piety, conscientiousness and understanding. [Sinaceur 1990, 294]

Only a year later, Riemann's gravesite (the plaque is shown in Fig. 4.8) received prominent attention in Dedekind's biographical essay. In fact, Elise's cover letter to Dedekind ended with the very words he used in conclusion [Dedekind 1876a, 526]:

> He rests in the churchyard of Biganzolo, to which Selasca belongs. His gravestone bears the inscription:
>
> Here rests in God
> GEORG FRIEDRICH BERNHARD RIEMANN, Prof. in Gött.
> born in Breselenz September 17, 1826, died in Selasca June 20, 1866
> For those who love God all things must work together for the good.

Those who have read Dedekind's biographical essay, "Riemann's Lebenslauf," have occasionally commented on how some passages sound quite out of character.[32] Both he and Weber clearly realized that such an account of Riemann's life could not be written without his widow's help and support, but Dedekind probably did not anticipate receiving such an emotionally charged letter from her. Given the abundance of source material she sent him along with it, he could hardly in good conscience withdraw from the task at hand. In writing about Riemann's life, Dedekind reconciled himself with the thought that this was, to a considerable extent, Elise Riemann's version of that story and not so much his own.

This becomes especially evident from her remarks about Riemann as a deeply religious personality, a side of his character she no doubt knew better than anyone:

> R[iemann] combined with his deep knowledge, the purest, noblest heart. The pious sensibility implanted in his father's house remained with him throughout his entire life; he served his God faithfully, albeit in a different form. I still remember a saying of R[iemann] that I often repeated to myself later: The main thing in religion is daily self-examination before the presence of God. With the greatest piety, R[iemann] avoided disturbing others in their faith.
>
> R[iemann]'s end was a very gentle one, I would like to say a going home without struggle or shudders of death. It seemed to me as if he was following with interest the separation of the soul from the body. I gave him bread and wine, and he asked me to send greetings to the loved ones at home and to kiss our child. I prayed the Lord's Prayer with him; he could no longer speak. At the words,'Forgive us our sins,' he turned his eyes up in faith and I felt his hand grow colder in mine. He took a few more breaths and then went up to God, where he will now see what his researches here failed to find.
>
> Should you, dear Professor, decide to draw up the biographical sketch and would like information about any point, I would of course be happy to provide it if it is in my power. [Scheel 2015, 311]

Dedekind did his best to accommodate her wishes, producing a portrait of Riemann that hewed closely to Elise's idealization of him. He incorporated practically everything in the passages above in his text. After completing a first draft in early September 1875, he took the text with him to Göttingen, where a number of those close to Riemann were given the chance to read it.

In a letter to Heinrich Weber from 10 November, he reported on their reactions:

> Immediately after my arrival in Göttingen on September 15th, I brought the biographical sketch to Frau Riemann and her sister-in-law,[33] ... Riemann's only living sister, to read through. Both expressed their satisfaction with it and, in the event that it were included in the publication, asked whether a number of separate copies of this biography could be made. Wilhelm Weber is also satisfied with it and gave me two more notes about the high opinion Gauss had of Riemann; I have also included these. Finally, I gave the text to my long-time friend, [Jakob] Henle, ... who has very fine literary taste. He

[32] Laugwitz rightly noted that the text was "influenced by consideration for Riemann's wife and sister" [Laugwitz 1996/1999, 3].

[33] At some point, Ida Riemann married Elise's father, who had become infirm after suffering a stroke; see Elise Riemann to Dedekind, 14 November 1874 [Scheel 2015, 308].

> openly declared that this sketch, although not entirely without warmth, was probably too plain and simple and lacking in adornment. [Scheel 2015, 79]

Henle's reaction only confirmed for Dedekind that he had achieved the rather dry, factual account that suited his own preferences. He had no desire to alter that tone and promised to send Heinrich Weber a new version soon with the final revisions. He then added these telling remarks: "I assure you once again that I am completely satisfied if it is not printed. You would be mistaken if you thought you were doing me a favor by accepting it" (*ibid.*).

Thus, even by this late date, Dedekind found it difficult to identify with the text he had written. One of the major reasons he felt ambivalent about it stemmed from his own personal relationship with Riemann, which had never been easy (see Section 3.6). This ambivalence can be clearly seen from the draft of a letter Dedekind wrote once he learned that Heinrich Weber definitely wanted to include his sketch of Riemann's life. For the title, Dedekind wanted simply "Bernhard Riemanns Lebenslauf" (the Course of his Life) without any addition. He had conferred with Jakob Henle about this, and both thought it best for Dedekind to suppress his name as author, while referring to himself in the third person in the text itself. He had asked Henle directly, whether to use:

> ... third person with the author named in a very distant place, namely in the preface – or first person with the author named in the title itself – whereupon he [Henle] immediately declared himself in favor of the former, and that seems to me better, too. To leave me out of the story entirely would be downright unnatural; but if I were to introduce myself in the first person, the reader would again notice that I don't even tell how I met Riemann, and other such things; and I would like to avoid anything disturbing. [Scheel 2015, 112]

Dedekind therefore requested that Weber add a brief remark in the preface, stating that "the biographical sketch was written at my request by R. Dedekind, mainly based on communications from the Riemann family." This Weber was more than happy to do.

On June 30 1876, Elise Riemann wrote to Dedekind, reporting that she had in the meantime received thirty offprints of his biographical essay. Teubner also sent her a copy of the book in which it appeared, about which she sent these words of appreciation:

> Could I only adequately express to you, dear Professor, as I did to Mr. Weber, how grateful I am for all the loyalty, diligence, and endless effort with which you have so beautifully brought this work to completion. I greeted the first copy of Riemann's Collected Works that came into my hands with tears of profound gratitude and emotion; the entire presentation of the book, the printing and everything else, is very beautiful, and it has indeed become a truly magnificent volume. [Scheel 2015, 318]

Three months earlier, Weber had gained Dedekind's assent that his name would appear as a supporting editor on the title page (see [Scheel 2015, 119]). Frau Riemann was apparently unaware of this decision up until she held the volume in her hands, since she mentioned this explicitly in the same letter:

> The biography that you were kind enough to add to the work was very pleasing for us to read again and can certainly be considered very successful. One feels how it flows from

> the pen of one who grasped with love and full understanding the spirit and entire being of the dearly deceased. For me, the biography and the philosophical aspects of the work are, of course, the most interesting, since unfortunately the mathematical aspects are not accessible to me. To my great joy, I hear how the work is greeted with the greatest interest and recognition from all sides and how happy I was to see your name on the title page. It would have really saddened me if you, who made such great contributions to the publication, had due to excessive modesty refused to allow mention of your name. [Scheel 2015, 319]

The following year, on 30 April 1877, celebrations took place in Braunschweig and Göttingen to mark the hundredth anniversary of Gauss's birth. Heinrich Weber had hoped to see Dedekind on that occasion, but the latter was committed to represent his home institution on that day. The Royal Society held a public meeting, during which various dignitaries representing European academies paid honor to Göttingen's greatest mathematician. The main speaker was Ernst Schering, who the year before had completed work on the first six volumes of the Gauss edition. As such, he had good reason to treat this ceremonial lecture as a special event in his still young scholarly life. As might be expected, Schering spoke with great pathos about his hero.[34] Richard Dedekind, who happened not to be present, was one of the few who knew that Schering owed his unique position as an authority on the works of Gauss to his vigilance in marginalizing all others who might have shared in this acclaim.

## 12.5 Epilogue: the Later Editions

The publication of the first edition of *Riemanns Werke* marked only the beginning of a scholarly pursuit that afterward led to three more editions. This closing section briefly describes the circumstances behind each of these. By the mid-1880s, Teubner reported that all remaining copies of the volume had been sold. Heinrich Weber, who was now in Marburg, still had Riemann's papers in his possession, and in June 1891 he contacted Elise Riemann to inform her that he planned to prepare a second edition. She answered him from Bremen, where she had been living with Riemann's sister Ida since 1890 [Scheel 2015, 328]. At that time, they joined the household of Elise's daughter, Ida , who had married in 1884. Her husband was Schwarz's former student, Carl David Schilling, who took his doctorate under Schwarz in 1880. At that time, Schwarz lived next door to Frau Riemann in Göttingen. Schilling eventually gained a teaching position at the Naval School in Bremen, where he later became the director.

For the second edition, Weber added significant new material to the final item from Roch's notebook (now item XXXI) [Riemann 1876/1892/1902, 1892: 491–504]. This clarified that Riemann's lectures included a thorough treatment of theta characteristics in the case $p = 3$ that were directly related to configurations

[34] A shorter version of Schering's speech appeared in the Society's *Nachrichten* (1877) (see [Schering 1909, 169–175]), whereas the entire speech was published in Volume 22 of the *Abhandlungen* ([Schering 1909, 176–214]).

of the 28 bitangents for a quartic curve. Weber also added a short manuscript (Nr. XXV) on the conduction of heat in an ellipsoid. Far more interesting than this note, however, were some of the others Weber either added or amended for this new edition. For example, earlier he had overlooked an important paper by Enrico Betti, which was directly related to the fragment on analysis situs (Nr. XXIX in the new edition). Henri Poincaré later referenced this paper [Betti 1871] when he introduced the term Betti numbers in algebraic topology. It took several decades, however, to establish a solid proof that these numbers were, indeed, topological invariants.

Another important change occurred with the notes Weber appended to Riemann's paper on the zeta function (Sections 4.2 and 8.2). Here, for the first time, Weber quoted from the letter Riemann sent to Weierstrass soon after visiting Berlin in the fall of 1859. This included the mysterious reference to a complicated expression, "which he had not yet simplified enough to be able to communicate" [Riemann 1876/1892/1902, 1892: 154]. More than three decades later, Carl Ludwig Siegel took up the challenge of recovering what Riemann meant when he wrote those words [Siegel 1932].

As noted earlier, Weber added various remarks that took account of the criticisms Schwarz had leveled against Hattendorff's text on minimal surfaces. He also wrote an entirely new commentary for Riemann's "Commentatio." Although he had struggled valiantly to explain its second part, his account in the first edition did not persuade everyone. Rather than trying to amend it, Weber decided to present a new and more comprehensive commentary, which he hoped would gain general approval. Dedekind also wrote a new commentary for his explanations relating to Riemann's two fragments (now Nr. XXVIII). At the outset, he noted that since the appearance of the first edition, the new independent theory of modular functions had been the subject of numerous investigations [Riemann 1876/1892/1902, 1892: 466]. This acknowledgment of Riemann as the pioneer for such studies would later become a standard theme for Felix Klein. This second edition served as the basis for those that would follow; thus, ever since 1892 it has served as the definitive version of *Riemanns Werke.*

After completing it, Heinrich Weber felt the time had come for him to deposit Riemann's scientific papers with the Göttingen Royal Society. As chance would have it, he had already accepted Schwarz's position in Göttingen, where he started teaching in the fall of 1892. The matter of transferring Riemann's papers turned out to be somewhat difficult, however, as Elise Riemann wanted to ensure that information pertaining to Riemann's private life would not be accessible to the general public. By the summer of 1892, though, she was apparently willing to accede to Weber's request, as on 22 July, Carl David Schilling wrote the following to Heinrich Weber:

> At first mother could not come to terms with the idea that Riemann's papers should no longer remain in private hands; to her, they are something sacred, and she doesn't like to think of them being made accessible to any student, who would then also be able to read the marginal notes, some of which are purely personal. [Neuenschwander 1988, 103]

This sounds as though someone must have persuaded Elise Riemann to grant Weber's request. At any rate, not long afterward he passed on Riemann's mathematical papers to the Royal Society. Since 1895, the Riemann papers have been housed at the Göttingen University Library.

There remained the question as to whether transcriptions of Riemann's lectures made by his former students might have survived. Of particular importance were those courses where he had elaborated on theories only sketched in his published papers. A few years before Heinrich Weber arrived in Göttingen, Felix Klein placed a written proposal before the Royal Society with the aim of conducting a systematic search for such *Nachschriften* [Gierl 2004, 180]. Klein was at the time the junior member in the Mathematics Section, whereas Ernst Schering was its senior member. Predictably, Schering opposed Klein's proposal, claiming that it would produce nothing of value. After a decisive meeting on July 5, 1890, Klein withdrew his proposal, though he apparently played a role in having Riemann's mathematical papers deposited at the library in 1895. After Schwarz's departure in 1892, followed by Schering's death in 1897, Klein held all the cards for pursuing his original goal. At the same time, he launched a similar plan for supplementing the original edition of Gauss's *Werke*, which had ended in the 1870s under Schering. This latter project got off to a dramatic start with the unanticipated recovery of Gauss's scientific diary, which resurfaced at the very time Schering passed away. Ernst Schering must have turned over in his grave when Klein first publicized this sensational find in [Klein 1903].

In 1900, Volume 8, the first of these supplemental volumes appeared, updating the first three volumes of the Gauss edition. Klein enlisted his nephew, Robert Fricke, to edit the portions on elliptic functions that Riemann had originally chosen to edit himself. As recounted in Section 12.1, Riemann had delegated this task to Karl Hattendorff, which opened the way for Schering to confiscate the relevant documents. Schering's editorial treatment, as described in [Gauss 1863–1933, 3: 492–494], went well beyond merely presenting the source material, and the same was true of Fricke's comments in [Gauss 1863–1933, 8: 102–105], which was framed as explaining three fragments on modular functions. Fricke's expertise in this field was well documented in [Klein/Fricke 1890/92], but a critical examination of the source material itself might well suggest a different interpretation. Part of Fricke's revision was discussed in [Bottazzini & Gray 2013, 64–65], where they compared two diagrams to make clear that Schering had not recognized Gauss's drawing of the modular figure. Comparing this with the figure Wilhelm Wirtinger later reconstructed from Riemann's 1858/59 lectures (see [Riemann 1876/1892/1902, 1902: 93], one might well wonder whether part of Riemann's inspiration came from encountering this figure in Gauss's posthumous papers.

Around the time of Schering's death in 1897, Klein began collecting new material for the Riemann estate. More decisive still, he won over two leading experts, Max Noether and Wilhelm Wirtinger, who agreed to prepare commentaries on selected lecture notes. Five years later, in 1902, Teubner published the *Nachträge* to the second edition, which consisted of 116 pages of additional material based on various lecture courses Riemann gave. Noether initiated this project, whereas Wirtinger devoted himself especially to two lecture courses on linear differential equations and hypergeometric series.

In [Wirtinger 1905/1990], a plenary lecture held at the 1904 ICM in Heidelberg, Wirtinger especially emphasized the significance of Riemann's 1858/59 lectures. These were transcribed in Gabelsberger shorthand by the physicist Wilhelm von Bezold, who was director of the Prussian Institute of Meteorology. Lazarus Fuchs learned about this and had a copy made for his own use in 1894. Especially remarkable are the last pages [Riemann 1876/1892/1902, 1902: 91–93], which strongly suggest that Dedekind's original interpretation of the material in XXIX was, in fact, tied to Riemann's understanding of the properties of modular functions; if so, the formulas he derived may have had nothing to do with his postdoctoral thesis.

Subsequent reprints of the *Werke* included these *Nachträge*; this was also the case after World War II, but under unusual circumstances. The Teubner firm remained in Leipzig as BSB B.G. Teubner Verlag, a partly state-owned business in the German Democratic Republic. Its former owners and managers mainly left, however, forming an independent company in Stuttgart, West Germany. When Dover Publications reprinted the third edition they did so with the permission of B.G. Teubner Stuttgart. Very likely, Richard Courant had a hand in this, as this Dover edition contains an interesting preface written in English. The author (unidentified in the version I own) was Courant's former student, Hans Lewy, who in 1935 joined the faculty at the University of California in Berkeley. Lewy's brief synopsis is noteworthy as he emphasized topics in analysis that were central interests in Göttingen during Courant's years as director. Of course, he did not fail to highlight the Riemann hypothesis, but mainly he stressed Riemann's contributions to differential equations (including shock waves), Dirichlet's principle, and mathematical physics.

The Leipzig branch of Teubner, but especially its editor Jürgen Weiss, had a strong interest in historical work focused on German mathematical literature, most of which had been published by the Teubner firm up until World War I. During the waning years of the Cold War, when relations between East and West Germany were normalizing, Weiss saw a new chance to undertake a project in cooperation with Springer in Heidelberg. The plan thus arose to publish a new fourth edition of *Riemanns Werke* as a novel East-West venture. This led to [Riemann 1990], which added more than two hundred pages to the third edition, beginning with its Preface written by the editor, Raghavan Narasimhan, from the University of Chicago. As an expert on complex analysis with strong interests in analytic number theory, Narasimhan brought those aspects of Riemann's intellectual legacy up to date in his impressive introduction to this volume.

This edition features reprints of three important texts: Wilhelm Wirtinger's lecture at the 1904 Heidelberg Congress on Riemann's lectures of hypergeometric series [Riemann 1990, 720–739]; Hermann Weyl's commentary on Riemann's 1854 lecture on foundations of geometry [Weyl 1923]; and Carl Ludwig Siegel's study of Riemann's work on analytical number theory [Siegel 1932]. In addition, two new essays discuss Riemann's two important papers on physical topics: in [Riemann 1990, 807–810], Peter Lax described the results in [Riemann 1860] on non-linear differential equations modeling shock waves in the atmosphere; the companion essay [Riemann 1990, 811–822] by S. Chandrasekhar and N. Lebovitz describes how Riemann examined the stability of ellipsoid figures in [Riemann 1861a/1876]. Beyond these five essays, the volume also contains biographical source material, a reprint of [Neuenschwander 1988], and detailed bibliographic information relating to both primary and secondary literature, compiled respectively by Walter Purkert and Erwin Neuenschwander. In short, this 1990 edition reflects that era's intense interest in Riemann's work and its ongoing influence.

In closing, one easily recognizes how Riemann's changing legacy has been part of a dynamical process, one that began when Richard Dedekind and Heinrich Weber began studying his unpublished papers. Although they barely knew one another when they joined hands to produce the first edition of *Riemanns Werke*, their mutual dedication to that project laid the groundwork for their lifelong friendship. Dedekind had the advantage of having known Riemann personally, but as he intimated to Weber from the beginning, he never felt really at home in Riemann's mathematical world. Weber had the advantage of youth and, as a product of the Königsberg tradition, a more intuitive understanding of the interplay between analysis and mathematical physics. Temperamentally, they harmonized completely as two generous personalities, who shared a deep appreciation for Riemann's intellectual accomplishments.

When Dedekind wrote to accept Weber's offer that both of their names appear on the title page, he expressed his appreciation in these words:

> You really wouldn't have needed to feel any pangs of conscience about appearing as the sole editor, for you not only did the main work on it but also guided the whole thing through with your complete mastery of Riemann's creative work, and so the world will say: it turned out well. That would have been quite impossible for me; I've told myself that countless times this winter, and only now, as your work actually progressed, have I truly understood everything that goes into it, and how little my own knowledge would have sufficed. I followed your work with the greatest interest, from which I learned a great deal, and the joy of having come into such close contact with you would alone have been ample reward for my contribution. Now, however, I have considered your renewed proposal, and I find it so tempting and honorable to appear in your company that I cannot resist ... [Scheel 2015, 109–110]

Richard Dedekind long outlived everyone from the Göttingen circle of his youth. When he left this world in 1916, his own intellectual legacy was still on the rise. Among those who promoted his name was the great Emmy Noether, not least by virtue of her role in editing his Collected Works [Dedekind 1930–1932]. One must step further back in time, though, to recognize some of the other sides of Dedekind's creative life. Only then can one appreciate how he and the young Heinrich Weber collaborated so successfully to produce the single volume that has ever since anchored Riemann's intellectual legacy.

# Photo and Figure Credits

Fig. 0.1: Wikipedia, Public Domain

Fig. 0.2: Used with permission of the Niedersächsische Staats- und Universitätsbibliothek Göttingen (SUB Göttingen)

Fig. 0.3: 2 Sammlung Voit: P. Lejeune-Dirichlet, used with permission of the SUB Göttingen

Fig. 0.4: Sammlung Voit: W. Weber, Nr. 7, used with permission of the SUB Göttingen

Fig. 0.5: Wikipedia, Public Domain

Fig. 0.6: Wikipedia, Public Domain

Fig. 1.1: Christian Fischer, 2015 – Creative Commons Attribution-Share Alike 3.0 Unported license

Fig. 1.2: Used with permission of the Göttingen University Archive

Fig. 1.3: Sammlung Voit: M. A. Stern, Nr. 1, used with permission of the SUB Göttingen

Fig. 1.4: Used with permission of the Berlin University Archive

Fig. 1.5: Wikipedia, Public Domain

Fig. 2.1: From [Sarkowski 1992, 16]

Fig. 2.2: Herzog August Bibliothek Wolfenbüttel: Graph. C: 223, Public Domain

Fig. 2.3: Wikipedia, Public Domain

Fig. 2.4: ETH Library Zurich, Image Archive, Public Domain

Fig. 3.1: Wikipedia, Public Domain

Fig. 3.2: Cod. Ms. Riemann 32c, p. 59, used with permission of the SUB Göttingen

Fig. 4.1: Cod. Ms. Riemann 32c, used with permission of the SUB Göttingen

Fig. 4.2: Wikipedia, Public Domain

Fig. 4.3: Sammlung Voit: W. Sartorius v. Waltershausen, Nr. 7, used with permission of the SUB Göttingen

Fig. 4.4: Ben Aveling, 2007 – Wikipedia, Public Domain

Fig. 4.5: Wikipedia, Public Domain

Fig. 4.6: Wikipedia, Public Domain

Fig. 4.7: Marta De Bortoli 1991, Creative Commons Attribution-Share Alike 4.0 International license



D. E. Rowe, *Bernhard Riemann: His Life and Wondrous Mathematical Legacy*,
https://doi.org/10.1007/978-3-032-25457-3

Fig. 4.8: Wikipedia, Public Domain

Fig. 5.1: Cod. Ms. B. Riemann 13, 100r, 100v, used with permission of the SUB Göttingen

Fig. 7.1: Wikipedia, Public Domain

Fig. 7.2: Cod. Ms. B. Riemann 32 : 1, p. 180, used with permission of the SUB Göttingen

Fig. 7.3: Cod. Ms. B. Riemann 32 : 1, p. 180, used with permission of the SUB Göttingen

Fig. 9.1: Gauss & Weber, *Resultate aus den Beobachtungen des magnetischen Vereins*, 1836, p. 210

Fig. 10.1: Gauss Bibl 1434: 1836, Taf. 1, used with permission of the SUB Göttingen

Fig. 11.1: MacTutor History of Mathematics Archive, Public Domain

Fig. 11.2: Sammlung Voit: A. Clebsch, Nr. 1, used with permission of the SUB Göttingen

Fig. 11.3: From [Klein 1921–23, 2: 92]

Fig. 11.4: From [Klein 1921–23, 2: 91]

Fig. 11.5: From [Klein 1921–23, 2: 94]

Fig. 11.6: From [Klein 1882/1923, 4]

Fig. 11.7: From [Klein 1882/1923, 38]

Fig. 11.8: From [Klein 1882/1923, 38]

Fig. 12.1: Cod. Ms. B. Riemann 16 Cim., used with permission of the SUB Göttingen

Fig. 12.2: Wikipedia, Public Domain

Fig. 12.3: Sammlung Voit: B. Riemann, Nr. 2, used with permission of the SUB Göttingen

# References


Abbe 1986. Abbe, Ernst: *Ernst Abbe: Briefe an seine Jugend- und Studienfreunde Carl Martin und Harald Schütz, 1858–1865*, Volker Wahl u. Joachim Wittig, Hrsg., Berlin: Akademie-Verlag.

Abel 1881. Abel, Niels Henrik: *Oeuvres Complètes de Niels Henrik Abel*, Ludvig Sylow et Sophus Lie, eds. Christiania: Grondahl.

Ahrens 1907. Ahrens, Wilhelm: *C.G.J. Jacobi als Politiker; ein Beitrag zu seiner Biographie*, Leipzig: Teubner.

Archibald 1989a. Archibald, Thomas: Physics as a Constraint on Mathematical Research: the Case of Potential Theory and Electrodynamics, 1840–1880, in [Rowe/McCleary 1989, 2: 29–76]Rowe-McCleary.

Archibald 1989b. Archibald, Thomas: Energy and the Mathematization of Electrodynamics in Germany, 1845–1875, *Archives internationales d'histoire des sciences* 39(123): 276–307.

Archibald 1991. Archibald, Thomas: Riemann and the Theory of Electrical Phenomena: Nobili's Rings, *Centaurus* 34: 247–271.

Archibald 1996. Archibald, Thomas: From Attraction Theory to Existence Theory: the Evolution of Potential-Theoretic Methods in the Study of Boundary-Value Problems, 1860–1890, *Revue d'histoire des mathématiques* 2: 101–128.

Ayoub 1974. Ayoub, Raymond: Euler and the Zeta Function, *American Mathematical Monthly* 81(10): 1067–1086.

Babbage 1826. Babbage, Charles: *A Comparative View of the Various Institutions for the Assurance of Lives*, London: Mawman.

Baeyer 1861. Baeyer, Johann Jacob: *Ueber die Grösse und Figur der Erde: eine Denkschrift zur Begründung einer mittel-europäischen Gradmessung; nebst einer Uebersichtskarte* Berlin : G. Reimer.

Begehr et al. 1998. Begehr, Heinrich, et al., eds.: *Mathematics in Berlin*, Berlin: Birkhäuser.

Beltrami 1868a. Beltrami, Eugenio: Saggio di interpretazione della geometria non-euclidea, *Giornale di Mathematiche* 6: 284–315; English translation in [Stillwell 1996, 7–39].

Beltrami 1868b. Beltrami, Eugenio: Teoria fondamentale degli spazii di curvatura costante, *Annali di Matematica Pura ed Applicata* 2: 232–255; English translation in [Stillwell 1996, 41–62].

Bessel 1839. Bessel, Friedrich Wilhelm: Ueber den Ausdruck einer Function $\phi x$ durch Cosinusse uud Sinusse der Vielfachen von $x$, *Astronomische Nachrichten* 16(374): 229–234.

Betti 1862. Betti, Enrico: Sopra le funzioni algebriche di una variabile complessa, *Annali delle Università Toscane* VII: 101–130.

Betti 1871. Betti, Enrico: Sopra Eli spazi di un numero qualunque di dimensioni, *Annali di Matematica Pura ed Applicata*, Ser. II, 4: 140–158.

Biermann 1959. Biermann, Kurt-R.: Johann Peter Gustav Lejeune Dirichlet. Dokumente für sein Leben und Wirken, *Abhandlungen der Deutschen Akademie der Wissenschaften zu Berlin, Klasse für Mathematik, Physik und Technik*, Jahrgang 1959, Nr. 2, pp. 2–88.





D. E. Rowe, *Bernhard Riemann: His Life and Wondrous Mathematical Legacy*, https://doi.org/10.1007/978-3-032-25457-3

Biermann 1980. Biermann, Kurt-R.: Eisenstein, Ferdinand Gotthold Max, *Dictionary of Scientific Biography* 4: 340–343.

Biermann 1988. Biermann, Kurt-R.: *Die Mathematik und ihre Dozenten an der Berliner Universität, 1810-1933*, Berlin: Akademie Verlag.

Biermann 2022. Biermann, Kurt-R.: *Vorschläge zur Wahl von Mathematikern in die Berliner Akademie, Ein Beitrag zur Gelehrten- und Mathematikgeschichte des 19. Jahrhunderts*, Berlin: DeGruyter.

Bonola 1906/1912. Bonola Roberto: *La geometria non-euclidea. Esposizione storico-critica del suo sviluppo*, Bologna: Zanichelli; *Non-Euclidean Geometry: a Critical and Historical Study of its Development*, trans. H.S. Carslaw, Chicago: Open Court.

Bottazzini 1977. Bottazzini, Umberto: Riemanns Einfluss auf E. Betti und F. Casorati, *Archive for History of Exact Sciences* 18: 27–37.

Bottazzini 2003. Bottazzini, Umberto: "Algebraic Truths" vs. "Geometric Fantasies": Weierstrass' Response to Riemann, https://arxiv.org/pdf/math/0305022.

Bottazzini & Gray 2013. Bottazzini, Umberto and Jeremy Gray: *Hidden Harmony – Geometric Fantasies. The Rise of Complex Function Theory.* New York: Springer.

Bottazzini & Tazzioli 1995. Bottazzini, Umberto and Rossana Tazzioli: Naturphilosophie and its Role in Riemann's Mathematics, *Revue d'histoire des mathématiques* 1: 3–38.

Brewster 1831. Brewster, David: *The Life of Sir Isaac Newton*, London: Murray.

Brieskorn/Knörrer 1986. Brieskorn, Egbert and Knörrer, Horst: *Plane Algebraic Curves*, trans. J. Stillwell, Basel: Birkhäuser.

Brill/Noether 1894. Brill, Alexander and Noether, Max: Die Entwickelung der Theorie der algebraischen Functionen in älterer und neuerer Zeit, *Jahresbericht der Deutschen Mathematiker-Vereinigung* 3: 107–566.

Briot/Bouquet 1859. Briot, Charles and Claude Bouquet: *Théorie des fonctions doublement périodiques et, en particulier, des fonctions elliptiques*, Paris: Mallet-Bachelier.

Bühler 1987. Bühler, Walter K.: *Gauss, eine biographische Studie*, New York: Springer, trans. from English edition of 1981.

Bullynck 2010. Bullynck Maarten: A History of Factor Tables with Notes on the Birth of Number Theory 1657–1817, *Revue d'histoire des mathématiques* 16(2): 133–216.

Caneva 2021. Caneva, Kenneth L.: *Helmholtz and the Conservation of Energy: Contexts of Creation and Reception*, Cambridge Mass.: MIT Press.

Casorati 1868. Casorati, Felice: *Teorica delle funzioni di variabili complesse*, vol. 1, Pavia: Fratelli Fusi.

Cayley 1865. Cayley, Arthur: On the Transformation of Plane Curves. *Proceedings of the London Mathematical Society* 1: 1–8.

Chapman 1962. Chapman, Sydney: Alexander von Humboldt and Geomagnetic Science, *Archive for History of Exact Sciences* 2: 41–51.

Christoffel 1869. Christoffel, E.B.: Ueber die Transformation der homogenen Differentialausdrücke zweiten Grades, *Journal für die reine und angewandte Mathematik* 70: 46–70.

Christoffel 1870a. Christoffel, E.B.: Ueber die Abbildung einer einblättrigen, einfach zusamenhängenden, ebenen Fläche auf einem Kreise, *Nachrichten der Göttingen Gesellschaft der Wissenschaften* 1870: 283–298.

Christoffel 1870b. Christoffel, E.B.: Ueber die Abbildung einer n-blättrigen, einfach zusamenhängenden, ebenen Fläche auf einem Kreise, *Nachrichten der Göttingen Gesellschaft der Wissenschaften* 1870: 359–369.

Clark 2006. Clark, William: *Academic Charisma and the Origins of the Research University*, Chicago. University of Chicago Press.

Clebsch 1864. Clebsch, Alfred: Über die Anwendung der Abelschen Functionen in der Geometrie, *Journal für die reine und angewandte Mathematik* 63: 189–243.

Clebsch/Gordan 1866. Clebsch, Alfred und Gordan, Paul: *Theorie der Abelschen Funktionen*, Leipzig: Teubner.

Clifford 1876. Clifford, William Kingdon: On the Space-Theory of Matter, *Proceedings of the Cambridge Philosophical Society* 2: 157–158.

Coen 2012. Coen, Salvatore: *Mathematicians in Bologna, 1861–1960*, Basel: Birkhäuser.

Cogliati 2014. Cogliati, Alberto: Riemann's Commentatio Mathematica, a Reassessment, *Revue d'histoire des mathématiques* 20: 73–94.
Confalonieri 2019. Confalonieri, Sara, Schmidt, Peter-Maximilian Schmidt, Volkert, Klaus, Hrsg.: *Der Briefwechsel von Wilhelm Fiedler mit Alfred Clebsch, Felix Klein und italienischen Mathematikern.* Siegen: Universitätsbibliothek der Universität Siegen.
Courant 1950. Courant, Richard: *Dirichlet's Principle, Conformal Mappings and Minimal Surfaces*, New York: Interscience.
Craig 1955. Craig, Gordon A.: *The Politics of the Prussian Army 1640–1945*, Oxford: Clarendon Press.
Courant/Hilbert 1924. Courant, Richard and Hilbert, David: *Methoden der mathematischen Physik*, Bd. 1, Berlin: Springer.
Darrigol 2000. Darrigol, Olivier: *Electrodynamics from Ampère to Einstein*, Oxford: Oxford University Press.
Darrigol 2015. Darrigol, Olivier: The Mystery of Riemann's Curvature, *Historia Mathematica* 42: 47–83.
Dauben 1979. Dauben, Joseph W.: *Georg Cantor: His Mathematics and Philosophy of the Infinite*, Boston: Harvard University Press.
De Benedetti 1864. De Benedetti, Salvatore: *Il terzo centenario di Galileo. Narrazione istorica*, Pisa: Tipografia Nistri.
Dedekind 1861. Dedekind, Richard: Zusatz zu der vorstehenden Abhandlung [[Dirichlet 1860/1861]], *Journal für die reine und angewandte Mathematik* 58: 217–228.
Dedekind 1872. Dedekind, Richard: *Stetigkeit und irrationale Zahlen*, Braunschweig: Vieweg.
Dedekind 1876a. Dedekind, Richard: Bernhard Riemanns Lebenslauf, in [Riemann 1876/1892/1902, 1876: 507–526].
Dedekind 1876b. Dedekind, Richard: Anzeige der ersten Auflage von Riemanns gesammelten Werken, *Göttingische Gelehrte Anzeigen*, Jahrgang 1876: 961–965; reprinted in [Riemann 1990, 716–718].
Dedekind 1877. Dedekind, Richard: Schreiben an Herrn Borchardt über die Theorie der elliptischen Modul-Functionen, *Journal für die reine und angewandte Mathematik* 83: 265–292.
Dedekind 1901. Dedekind, Richard: Gauss in seiner Vorlesung über die Methode der kleinsten Quadrate, *Festschrift zur Feier des hundertfünfzigenjährigen Bestehens der Königlichen Gesellschaft der Wissenschaften zu Göttingen*, pp. 45–59; reprinted in [Dedekind 1930–1932, 2: 293–306].
Dedekind 1930–1932. Dedekind, Richard: *Richard Dedekind, Gesammelte mathematische Werke*, Emmy Noether, Robert Fricke, Oystein Ore, eds., 3 Bde., Braunschweig, Vieweg.
Dedekind/Weber 1882/2012. Dedekind, Richard Und Weber, Heinrich: Theorie der algebraischen Funktionen einer Veränderlichen. *Journal für die reine und angewandte Mathematik* 92: 181–290; English translation and commentary by John Stillwell, Providence, RI: American Mathematical Society.
de Paz/Ferreirós 2020. de Paz, Maria, and José Ferreirós: From Gauss to Riemann Through Jacobi: Interactions Between the Epistemologies of Geometry and Mechanics? *Journal for General Philosophy of Science* 51: 147–172.
Dirichlet 1828. Dirichlet, P. G. Lejeune: Recherches sur les diviseurs premier d'une classe de formules du quatrième degré, *Journal für die reine und angewandte Mathematik* 3: 35–69.
Dirichlet 1829. Dirichlet, P. G. Lejeune: Sur la convergence des séries trigonométriques qui servent à représenter une fonction arbitraire entre des limites données, *Journal für die reine und angewandte Mathematik* 4: 157–169.
Dirichlet 1839. Dirichlet, P. G. Lejeune: Beweis des Satzes, dass jede unbegrenzte arithmetische Progression, deren erstes Glied und Differenz ganze Zahlen ohne gemeinschaftlichen Factor sind, unendlich viele Primzahlen enthält, *Abhandlungen der Preussischen Akademie der Wissenschaften zu Berlin aus dem Jahre 1837*, 45–71.
Dirichlet 1839–40. Dirichlet, P. G. Lejeune: Recherches sur diverses applications de l'analyse infinitésimale a la théorie des nombres, *Journal für die reine und angewandte Mathematik* 19: 324–369, 21: 1–12, 134–155; reprinted in [Dirichlet 1889/1897, 1: 413–496].

Dirichlet 1857. Dirichlet, P. G. Lejeune: Untersuchungen über ein Problem der Hydrodynamik, *Nachrichten der Göttingen Gesellschaft der Wissenschaften* 1857: 205–207.

Dirichlet 1860/1861. Dirichlet, P. G. Lejeune: Untersuchungen über ein Problem der Hydrodynamik, bearb. R. Dedekind, *Abhandlungen der Königlichen Gesellschaft der Wissenschaften zu Göttingen* Band 8: 3–42; reprinted in *Journal für die reine und angewandte Mathematik* 58: 181–216.

Dirichlet 1863. Dirichlet, P.G. Lejeune: *Vorlesungen über Zahlentheorie*, R. Dedekind, Hrsg., Braunschweig: Vieweg.

Dirichlet 1871. Dirichlet, P.G. Lejeune: *Vorlesungen über Zahlentheorie* 2. Aufl., R. Dedekind, Hrsg., Braunschweig: Vieweg.

Dirichlet 1879. Dirichlet, P.G. Lejeune: *Vorlesungen über Zahlentheorie* 3. Aufl., R. Dedekind, Hrsg., Braunschweig: Vieweg.

Dirichlet 1889/1897. Dirichlet, P. G. Lejeune: *G. Lejeune Dirichlet's Werke*, 2 Bde., Berlin: Georg Reimer.

Dugac 1973. Dugac, Pierre: Eléments d'analyse de Karl Weierstrass, *Archive for History of Exact Sciences* 10: 41–176.

Dugac 1976. Dugac, Pierre: *Richard Dedekind et les fondements des mathématiques*, Paris: Vrin.

Durège 1864. Durège, Heinrich: *Elemente der Theorie der Functionen einer complexen Veränderlichen. Mit besonderer Berücksichtigung der Schöpfungen Riemanns*, Leipzig: Teubner.

Du Sautoy 2003. Du Sautoy, Marcus: *The Music of the Primes*, New York HarperCollins.

Edwards 1974. Edwards, Harold M.: *Riemann's Zeta Function*, New York: Academic Press.

Edwards 2020. Edwards, Harold M.: The Role of History in the Study of Mathematics, *The Mathematical Intelligencer* 42(1): 66–69.

Einstein 1916. Einstein, Albert: Grundlagen der allgemeinen Relativitätstheorie, *Annalen der Physik*, 49: 769–822 ; reprinted in [Einstein 1996, 283–339].

Einstein 1917. Einstein, Albert: Kosmologische Betrachtungen zur allgemeinen Relativitätstheorie, *Sitzungsberichte der Königlichen Preußischen Akademie der Wissenschaften*, 142–152; reprinted in [Einstein 1996, 540–552].

Einstein 1989. Einstein, Albert: *Collected Papers of Albert Einstein, Vol. 2: The Swiss Years: Writings, 1900–1909*, John Stachel, ed., Princeton: Princeton University Press.

Einstein 1995. Einstein, Albert: *Collected Papers of Albert Einstein, vol. 4: The Swiss Years: Writings, 1912–1914*, Martin J. Klein, et al., eds., Princeton: Princeton University Press.

Einstein 1996. Einstein, Albert: *Collected Papers of Albert Einstein, Vol. 6: The Berlin Years: Writings, 1914–1917*, A.J. Kox, et al., eds., Princeton: Princeton University Press.

Einstein/Grossmann 1913. Einstein, Albert and Grossmann, Marcel: *Entwurf einer verallgemeinerten Relativitätstheorie und einer Theorie der Gravitation*, Leipzig: Teubner; reprinted in [Einstein 1995, 302–343].

Eisenstein 1847. Eisenstein, Gotthold: *Mathematische Abhandlungen besonders aus dem Gebiete der höhern Arithmetik und der elliptischen Functionen*, Berlin: Reimer.

Elstrodt 2007. Elstrodt, Jürgen: The Life and Work of Gustav Lejeune Dirichlet (1805–1859), in William Duke, Yuri Tschinkel, eds.: *Analytic Number Theory: a Tribute to Gauss and Dirichlet*, pp. 1–37.

Elstrodt/Ullrich 1999. Elstrodt, Jürgen and Peter Ullrich: A Real Sheet of Complex Riemannian Function Theory: A Recently Discovered Sketch in Riemann's Own Hand, *Historia Mathematica* 26: 268–288.

Epple 1999. Epple, Moritz: *Die Entstehung der Knotentheorie*, Braunschweig: Vieweg.

Euler 1748. Euler, Leonhard: *Introductio in analysin infinitorum*, T. 1, Lausanne.

Euler 1750. Euler, Leonhard: De seriebus quibusdam considerationes, *Commentarii academiae scientiarum Petropolitanae* 12: 53-96; English trans. Alexander Aycock for the Euler-Kreis Mainz, E130: uni-mainz.de/mathematik/Algebraische Geometrie/Euler-Kreis Mainz.

Farwell/Knee 1990. Farwell, Ruth and Knee, Christopher, The Missing Link: Riemann's "Commentatio," Differential Geometry and Tensor Analysis, *Historia Mathematica* 17: 223–255.

Fechner 1845. Fechner, Gustav T.: Ueber die Verknüpfung der Faraday'schen Inductions-Erscheinungen mit den Ampère'schen elektro-dynamischen Erscheinungen, *Annalen der Physik und Chemie* 64: 337–345.

Ferreirós 2006. Ferreirós, José: Riemann's Habilitationsvortrag at the Crossroads of Mathematics, Physics, and Philosophy, in [Ferreirós/Gray 2006, 47–96].

Ferreirós 2007. Ferreirós, José: *Labyrinth of Thought. A History of Set Theory and Its Role in Modern Mathematics*, 2nd ed., Basel/Boston/Berlin: Birkhäuser.

Ferreirós/Gray 2006. Ferreirós, José, and Gray, Jeremy: *The Architecture of Modern Mathematics: Essays in History and Philosophy*, Oxford: Oxford Academic.

Folkerts 2024. Folkerts, Menso: Die Vorgeschichte und die Gründung des mathematisch-physikalischen Seminars an der Universität Göttingen (1850), in A. Verdugo Rohrer and J. Zender, eds., *History of Mathematics and Its Contexts*, Basel: Birkhäuser, pp. 103–118.

Fontane 1898. Fontane, Theodor: *Von Zwanzig bis Dreißig. Autobiographisches*, Berlin: Fontane.

Frei/Stammbach 1994. Frei, Gunther and Stammbach, Urs: Die Mathematiker an den Zürcher Hochschulen, Basel: Birkhäuser.

Freudenthal 1964. Freudenthal, Hans: Lie Groups in the Foundations of Geometry, *Advances in Mathematics* 1(2): 145–190.

Friedman 1992. Friedman, Michael: *Kant and the Exact Sciences*, Cambridge, MA: Harvard University Press.

Fuchs 1877. Fuchs, Lazarus: Sur quelques propriétés des intégrales des équations différentielles ..., Extrait d'une lettre adressée à M. Hermite, *Journal für die reine und angewandte Mathematik* 83: 13–37.

Gabke 2016. Gabke, Wolfgang: Transcriptions of six Letters of Bernhard Riemann preserved in the Smithsonian Libraries, arXiv:1602.08264v1 [math.HO]

Gardner/Wilson 1993. Gardner, Helen J. and Wilson, Robin J.: Thomas Archer Hirst – Mathematician Xtravagant III. Berlin and Göttingen, *American Mathematical Monthly* 100: 619–625.

Gauss 1799. Gauss, Carl Friedrich: *Demonstratio nova theorematis omnem functionem algebraicam rationalem integram unius variabilis in factores reales primi vel secundi gradus resolvi posse*, Helmstedt; available online at https://doi.org/10.18452/80; reprinted in [Gauss 1863–1933, 3: 1–30].

Gauss 1801/1966. Gauss, Carl Friedrich: *Disquisitiones arithmeticae*, English translation, Arthur A. Clarke, New York & London: Yale University Press.

Gauss 1809/1857. Gauss, Carl Friedrich: *Theoria motus corporum coelestium in sectionibus conicis solem ambientum*, Hamburg; English trans. by Charles Henry Davis, Little & Brown, 1857.

Gauss 1812. Gauss, Carl Friedrich: Disquisitiones generales circa seriem infinitam ..., Pars I, *Commentationes societatis regiae scientiarum Gottingensis* 2: 1–46.

Gauss 1815. Gauss, Carl Friedrich: Anzeige von Pfaff, Methodus generalis, etc., *Göttingische Gelehrte Anzeigen*, 1 July 1815; reprinted in [Gauss 1863–1933, 3: 231–240].

Gauss 1825a. Gauss, Carl Friedrich: Allgemeine Auflösung der Aufgabe: Die Theile einer gegebenen Fläche auf einer anderen Fläche so abzubilden, dass die Abbildung dem Abgebildeten in den kleinsten Theilen ähnlich wird, *Astronomische Nachrichten* 3: 1–30; reprinted in [Gauss 1863–1933, 4: 189–216].

Gauss 1825b. Gauss, Carl Friedrich: Theoria Residuorum Biquadraticorum, Commentatio prima, *Göttingische Gelehrte Anzeigen*, 11 April 1825; reprinted in [Gauss 1863–1933, 2: 165–168].

Gauss 1828a. Gauss, Carl Friedrich: Theoria Residuorum Biquadraticorum, Commentatio prima, *Commentationes societatis regiae scientiarum Gottingensis* 6: 27–56; reprinted in [Gauss 1863–1933, 2: 65–92].

Gauss 1828b. Gauss, Carl Friedrich: Disquisitiones generales circa superficies curvas, *Commentationes societatis regiae scientiarum Gottingensis* 6: 99–146; reprinted in [Gauss 1863–1933, 4: 217–258].

Gauss 1828c. Gauss, Carl Friedrich: *Bestimmung des Breitenunterschiedes zwischen den Sternwarten von Göttingen und Altona durch Beobachtungen am Ramsdenschen Zenithsector*, Göttingen: Vandenhoeck & Ruprecht.

Gauss 1831. Gauss, Carl Friedrich: Selbst-Anzeige der Theoria Residuorum Biquadraticorum, Commentatio secunda, *Göttingische Gelehrte Anzeigen*, 23 April 1831; reprinted in [Gauss 1863–1933, 2: 169–180].

Gauss 1832a. Gauss, Carl Friedrich: Theoria Residuorum Biquadraticorum, Commentatio secunda *Commentationes societatis regiae scientiarum Gottingensis* 7: 89–148; reprinted in [Gauss 1863–1933, 2: 93–150].

Gauss 1832b. Gauss, Carl Friedrich: Selbst-Anzeige der Intensitas vis magneticae terrestris ad mensuram absolutam revocata, *Göttingische Gelehrte Anzeigen*, December 1832; reprinted in [Gauss 1863–1933, 5: 293–304].

Gauss 1833/1995. Gauss, Carl Friedrich: *Intensitas vis magneticae terrestris ad mensuram absolutam revocata, Commentationes Societatis Regiae Scientiarum Gottingensis recentiores*, 8: 3–44, reprinted in [Gauss 1863–1933, 5: 81–118]; English trans., Susan P. Johnson (1995), https://21sci-tech.com/translations/gaussMagnetic.pdf.

Gauss 1839. Gauss, Carl Friedrich: Allgemeine Theorie des Erdmagnetismus, *Resultate aus den Beobachtungen des magnetischen Vereins im Jahre 1838*: 1–57; reprinted in [Gauss 1863–1933, 5: 119–193].

Gauss 1840a. Gauss, Carl Friedrich: Allgemeine Lehrsätze in Beziehung auf die im verkehrten Verhältnis des Quadrats der Entfernung wirkenden Anziehungs- und Abstoßungskräfte, *Resultate aus den Beobachtungen des magnetischen Vereins im Jahre 1839*: 1–51; reprinted in [Gauss 1863–1933, 5: 195–242].

Gauss 1840b. Gauss, Carl Friedrich: Zur Bestimmung der Constanten des Bifilarmagnetometers, *Resultate aus den Beobachtungen des magnetischen Vereins im Jahre 1840*, 1–25; reprinted in [Gauss 1863–1933, 5: 405–421].

Gauss 1844. Gauss, Carl Friedrich: Untersuchungen über Gegenstände der höheren Geodäsie, erste Abhandlung, *Abhandlungen der Königlichen Gesellschaft der Wissenschaften in Göttingen* 2: 3–45; reprinted in [Gauss 1863–1933, 4: 259–300].

Gauss 1847. Gauss, Carl Friedrich: Untersuchungen über Gegenstände der höheren Geodäsie, zweite Abhandlung, *Abhandlungen der Königlichen Gesellschaft der Wissenschaften in Göttingen* 3: 3–43; reprinted in [Gauss 1863–1933, 4: 301–334].

Gauss 1863–1933. Gauss, Carl Friedrich: *Werke*, 12 Bände, Göttingen.

Gauss/Weber 1837–43. Gauss, Carl Friedrich und Wilhelm Weber: *Resultate aus den Beobachtungen des Magnetischen Vereins aus den Jahren 1836–1841*, 6 Bde., Göttingen: Dieterich.

Gauss/Weber 1840. Gauss, Carl Friedrich und Wilhelm Weber: *Atlas des Erdmagnetismus nach den Elementen der Theorie entworfen. Supplement zu den Resultaten aus den Beobachtungen des magnetischen Vereins unter Mitwirkung von C. W. B. Goldschmidt*, Leipzig; reprinted in [Gauss 1863–1933, 12: 337–408].

Geyer 1981. Geyer, Wulf-Dieter: Die Theorie der algebraischen Funktionen einer Veränderlichen nach Dedekind und Weber, in [Scharlau 1981, 109–133].

Gierl 2004. Gierl, Martin: *Geschichte und Organisation. Institutionalisierung als Kommunikationsprozess am Beispiel der Wissenschaftakademien um 1900*, Göttingen: Vandenhoeck & Ruprecht.

Goldstein et al. 2007. Goldstein, Catherine, Schappacher, Norbert, and Schwermer, Joachim: *The Shaping of Arithmetic after C.F. Gauss's Disquisitiones Arithmeticae*, Heidelberg: Springer.

Gray 1998. Gray, Jeremy: The Riemann-Roch Theorem and Geometry, 1854–1914, *Documenta Mathematica, Extra Volume ICM 1998* III: 811–822.

Gray 2006. Gray, Jeremy: Gauss and non-Euclidean Geometry, in A. Prékopa and E. Molnar, eds., *Non-Euclidean Geometries: János Bolyai Memorial Volume*, New York: Springer, pp. 61–80.

Gray 2008. Gray, Jeremy: *Linear Differential Equations and Group Theory from Riemann to Poincaré*, 2nd ed., Basel: Birkäuser.

Gray 2015. Gray, Jeremy: *The Real and the Complex: A History of Analysis in the 19th Century*, Cham, Switzerland: Springer.

Gray 2021. Gray, Jeremy: *Change and Variations: A History of Differential Equations to 1900*, Cham, Switzerland: Springer.

Green 1850/52/54. Green, George: An Essay on the Application of Mathematical Analysis to the Theories of Electricity and Magnetism, *Journal für die reine und angewandte Mathematik* 39: 73–89; 44: 356–374; 47: 161–221.

Günther 1897. Günther, Siegmund: Westphal, Justus Georg *Allgemeine Deutsche Biographie* 42: 203–204.

Haffner 2021. Haffner, Emmylou: The Edition of Bernhard Riemann's Collected Works: Then and now, *European Mathematical Society Magazine* 120: 29–39.

Hall 1980. Hall, A. Rupert: *Philosophers at War. The Quarrel between Newton and Leibniz*, Cambridge: Cambridge University Press.

Hattendorff 1869. Hattendorff, Karl: *Die elliptischen Functionen in dem Nachlasse von Gauss*, Hannover: Schmorl & von Seefeld.

Hawkins 2013. Hawkins, Thomas: *The Mathematics of Frobenius in Context: A Journey Through 18th to 20th Century Mathematics*, New York: Springer.

Heidelberger 2004. Heidelberger, Michael: *Nature from Within: Gustav Theodor Fechner and His Psychophysical Worldview*, Pittsburgh: University of Pittsburgh Press.

Heinrich 1985. Heinrich, Gerd, Hrsg.: *Berlin 1848. Das Erinnerungswerk des Generalleutnants Karl Ludwig von Prittwitz und andere Quellen zur Berliner Märzrevolution und zur Geschichte Preußens um die Mitte des 19. Jahrhunderts*, Berlin: de Gruyter.

Helmholtz 1853. Helmholtz, Hermann von: Ueber einige Gesetze der Vertheilung elektrischer Ströme in körperlichen Leitern, mit Anwendung auf die thierischen elektrischen Versuche, *Annalen der Physik und Chemie* 89: 211–233, 353–377.

Helmholtz 1858. Helmholtz, Hermann von: Ueber Integrale der hydrodynamischen Gleichungen welche den Wirbelbewegungen entsprechen, *Journal für die reine und angewandte Mathematik* 55: 25–55.

Helmholtz 1863. Helmholtz, Hermann von: *Die Lehre von den Tonempfindungen als physiologische Grundlage für die Theorie der Musik*, Braunschweig: Vieweg.

Helmholtz 1868/1869. Helmholtz, Hermann von: Ueber die thatsächlichen Grundlagen der Geometrie, *Verhandlungen des naturhistorisch-medicinischen Vereins zu Heidelberg* IV: 197–202. 22; Zusatz ebenda. V: 31–32.

Helmholtz 1868. Helmholtz, Hermann von: Ueber die Thatsachen, die der Geometrie zum Grunde liegen, *Nachrichten der Königlichen Gesellschaft der Wissenschaften zu Göttingen. Math.-phys. Klasse*, 1868: 193–221.

Helmholtz 1869. Helmholtz, Hermann von: Die Mechanik der Gehörknöchelchen und des Trommelfells, Separatdruck aus *Pflügers Archiv der gesamten Physiologie*, Band 1.

Helmholtz 1870/1883. Helmholtz, Hermann von: Ueber den Ursprung und die Bedeutung der geometrischen Axiome, in *Hermann von Helmholtz, Vorträge und Reden*, 4te Aufl., Bd. 2, Braunschweig: Vieweg, pp. 1–31.

Hensel 1911. Hensel, Sebastian: *Die Familie Mendelssohn, 1729–1847: Nach Briefen Und Tagebüchern*, Bd. 2, Berlin: Reimer.
it

Hentschel 2005. Hentschel, Klaus: *Gaußens unsichtbare Hand: Der Universitäts-Mechanicus und Maschinen-Inspector Moritz Meyerstein: Ein Instrumentenbauer im 19. Jahrhundert*, Göttingen: Vandenhoeck & Ruprecht.

Herbart 1851. Herbart, Johann Friedrich: *Johann Friedrich Herbart's Sämmtliche Werke*, G. Hartenstein, Hrsg., Leipzig: Voss.

Hermite 1862/1905. Hermite, Charles: Note sur la théorie des fonctions elliptiques (Extrait de la 6e edition du calcul différentiel et calcul integral de Lacroix, Paris, Mallet- Bachelier); reprinted in *Oeuvres de Charles Hermite*, Tome II, pp. 125–238.

Hilbert 1900/1935. Hilbert, David: Mathematische Probleme. Vortrag, gehalten auf dem Internationalen Mathematikerkongress zu Paris, *Nachrichten der Königlichen Gesellschaft der Wissenschaften zu Göttingen. Math.-phys. Klasse* 253–297; in [Hilbert 1932–1935, 3: 290–329].

Hilbert 1915. Hilbert, David: Die Grundlagen der Physik I, *Nachrichten der Göttingen Gesellschaft der Wissenschaften*, 1915: 395-407; reprinted in [Sauer/Majer 2009, 28–46].

Hilbert 1924. Hilbert, David: Grundlagen der Physik, *Mathematische Annalen* 92: 1–32.

Hilbert 1932–1935. Hilbert, David: *Gesammelte Abhandlungen*, 3 vols., Berlin: Springer.

Houzel 1978. Houzel, Christian: Fonctions elliptiques et intégrales abéliennes, *Abrégé d'histoire des mathématiques 1700-1900*, tome II, Paris: Hermann.

Houzel 2004. Houzel, Christian: The Work of Niels Henrik Abel, in Laudal, Olav Arnfinn and Piene, Ragni, eds., *The Legacy of Niels Henrik Abel, The Abel Bicentennial, Oslo, 2002*, Heidelberg: Springer, pp. 21–177.

Hudson 1905/1990. Hudson, Ronald W. H. T.: *Kummer's Quartic Surface.* Cambridge: Cambridge University Press; reprinted in 1990.

Hurwitz/Rudio 1895. Hurwitz, Adolf and Ferdinand Rudio, eds.: Briefe von G. Eisenstein and M.A. Stern, *Zeitschrift für Mathematik und Physik* 40: 169–204.

Jacobi 1829. Jacobi, C.G.J.: *Fundamenta nova theoriae functionum ellipticarum*, Königsberg: Borntraeger, engl. trans. by Alexander Aycock, available online from Euler-Kreis Mainz.

Jacobi 1846. Jacobi, C.G.J.: Über die Kreistheilung und ihre Anwendung auf die Zahlentheorie, *Journal für die reine und angewandte Mathematik* 30: 166–182.

Jacobi 1881–91. Jacobi, C.G.J.: *C.G.J. Jacobi's Gesammelte Werke*, 7 Bde., C.W. Borchardt, K. Weierstrass, Hrsg., Berlin: Reimer.

Jacobi 1996. Jacobi, C.G.J.: *Vorlesungen über analytische Mechanik, Berlin 1847/48*, Helmut Pulte, Hrsg., Braunschweig: Vieweg.

Janssen/Renn 2022. Janssen, Michel, and Renn, Jürgen: *How Einstein Found His Field Equations. Sources and Interpretations*, Basel: Birkhäuser.

Jost 2025. Jost, Jürgen: *Bernhard Riemann, On the Hypotheses Which Lie at the Bases of Geometry*, Basel: Birkhäuser.

Jungnickel/McCormmach 1986. Jungnickel, Christa and Russell McCormmach, *Intellectual Mastery of Nature-Theoretical Physics from Ohm to Einstein*, 2 vols., 1: The Torch of Mathematics 1800–1870, 2: The Now Mighty Theoretical Physics 1870–1925, Chicago: The University of Chicago Press.

Kirchhoff 1848. Kirchhoff, Gustav: Ueber die Anwendbarkeit der Formeln für die Intensitäten der galvanischen Ströme in einem Systeme linearer Leiter auf Systeme, die zum Theil aus nicht linearen Leitern bestehen, *Annalen der Physik und Chemie* 75: 189–205.

Kirchhoff 1854. Kirchhoff, Gustav: Ueber den inducirten Magnetismus eines unbegrenzten Cylinders von weichem Eisen, *Journal für die reine und angewandte Mathematik* 48: 348–376.

Klein 1872/1893. Klein, Felix: *Vergleichende Betrachtungen über neuere geometrische Forschungen*, Erlangen: A. Deichert; reprinted in *Mathematische Annalen*, 43: 63–100; reprinted in [Klein 1921–23, 1: 460–496].

Klein 1874. Klein, Felix: Ueber eine neue Art der Riemannschen Flächen (Erste Mitteilung), *Mathematische Annalen* 7: 558–566; reprinted in [Klein 1921–23, 2: 89–98].

Klein 1878. Klein, Felix: Ueber die Transformation der elliptischen Functionen und die Auflösung der Gleichungen fünften Grades, *Mathematische Annalen* 14: 111–172; reprinted in [Klein 1921–23, 3: 13–75].

Klein 1882/1923. Klein, Felix: *Über Riemanns Theorie der algebraischen Functionen und ihrer Integrale*, Leipzig: Teubner; reprinted in [Klein 1921–23, 3: 499–573].

Klein 1892/1986. Klein, Felix: *Riemannsche Flächen, Vorlesungen, gehalten in Göttingen, 1891/92*, G. Eisenreich und W. Purkert, Hrsg., Leipzig: BSB Teubner.

Klein 1894. Klein, Felix: *Ueber die hypergeometrische Funktion. Vorlesung, gehalten im Wintersemester 1893/94*, ausg. Ernst Ritter, Göttingen.

Klein 1895. Klein, Felix: Riemann und seine Bedeutung für die Entwicklung der modernen Mathematik, *Jahresbericht der Deutschen Mathematiker-Vereinigung* 4: 71–87; reprinted in [Klein 1921–23, 3: 482–498].

Klein 1903. Klein, Felix: Gauss' wissenschaftliches Tagebuch 1796–1814, *Mathematische Annalen* 57: 1–34.

Klein 1921–23. Klein, Felix: *Gesammelte Mathematische Abhandlungen*, 3 Bde., Berlin: Julius Springer.

Klein 1926. Klein, Felix: *Vorlesungen über die Entwicklung der Mathematik im 19. Jahrhundert*, Bd. 1, Berlin: Julius Springer.

Klein 1927. Klein, Felix: *Vorlesungen über die Entwicklung der Mathematik im 19. Jahrhundert*, vol. 2, Berlin: Julius Springer.

Klein/Fricke 1890/92. Klein, Felix and Robert Fricke: *Vorlesungen über die Theorie der elliptischen Modulfunctionen*, 2 Bde., Leipzig: Teubner.

Koenigsberger 1902. Koenigsberger, Leo: *Hermann von Helmholtz*, Bd. 2, Braunschweig: Vieweg.

Koenigsberger 1904. Koenigsberger, Leo: *Carl Gustav Jacob Jacobi: Festschrift zur Feier der 100. Wiederkehr seines Geburtstages*, Leipzig: Teubner.

Kohlrausch/Weber 1857. Kohlrausch, Rudolf and Weber, Wilhelm: Elektrodynamische Maassbestimmungen insbesondere Zurückführung der Stromintensitäts-Messungen auf mechanisches Maass, *Abhandlungen der königlich sächsischen Gesellschaft der Wissenschaften* 5: 219–292.

Koyré 1957. Koyré, Alexander: *From the Closed World to the Infinite Universe*, Baltimore: Johns Hopkins University Press.

Krazer 1903. Krazer, Adolf: *Lehrbuch der Thetafunktionen*, Leipzig: Teubner.

Krazer 1917. Krazer, Adolf: Friedrich Prym, *Jahresbericht der Deutschen Mathematiker-Vereinigung* 25: 1–15.

Kronecker 1882. Kronecker, Leopold: Grundzüge einer arithmetischen Theorie der algebraischen Grössen, *Journal für die reine und angewandte Mathematik* 92: 1–122.

Küssner 1979. Küssner, Martha: *Carl Friedrich Gauss und seine Welt der Bücher*, Göttingen: Musterschmidt.

Kummer 1836. Kummer, Ernst Eduard: Ueber die hypergeometrische Reihe, *Journal für die reine und angewandte Mathematik* 15: 39–83.

Kummer 1860. Kummer, Ernst Eduard: Gedächtnissrede auf Gustav Peter Lejeune-Dirichlet, Berlin: Buchdruckerei der Königlichen Akademie der Wissenschaften; reprinted in [Dirichlet 1889/1897, 2: 311–344].

Lamé 1852. Lamé, Gabriel: *Leçons sur la théorie mathématique de l'élasticité des corps solides*, Paris: Bachelier.

Laßwitz 1883. Laßwitz, Kurd: *Die Lehre Kants von der Idealität des Raumes und der Zeit im Zusammenhange mit seiner Kritik des Erkennens allgemeinverständlich dargestellt*, Berlin: Weidmannsche Buchhandung.

Laßwitz 1890. Laßwitz, Kurd: *Geschichte der Atomistik vom Mittelalter bis Newton, Band 1: Die Erneuerung der Korpuskulartheorie, Band 2: Höhepunkt und Verfall der Korpuskulartheorie des siebzehnten Jahrhunderts*, Hamburg u. Leipzig: Voss.

Laßwitz 1896. Laßwitz, Kurd: *Gustav Theodor Fechner*, Stuttgart: Frommanns.

Laugwitz 1996/1999. Laugwitz, Detlef: *Bernhard Riemann 1826–1866*; Engl. trans. Abe Schenitzer, Basel: Birkhäuser.

Lê 2020. Lê, François: Are the genre and the Geschlecht one and the same number? An inquiry into Alfred Clebsch's Geschlecht. *Historia mathematica* 53: 71–107.

Legendre 1825/26. Legendre, Adrien Marie: *Traité des fonctions elliptiques et des intégrales eulériennes*, T. 1,2, Paris: Huzard-Courcier.

Legendre 1830. Legendre, Adrien Marie: *Théorie des nombres*, 3rd ed., 2 vols., Paris: Didot.

Leibniz 1920. Leibniz, Gottfried Wilhelm: *The Early Mathematical Manuscripts of Leibniz*, Translated from the Latin Texts Published by Carl Immanuel Gerhardt With Critical and Historical Notes by J.M. Child, Chicago: Open Court.

Liebmann 1921. Liebmann, Heinrich: Johannes Thomae, *Jahresbericht der Deutschen Mathematiker-Vereinigung* 30: 133–144.

Lipschitz 1869. Lipschitz, Rudolf: Untersuchungen in Betreff der ganzen homogenen Functionen von n Differentialen, *Journal für die reine und angewandte Mathematik* 70: 7–102.

Lipschitz 1870. Lipschitz, Rudolf: Fortgesetzte Untersuchungen in Betreff der ganzen homogenen Functionen von n Differentialen, *Journal für die reine und angewandte Mathematik* 72: 1–56.

Lipschitz 1876. Lipschitz, Rudolf: Bemerkungen zu dem Prinzip des kleinsten Zwanges, *Journal für die reine und angewandte Mathematik* 82: 316–342.

Lorey 1916. Lorey, Willhelm: *Das Studium der Mathematik an den deutschen Universitäten selt Anfang des 19. Jahrhunderts*, Abhandlungen über den mathematischen Unterricht in Deutschland, Leipzig: Teubner.

Mayer 1880. Mayer, Adolf: Zur Pfaffschen Lösung des Pfaff'schen Problems, *Mathematische Annelen* 17: 523–530.

Mazur/Stein 2016. Mazur, Barry; Stein, William: *Prime Numbers and the Riemann Hypothesis*, New York: Cambridge University Press.

Merzbach 2018. Merzbach, Uta C.: *Dirichlet. A Mathematical Biography*, Cham: Switzerland: Birkhäuser.

Monna 1975. Monna, Antonie F.: *Dirichlet's Principle: A Mathematical Comedy of Errors and Its Influence on the Development of Analysis*, Utrecht: Oosthoek, Scheltema & Holkema, 1975.

Mossotti 1838. Mossotti, Ottaviano-Fabrizio: *Sur les forces qui régissent la constitution intérieure des corps, aperçu pour servir à la détermination de la cause et des lois de l'action moléculaire*, Turin.

Neuenschwander 1978. Neuenschwander, Erwin: Der Nachlass von Casorati (1835–1890) in Pavia, *Archive for History of Exact Sciences* 19: 1–89.

Neuenschwander 1981a. Neuenschwander, Erwin: Lettres de Bernhard Riemann à sa famille, *Cahiers du séminaire d'histoire des mathématiques* 2: 85–131.

Neuenschwander 1981b. Neuenschwander, Erwin: Über die Wechselwirkungen zwischen der französischen Schule, Riemann und Weierstrass. Eine Übersicht mit zwei Quellenstudien, *Archive for History of Exact Sciences* 24: 221–255.

Neuenschwander 1981c. Neuenschwander, Erwin: Studies in the History of Complex Function Theory II: Interactions among the French School, Riemann, and Weierstrass, *Bulletin of the American Mathematical Society* 5(2): 87–105.

Neuenschwander 1988. Neuenschwander, Erwin: A Brief Report on a Number of Recently Discovered Sets of Notes on Riemann's Lectures and on the Transmission of the Riemann Nachlass, Historia Mathematica 15: 101–113; reprinted in [Riemann 1990, 855–868].

Neuenschwander 1996. Neuenschwander, Erwin, Hrsg.: *Riemanns Einführung in die Funktionentheorie: Eine quellenkritische Edition seiner Vorlesungen mit einer Bibliographie zur Wirkungsgeschichte der Riemannschen Funktionentheorie*, Göttingen: Vandenhoeck & Ruprecht, 1996.

Neuenschwander 2022. Neuenschwander, Erwin: Die Ausleihjournale der Göttinger Universitätsbibliothek: eine bisher kaum benutzte Quelle zur Analyse von Riemanns bahnbrechenden Ideen, Max Planck Institute for History of Science Berlin. Preprint no. 512.

Neumann 1865/1884. Neumann, Carl: *Vorlesungen über Riemann's Theorie der Abel'schen Integrale*; 2nd revised ed., Leipzig: Teubner.

Neumann 1865. Neumann, Carl: *Das Dirichlet'sche Princip in seiner Anwendung auf die Riemannschen Flächen*, Leipzig: Teubner.

Newton 1713/1934. Newton, Isaac: *Sir Newton's Principles of Natural Philosophy and his System of the World*, Motte's Translation, revised by Florian Cajori, Berkeley: University of California Press.

Newton 1756. Newton, Isaac: *Four Letters from Sir Isaac Newton to Doctor Bentley, Containing some Arguments in Proof of a Deity*, London: R. and J. Dodsley.

Noether 1882. Noether, Max: Zu F. Kleins Schrift "Ueber Riemanns Theorie der algebraischen Functionen", *Zeitschrift für Mathematik und Physik* 27: 210–216.

Oesterley 1838. Oesterley, Georg Heinrich: *Johann Stephan Pütter: Versuch einer academischen Gelehrten-Geschichte von der Georg-Augustus Universität zu Göttingen*, Theil 4, 1820–1837, fortgesetzt vom Dr. Oesterley.

Olesko 1991. Olesko, Kathryn M.: *Physics as a Calling: Discipline and Practice in the Königsberg Seminar for Physics*, Ithaca, NY: Cornell University Press.

Olesko 1997. Olesko, Kathryn M.: The Meaning of Precision, in [Wise 1997, 103–134].

Parshall 1989. Parshall, Karen Hunger. 1989. Toward a History of Nineteenth-Century Invariant Theory, in [Rowe/McCleary 1989, 1: 157–206].

Parshall/Rowe 1994. Parshall, Karen Hunger and Rowe, David E.: *The Emergence of the American Mathematical Research Community, 1876–1900: James Joseph Sylvester, Felix Klein, and E. H. Moore*, Providence: American Mathematical Society and London: London Mathematical Society.

Patterson 2017. Patterson, S.J.: Reading Riemann, in H. Montgomery, et al., eds., *Exploring the Riemann Zeta Function: 190 years from Riemann's Birth*, New York: Springer, pp. 265–285.

Prym 1882. Prym, Friedrich: *Untersuchungen ueber die Riemann'sche Thetaformel und die Riemann'sche Charakteristikentheorie*, Leipzig: Teubner.

Prym 1883. Prym, Friedrich: Ein neuer Beweis für die Riemann'sche Thetaformel, *Acta Mathematica* 3: 201–215.

Purkert/Ilgauds 1987. Walter Purkert, and H.J. Ilgauds, *Georg Cantor 1845–1918*, Basel: Birkhäuser.

Purkert 1989. Walter Purkert, Cantor's Views on the Foundations of Mathematics, in [Rowe/McCleary 1989, 49–65].

Reich 1994. Reich, Karin: *Die Entwicklung des Tensorkalküls: Vom absoluten Differentialkalkül zur Relativitätstheorie*, Basel: Birkhäuser.

Reiff/Sommerfeld 1902. Reiff, Richard und Sommerfeld Arnold: Elektrizität und Optik: Standpunkt der Fernwirkung, Elementargesetze, *Encyklopädie der mathematischen Wissenschaften*, 5–2, pp. 4–66.

Remmert 1993. Remmert, Reinhold: The Riemann-file Nr. 135 of the Philosophische Fakultät of the Georgia Augusta at Göttingen, *The Mathematical Intelligencer* 15(3): 44–48.

Ricci/Levi-Civita 1901. Ricci, Gregorio and Levi-Civita, Tullio: Méthodes de calcul différentiel absolu et leurs applications, *Mathematische Annalen* 54: 125–201.

Riemann 1847/1876. Riemann, Bernhard: Versuch einer allgemeinen Auffassung der Integration und Differentiation, first printed in [Riemann 1876/1892/1902, 331–345].

Riemann 1851/1876. Riemann, Bernhard: Grundlagen für eine allgemeine Theorie der Functionen einer veränderlichen complexen Grösse, Inauguraldissertation, Göttingen; reprinted in [Riemann 1876/1892/1902, 3–45].

Riemann 1854. Riemann, Bernhard: Ueber die Gesetze der Vertheilung von Spannungselectricität in ponderabeln Körpern, wenn diese nicht als vollkommene Leiter oder Nichtleiter, sondern als dem Enthalten von Spannungselectricität mit endlicher Kraft widerstrebend betrachtet werden, *Amtlicher Bericht über die 31. Versammlung deutscher Naturforscher und Aerzte zu Göttingen im September 1854*; reprinted in [Riemann 1876/1892/1902, 49–62].

Riemann 1855. Riemann, Bernhard: Zur Theorie der Nobili'schen Farbenringe, *Annalen der Physik und Chemie* 95: 130–138.

Riemann 1857a. Riemann, Bernhard: Beiträge zur Theorie der durch die Gauss'sche Reihe $F(\alpha, \beta, \gamma, x)$ darstellbaren Functionen by Bernhard Riemann, *Abhandlungen der Königlichen Gesellschaft der Wissenschaften zu Göttingen* Band 7: 3–22; reprinted in [Riemann 1876/1892/1902, 1876: 62–78].

Riemann 1857b. Riemann, Bernhard: Theorie der Abel'schen Functionen, *Journal für die reine und angewandte Mathematik* 54: 101–155; reprinted in [Riemann 1876/1892/1902, 1876: 81–144].

Riemann 1859/1876. Riemann, Bernhard: Ueber die Anzahl der Primzahlen unter einer gegebenen Grösse, *Monatsberichte der Berliner Akademie* 671–680; reprinted in [Riemann 1876/1892/1902, 145–153].

Riemann 1859/1990. Riemann, Bernhard: Entwurf eines Briefes von B. Riemann an K. Weierstrass, 29. Oktober 1859, in [Riemann 1990, 822–825].

Riemann/Betti 1859. Riemann, Bernhard: Fondamenti di una teorica generale delle funzioni di una variable complessa (trad. di E. Betti), *Annali di Matematica Pura ed Applicata* 2: 288–304; 337–356.

Riemann 1860. Riemann, Bernhard: Ueber die Fortpflanzung ebener Luftwellen von endlicher Schwingungsweise, *Abhandlungen der Königlichen Gesellschaft der Wissenschaften zu Göttingen* Band 8: 43–66; reprinted in [Riemann 1876/1892/1902, 1876: 145–164].

Riemann 1861a/1876. Riemann, Bernhard: Ein Beitrag zu den Untersuchungen über die Bewegungen eines flüssigen gleichartigen Ellipsoides, *Abhandlungen der Königlichen Gesellschaft der Wissenschaften zu Göttingen* Band 9: 3–36; reprinted in [Riemann 1876/1892/1902, 182–211].

Riemann 1861b/1876. Riemann, Bernhard: Commentatio mathematica, qua respondere tentatur quaestioni ab Ill Academia Parisiensi propositae; in [Riemann 1876/1892/1902, 1876: 370–383].

Riemann 1861*. Riemann, Bernhard: Theorie complexer Functionen. Im Sommersemester 1861, 4stündig vorgetragen von B. Riemann. Lecture notes edited by H. Hankel, from the estate of Georg Thieme. Cod. Ms. Riemann 32g, Niedersächsische Staats- und Universitätsbibliothek Göttingen.

Riemann 1861/1862*. Riemann, Bernhard: Theorie der Functionen complexer Variabeln: Vorlesung des Prof. Riemann, Göttingen, Sommersemester 1861. Wintersemester 1861/62. Lecture notes taken by Ernst Abbe, from the estate of Georg Thieme. Cod. Ms. Riemann 32c, Niedersächsische Staats- und Universitätsbibliothek Göttingen.

Riemann 1866/1876. Riemann, Bernhard: Ueber das Verschwinden der Theta-Functionen, *Journal für die reine und angewandte Mathematik* 65: 161–172; reprinted in [Riemann 1876/1892/1902, 212–224].

Riemann 1867a/1876. Riemann, Bernhard: Ein Beitrag zur Elektrodynamik, *Annalen der Physik und Chemie* 131: 237–243; reprinted in [Riemann 1876/1892/1902, 1876: 270–275].

Riemann 1867b/1876. Riemann, Bernhard: Mechanik des Ohres, *Zeitschrift für rationelle Medicin* 29; reprinted in [Riemann 1876/1892/1902, 1876: 338–350].

Riemann 1868/1876. Riemann, Bernhard: Ueber die Darstellbarkeit einer Function durch eine trigonometrische Reihe, *Abhandlungen der Königlichen Gesellschaft der Wissenschaften in Göttingen* 13: 87–132; reprinted in [Riemann 1876/1892/1902, 1876: 213–251]

Riemann 1868/1876. Riemann, Bernhard: Ueber die Hypothesen, welche der Geometrie zu Grunde liegen, *Abhandlungen der Königlichen Gesellschaft der Wissenschaften in Göttingen* 13: 133–152; reprinted in [Riemann 1876/1892/1902, 1876: 254–269]; reprinted with historical and mathematical commentary in [Jost 2025]; translated by William K. Clifford, *Nature* 8(1873): 14–17, 36–37.

Riemann 1868/1876. Riemann, Bernhard: Ueber die Fläche vom kleinsten Inhalt bei gegebener Begrenzung. Hrsg. von Karl Hattendorff, *Abhandlungen der Königlichen Gesellschaft der Wissenschaften in Göttingen* 13: 3–52; reprinted in revised form in [Riemann 1876/1892/1902, 1876: 283–315].

Riemann 1869/1876. Riemann, Bernhard: Beweis des Satzes, dass eine einwerthige mehr als $2n$-fach periodische Function von $n$ Veränderlichen unmöglich ist, *Journal für die reine und angewandte Mathematik* 71: 197–200; reprinted in [Riemann 1876/1892/1902, 294–297].

Riemann/Hattendorff 1869. Riemann, Bernhard: *Partielle Differenzialgleichungen und deren Anwendung auf physicalische Fragen. Vorlesungen von Berh. Riemann*, Karl Hattendorff, Hrsg., Braunschweig: Vieweg.

Riemann 1876a. Riemann, Bernhard: Zur Theorie der der Abel'schen Functionen (based on a text of Gustav Roch), in [Riemann 1876/1892/1902, 487–508].

Riemann 1876b. Riemann, Bernhard: Commentatio mathematica, qua respondere tentatur quaestioni ab Illma Academia Parisiensi propositae, in [Riemann 1876/1892/1902, 370–383–508].

Riemann/Hattendorff 1876. Riemann, Bernhard: *Schwere, Elektrizität und Magnetismus*, Karl Hattendorff, Hrsg., Hannover: Rümpler.

Riemann 1876/1892/1902. Riemann, Bernhard: *Bernhard Riemanns Gesammelte Mathematische Werke und wissenschaftlicher Nachlass*, Heinrich Weber, Hrsg., 2. Aufl. 1892; reprinted with Nachträge edited by M. Noether and W. Wirtinger in 1902, Teubner: Leipzig.

Riemann 1899. Riemann, Bernhard: *Elliptische Functionen. Vorlesungen von Bernhard Riemann, mit Zusätzen herausgegeben von Hermann Stahl*, Leipzig: Teubner.

Riemann 1990. Riemann, Bernhard: *Riemanns Gesammelte Werke*, Raghavan Narasimhan, ed., Leipzig: Teubner/Springer.

Roch 1863a. Roch, Gustav: Ueber Functionen complexer Grössen, *Zeitschrift für Mathematik und Physik* 8: 12–26, 183–203.

Roch 1863b. Roch, Gustav: De theoremate quodam circa functiones Abelianas, Habilitationsschrift Halle, Leipzig: Teubner.
Roch 1865a. Roch, Gustav: Ueber Functionen complexer Grössen, *Zeitschrift für Mathematik und Physik* 10: 169–194.
Roch 1865b. Roch, Gustav: Ueber die Ausdrücke elliptischer Integrale zweiter und dritter Gattung durch Theta-Functionen. *Zeitschrift für Mathematik und Physik* 10: 317–320.
Roch 1865c. Roch, Gustav: Ueber die Anzahl der willkürlichen Constanten in algebraischen Functionen, *Journal für die reine und angewandte Mathematik* 64: 372–376.
Roch 1866a. Roch, Gustav: Ueber Integrale zweiter Gattung und die Werthermittelung der Theta-Functionen, *Zeitschrift für Mathematik und Physik* 11: 53–63.
Roch 1866b. Roch, Gustav: Ueber specielle vierfach periodische Functionen, *Zeitschrift für Mathematik und Physik* 11: 463–474.
Roch 1866c. Roch, Gustav: Ueber die dritte Gattung Abel'scher Integrale erster Ordnung, *Journal für die reine und angewandte Mathematik* 65: 42–51.
Roch 1866d. Roch, Gustav: Ueber die Doppeltangenten an Curven vierter Ordnung, *Journal für die reine und angewandte Mathematik* 66: 97–120.
Roch 1868. Roch, Gustav: Ueber Abel'sche Integrale dritter Gattung, *Journal für die reine und angewandte Mathematik* 68: 170–175.
Rosenhain 1851/1895. Rosenhain, Georg: *Mémoire sur les fonctions de deux variables et à quatre périodes, qui sont les inverses des intégrales ultra-elliptiques de la première classe*, Paris; *Abhandlung über die Functionen zweier Variabler mit vier Perioden, welche die Inversen sind der ultra-elliptischen Integrale erster Klasse*, H. Weber, Hrsg., Leipzig: Engelmann.
Rowe 1986. Rowe, David E.: "Jewish Mathematics" at Göttingen in the Era of Felix Klein, *Isis* 77: 422–449.
Rowe 1998. Rowe, David E.: Mathematics in Berlin, 1810–1933, in [Begehr et al. 1998, 9–26].
Rowe 2004. Rowe, David E.: Making Mathematics in an Oral Culture: Göttingen in the Era of Klein and Hilbert, *Science in Context* 17: 85–129.
Rowe 2018. Rowe, David E.: *A Richer Picture of Mathematics: The Göttingen Tradition and Beyond*, New York: Springer.
Rowe 2022. Rowe, David E.: Felix Klein and Emmy Noether on Invariant Theory and Variational Principles, in *The Philosophy and Physics of Noether's Theorems*, James Read and Nicholas Teh, eds., Cambridge: Cambridge University Press, pp. 25–50.
Rowe 2023. Rowe, David E.: Mathematiker als Schriftsteller und Dichter: Geometrie und Naturphilosophie, 1700–1900, in [Rowe/Volkert 2023, 101–174].
Rowe 2024. Rowe, David E.: Felix Klein's Early Contributions to *anschauliche Geometrie*, *Archive for History of Exact Sciences* 78: 401–477.
Rowe 2025. Rowe, David E.: *Felix Klein. The Erlangen Program*, Cham, Switzerland: Springer Nature.
Rowe/McCleary 1989. Rowe, David E. and McCleary, John, eds.: *The History of Modern Mathematics*, 2 vols., Boston: Academic Press.
Rowe/Volkert 2023. Rowe, David E. and Klaus Volkert: *Jenseits von Flachland: Mathematische Grenzüberschreitungen und ihre Auswirkungen*, Heidelberg: Springer.
Sarkowski 1992. Sarkowski, Heinz: *Der Springer-Verlag. Stationen seiner Geschichte. Teil I: 1842 bis 1945*, New York: Springer.
Sartorius 1856/2012. Sartorius von Waltershausen, Wolfgang: *Gauss zum Gedächtniss*, Leipzig: Hirzel; Neudruck mit Kommentar von Karin Reich, Leipzig: Eagle.
Sauer/Majer 2009. Sauer, Tilman and Majer, Ulrich, eds.: *David Hilbert's Lectures on the Foundations of Physics, 1915-1927*, Heidelberg: Springer.
Schappacher 1998. Schappacher, Norbert: Gotthold Eisenstein, 16 April 1823 – 11 October 1852, in [Begehr et al. 1998, 55–60].
Schappacher 2016. Schappacher, Norbert: Bernhard Riemann as a Puzzle for the Historian, The Riemann Conference, Münster, unpublished lecture, 6 October 2016.
Scharlau 1981. Scharlau, Winfried, Hrsg.: *Richard Dedekind: 1831–1981, eine Würdigung zu seinem 150. Geburtstag*, Braunschweig: Vieweg.

Scharlau 1986. Scharlau, Winfried, Hrsg.: *Rudolf Lipschitz: Briefwechsel mit Cantor, Dedekind, Helmholtz, Kronecker, Weierstrass und Anderen*, Wiesbaden: Vieweg.

Scharlau 1990. Scharlau, Winfried, Hrsg.: *Mathematische Institute in Deutschland, 1800–1945*, Braunschweig: Vieweg.

Scheel 2015. Scheel, Katrin, Hrsg.: *Der Briefwechsel Richard Dedekind – Heinrich Weber*, Berlin: De Gruyter.

Scheibner 1860. Scheibner, Wilhelm: Ueber die Anzahl der Primzahlen unter einer beliebigen Grenze, *Zeitschrift für Mathematik und Physik* 5: 233–252.

Schering 1867. Schering, Ernst: Bernhard Riemann zum Gedächtniss, *Nachrichten der Göttingen Gesellschaft der Wissenschaften*, 1867: 305–314; reprinted in [Schering 1909, 161–168].

Schering 1909. Schering, Ernst: *Gesammelte mathematische Werke von Ernst Schering*, Bd. 2, Robert Haussner & Karl Schering, Hrsg., Berlin: Mayer & Müller.

Schering 1909/1990. Schering, Ernst: Zum Gedächtniss an B. Riemann; in [Schering 1909, 367–392]; reprinted in [Riemann 1990, 828–848].

Schering, K. 1909. Schering, Karl: Lebenslauf von Ernst Schering, in [Schering 1909, 449–472].

Scholz 1980. Scholz, Erhard: *Geschichte des Mannigfaltigkeitsbegriffs von Riemann bis Poincaré*, Boston: Birkhäuser.

Scholz 1982a. Scholz, Erhard: Herbart's Influence on Riemann, *Historia Mathematica* 9: 413–440.

Scholz 1982b. Scholz, Erhard: Riemanns frühe Notizen zum Mannigfaltigkeitsbegriff und zu den Grundlagen der Geometrie, *Archive for History of Exact Sciences* 27: 213–232.

Scholz 1992. Scholz, Erhard: Gaußund die Begründung der höheren Geodäsie, in Demidov, S.S., Rowe, D., Folkerts, M., Scriba, C.J., eds., *Amphora*, Basel: Birkhäuser, pp. 631–647.

Scholz 1999. Scholz, Erhard: The Concept of Manifold, 1850–1950, in *History of Topology*, I.M. James, ed., Dordrecht: Kluwer, pp. 25–64.

Scholz 2004. Scholz, Erhard: C.F.Gauß' Präzisionsmessungen terrestrischer Dreiecke und seine Überlegungen zur empirischen Fundierung der Geometrie in den 1820er Jahren, in Menso Folkerts, Ulf Hashagen, Rudolf Seising, eds., *Form, Zahl, Ordnung. Studien zur Wissenschafts- und Technikgeschichte. Ivo Schneider zum 65. Geburtstag*, Stuttgart: Franz Steiner, 355–380, http:// arxiv.org/abs/math/0409578.

Schramm 1985. Schramm, Matthias: *Natur Ohne Sinn? Das Ende des Teleologischen Weltbildes*, Graz: Verlag Styria.

Schubring 1984. Schubring, Gert: Die Promotion von P.G. Lejeune Dirichlet. Biographische Mitteilungen zum Werdegang Dirichlets, *NTM (Schriftenreihe für Geschichte der Naturwissenschaften, Technik und Medizin)* 21(1): 45–65.

Schubring 1986. Schubring, Gert: The Three Parts of the Dirichlet Nachlass, *Historia Mathematica* 13: 52–56.

Schubring 1990. Schubring, Gert: Zur strukturellen Entwicklung der Mathematik an den deutschen Hochschulen 1800–1945, in [Scharlau 1990, 264–279].

Schubring 1992. Schubring, Gert: Zur Modernisierung des Studiums der Mathematik in Berlin, 1820–1840, in *Amphora. Festschrift für Hans Wussing zu seinem 65. Geburtstag*, Sergei Demidov, Menso Folkerts, David E. Rowe, and Christoph Scriba, eds., Basel: Birkhäuser, pp. 649–675.

Schubring 2012. Schubring, Gert: Antagonisms between German States regarding the Status of Mathematics Teaching during the 19th Century: Processes of Reconciling Them, *ZDM - The International Journal on Mathematics Education* 44(4): 525–535.

Schubring 2015. Schubring, Gert: The Emergence of the Profession of Mathematics Teachers – an International Analysis of Characteristic Patterns, in *"Dig Where you Stand": Proceedings of the Third International Conference on the History of Mathematics Education*, Uppsala, September 25–28, 2013, eds. KristÃn Bjarnadottir, Fulvia Furinghetti, Johan Prytz, Gert Schubring, Uppsala: Uppsala University, Department of Education, 389–403.

Schubring 1892/1973. Schubring, Julius, Hrsg.: *Briefwechsel zwischen Felix Mendelssohn Bartholdy und Julius Schubring: zugleich ein Beitrag zur Geschichte und Theorie des Oratoriums*, Walluf bei Wiesbaden: Sändig.

Schwarz 1869. Schwarz, Hermann Amandus: Ueber einige Abbildungsaufgaben, *Journal für die reine und angewandte Mathematik* 70: 105–120.

Schwarz 1872. Schwarz, Hermann Amandus: Ueber diejenigen Fälle, in welchen die Gaussische hypergeometrische Reihe eine algebraische Function ihres vierten Elementes darstellt, *Journal für die reine und angewandte Mathematik* 75: 292–335.

Shafarevich 1983. Shafarevich, Igor: Zum 150. Geburtstag von Alfred Clebsch. *Mathematische Annalen* 266: 135–140.

Siegel 1932. Siegel, Carl Ludwig: Über Riemanns Nachlass zur analytischen Zahlentheorie, *Quellen und Studien zur Geschichte der Mathematik, Astronomie und Physik*, Abt. B: Studien 2: 45–80; reprinted in [Riemann 1990, 770–806].

Siegmund-Schultze 2007. Siegmund-Schultze, Reinhard: Philipp Frank, Richard von Mises, and the Frank-Mises, *Physics in Perspective* 9: 26–57.

Sinaceur 1990. Sinaceur, Mohammed Allal: Dedekind et le programme de Riemann, *Revue d'histoire des sciences* 43: 221–296.

Spivak 1979. Spivak, Michael: *A Comprehensive Introduction to Differential Geometry*, vol. 2, Berkeley, CA: Publish or Perish.

Springer 2001. Springer, George: *Introduction to Riemann Surfaces*, 3rd ed., Providence RI, AMS Chelsea Publishing.

Stäckel 1922. Stäckel, Paul: Gauss als Geometer, in [Gauss 1863–1933, 10(2), 4:1–121].

Stern 1860. Stern, M.A.: *Lehrbuch der algebraischen Analysis*, Leipzig: C.F. Winter.

Stigler 1981. Stigler, Stephen M.: Gauss and the Invention of Least Squares, *Annals of Statistics* 9(3): 465–474.

Stigler 1986. Stigler, Stephen M.: *The History of Statistics: The Measurement of Uncertainty Before 1900*, Cambridge, MA: Harvard University Press.

Stillwell 1996. Stillwell, John: *Sources of Hyperbolic Geometry*, Providence, RI: American Mathematical Society.

Stöltzner 2002. Stöltzner, Michael: The Principle of Least Action as the Logical Empiricist's Shibboleth, *Studies in History and Philosophy of Science Part B: Studies in History and Philosophy of Modern Physics* 34(2): 285–318.

Tazzioli 2012. Tazzioli, Rossana: New Perspectives on Beltrami's Life and Work - Considerations Based on his Correspondence, in [Coen 2012, 465–518].

Tazzioli 2025. Tazzioli, Rossana: *From Differential Geometry to Relativity. Levi-Civita's Lectures on the Absolute Differential Calculus, 1925–1928*, EMS Press.

Tobies 2021. Tobies, Renate: *Felix Klein: Visions for Mathematics, Applications, and Education*, Heidelberg: Springer.

Tschinkel 2005. Tschinkel, Yuri: About the Cover: On the Distribution of Primes – Gauss' Tables, *Bulletin of the American Mathematical Society* 43(1): 89–91.

Turner 1980. Turner, R. Steven: The Prussian Universities and the Concept of Research, *Internationales Archiv für Sozialgeschichte der deutschen Literatur*, 5: 68–86.

Ullrich 1990. Ullrich, Peter: Wie man beim Weierstrasschen Aufbau der Funktionentheorie das Cauchysche Integral vermeidet, *Jahresbericht der Deutschen Mathematiker-Vereinigung* 92: 89–110.

Ullrich 2003. Ullrich, Peter: Die Weierstraßschen "analytischen Gebilde": Alternative zu Riemanns "Flächen" und Vorboten der komplexen Räume, *Jahresbericht der Deutschen Mathematiker-Vereinigung* 105: 30–59.

Volkert 2013. Volkert, Klaus: *Das Undenkbare denken. Die Rezeption der nicht-euklidischen Geometrie im deutschsprachigen Raum (1860–1900)*, Heidelberg: Springer.

Volkert 2017. Volkert, Klaus: Dedekind goes Zürich, *Mathematische Semesterberichte* 64: 1–12.

Volkert 2023. Volkert, Klaus: Im Reich der unbegrenzten Möglichkeiten – Die vierte Dimension in Mathematik, Kunst und Philosophie, in [Rowe/Volkert 2023, 1–99].

H.M. Weber 1869. Weber, Heinrich Martin: Note zu Riemanns Beweis des Dirichletschen Princips, *Journal für die reine und angewandte Mathematik* 71: 29–39.

H.M. Weber 1876. Weber, Heinrich Martin: *Theorie der Abelschen Functionen vom Geschlecht 3*, Berlin: Reimer.

H.M. Weber 1877. Weber, Heinrich Martin: Rezension von [Riemann 1876/1892/1902, 1876], *Repertorium der literarischen Arbeiten aus dem Gebiete der reinen und angewandten Mathematik* 1: 145–154; in Heidelberger Texte zur Geschichte der Mathematik, www.ub.uni-heidelberg.de/archiv/13225.

H.M. Weber 1900. Weber, Heinrich Martin: *Die partiellen Differentialgleichungen der mathematischen Physik, nach Riemanns Vorlesungen*, Braunschweig: Vieweg.

H. Weber 1893. Weber, Heinrich: *Wilhelm Weber: eine Lebensskizze*, Breslau: Trewendt; Engl. trans. in Andre Koch Torres Assis, ed. *Wilhelm Weber's Main Works on Electrodynamics Translated into English*, vol. 5, pp. 203–256. Available online at https://www.ifi.unicamp.br/ assis/Weber-in-English-Vol-5.pdf.

W. Weber 1846. Weber, Wilhelm: Elektrodynamische Maassbestimmungen – Über ein allgemeines Grundgesetz der elektrischen Wirkung, *Abhandlungen bei Begründung der Königlich Sächsischen Gesellschaft der Wissenschaften am Tage der zweihundertjährigen Geburtstagfeier Leibnizen's*, pp. 211–378.

W. Weber 1848. Weber, Wilhelm: Elektrodynamische Maassbestimmungen, *Annalen der Physik und Chemie* 73: 193–240.

W. Weber 1852. Weber, Wilhelm: Elektrodynamische Maassbestimmungen insbesondere über Diamagnetismus, *Abhandlungen der Königlich Sächsischen Gesellschaft der Wissenschaften zu Leipzig, mathematisch-physische Classe* 1: 485–577, 1852.

Weierstrass 1854. Weierstrass, Karl: Theorie der Abel'schen Functionen, *Journal für die reine und angewandte Mathematik* 47: 289–306.

Weierstrass 1902. Weierstrass, Karl: Vorlesungen über die Theorie der Abelschen Transcendenten von Karl Weierstrass. Bearbeitet von G. Hettner und J. Knobloch, in *Mathematische Werke*, Bd. 4, Berlin: Mayer & Müller.

Weierstrass 1988. Weierstrass, Karl: *Einleitung in die Theorie der analytischen Funktionen*, Peter Ullrich, ed., Heidelberg: Springer.

Weil 1979. Weil, André: Riemann, Betti and the Birth of Topology, *Archive for History of Exact Sciences* 20: 91–96.

Weil 1989. Weil, André: On Eisenstein's Copy of the Disquisitiones, *Algebraic Number Theory*, (Advanced Studies in Pure Mathematics, vol. 17), Boston: Academic Press, pp. 463–469.

Weyl 1913. Weyl, Hermann: *Die Idee der Riemannschen Fläche*, Leipzig: Teubner.

Weyl 1923. Weyl, Hermann: Kommentar zu [Riemann 1868/1876]; reprinted in [Riemann 1990, 740–769] and in [Jost 2025, 45–73].

Weyl 1988. Weyl, Hermann: *Riemanns geometrische Ideen, ihre Auswirkung und ihre Verknüpfung mit der Gruppentheorie*, Heidelberg: Springer.

Wiescher 2025. Wiescher, Michael: Two Jewish Mathematicians at the Time of Emancipation in Prussia, preprint.

Wiescher 2026. Wiescher, Michael: *From Crystal Symmetries to the Geometry of Space: The Life of Julius Plücker*, Oxford: Oxford University Press.

Wirtinger 1895. Wirtinger, Wilhelm: Zur Theorie der 2n-fach periodischen Functionen, *Monatshefte für Mathematik und Physik* 6: 69–98.

Wirtinger 1905/1990. Wirtinger, Wilhelm: Riemanns Vorlesungen über die hypergeometrische Reihe und ihre Bedeutung; reprinted in [Riemann 1990, 720–738].

Wise 1981. Wise, M. Norton: German Concepts of Force, Energy, and the Electromagnetic Ether: 1845–1880, in *History of Ether Theories in Modern Science*, ed. G.N. Cantor and M.J.S. Hodge, Cambridge, UK, pp. 269–307.

Wise 1997. Wise, M. Norton, ed.: *The Values of Precision*, Princeton: Princeton University Press.

Wittmann 2018. Wittmann, Axel, Hrsg.: *Obgleich und indeßen: Der Briefwechsel zwischen Carl Friedrich Gauß und Johann Franz Encke, 1813-1854*, 2 Bde., Kassel: Kessel.

Wolter 1987. Wolter, Andrä: *Das Abitur. Eine bildungssoziologische Untersuchung zur Entstehung und Funktion der Reifeprüfung*, Oldenburg: Holzberg.

Zagier 1977. Zagier, Don: The First 50 Million Prime Numbers, *The Mathematical Intelligencer* 1: 7–19.

Zöllner 1876. Zöllner, Friedrich: *Principien einer elektrodynamischen Theorie der Materie*, Leipzig: Engelmann.

# Name Index





D. E. Rowe, *Bernhard Riemann: His Life and Wondrous Mathematical Legacy*,
https://doi.org/10.1007/978-3-032-25457-3